Generative AI and Internet of Medical Things

Mathematical Methods in the Digital Age

Edited by

Abhishek Kumar, Satya Prakash Yadav, Ashutosh Kumar Dubey, Andrei Chtcheprov, Sayavur Bakhtiyarov

VOLUME 10

Generative AI and Internet of Medical Things

Artificial Intelligence, Biomedical Engineering, Wearable Sensors

Edited by

Pethuru Raj Chelliah, Satya Prakash Yadav,
Victor Hugo C. de Albuquerque, H. Herbert Song

DE GRUYTER

Editors
Dr Pethuru Raj Chelliah
Reliance Jio Platforms Ltd.
Avana Building
Sarjapur Road
Bengaluru 560103, Karnataka, India
peterindia@gmail.com

Dr. Satya Prakash Yadav
Department of Computer Science and Engineering
Madan Mohan Malaviya University of Technology
Gorakhpur, Uttar Pradesh, India
prakashyadav.satya@gmail.com, spycs@mmmut.ac.in

Prof. Victor Hugo C. de Albuquerque
Department of Teleinformatics Engineering
Federal University of Ceará
60455-970 Fortaleza, Brazil
victor120585@yahoo.com.br

Prof. (Dr) H. Herbert Song
Department of Information Systems
University of Maryland Balitmore Country (UMBC)
21250 Baltimore, MD, USA
song@computer.org

ISSN 2943-2731
ISBN 978-3-11-223707-6
ISBN 978-3-11-223708-3 (PDF)
ISBN 978-3-11-223709-0 (E-PUB)
DOI https://doi.org/10.1515/9783112237083

Library of Congress Control Number: 2026940651

Bibliographic information published by the Deutsche Nationalbibliothek
The Deutsche Nationalbibliothek lists this publication in the Deutsche Nationalbibliografie; detailed bibliographic data are available on the Internet at http://dnb.dnb.de.

De Gruyter and Walter de Gruyter GmbH are part of De Gruyter Brill.
www.degruyterbrill.com

Questions about General Product Safety Regulation: productsafety@degruyterbrill.com

Printing: CPI books GmbH, Leck

MIX
Papier | Fördert gute Waldnutzung
FSC® C083411

Contents

Preface

The rapid convergence of generative artificial intelligence (AI), the Internet of Medical Things (IoMT), and biomedical engineering is transforming the landscape of modern healthcare. This book presents a comprehensive and forward-looking exploration of how intelligent technologies are reshaping clinical practice, remote monitoring, and personalized medicine.

In recent years, the IoMT has emerged as a powerful paradigm for connecting medical devices, wearable sensors, and healthcare systems. This interconnected ecosystem empowers medical professionals, caregivers, and surgeons with real-time insights, enabling timely diagnosis, continuous monitoring, and data-driven decision-making. This book discusses the latest trends and technological advancements in IoMT, highlighting its critical role in improving healthcare accessibility, operational efficiency, and patient outcomes.

A key focus of this book is the acquisition and analysis of physiological and environmental data. Various physical parameters – such as heart rate, blood glucose levels, blood pressure, body temperature, and air quality – are captured through advanced sensor technologies. These sensors, embedded in wearable devices and medical instruments, generate vast volumes of data that require intelligent processing and interpretation.

This book further explores how modern medical devices – including scanners, robotic systems, smart appliances, imaging systems, handheld devices, and wearable technologies – can be seamlessly integrated with cloud-based platforms, digital twins, databases, and advanced analytics frameworks. By leveraging generative AI and machine learning models, these systems can derive meaningful insights, generate realistic synthetic medical data, and enhance predictive capabilities.

Another significant contribution of this book is its demonstration of how generative AI can synthesize highly realistic datasets from limited input samples, thereby addressing data scarcity challenges in healthcare. These capabilities support improved training of AI models and facilitate the extraction of actionable knowledge from raw healthcare data. As a result, medical professionals can perform real-time monitoring, intelligent diagnostics, and personalized treatment planning with greater accuracy and confidence.

Overall, this book serves as a valuable resource for researchers, practitioners, academicians, and industry professionals working at the intersection of AI, IoMT, and healthcare technologies. It aims to bridge the gap between theoretical advancements and practical implementations, thereby fostering innovation in smart, connected, and intelligent healthcare systems.

 | HTTPS://DOI.ORG/10.1515/9783112237083-203

List of contributors

Manivannan Karunakaran
Department of Information Science and Engineering, JAIN (Deemed-to-be University) Bengaluru, Karnataka, India, manivannan.k@jainuniversity.ac.in

Chapters 1, 2, 6, 7, 10, 12, 16, 20

Suhel Khan
Department of Information Science, Presidency University, Bengaluru, Karnataka, India, suhelkhan@presidencyuniversity.in

Chapter 1

Akash Kumar Bhagat
Department of Computer Science and IT, Arka Jain University, Jamshedpur, Jharkhand, India, akash.b@arkajainuniversity.ac.in

Chapter 1

Kalpesh Jadav
Department of Electronics and Communication Engineering, Faculty of Engineering and Technology, Parul Institute of Engineering and Technology, Parul University, Vadodara, Gujarat, India, kalpesh.jadav@paruluniversity.ac.in

Chapter 2

Shet Reshma Prakash
Department of Computer Science and Engineering, Presidency University, Bengaluru, Karnataka, India, reshma.prakash@presidencyuniversity.in

Chapter 2

Vijay Jagdish Upadhye
Department of Microbiology, Research and Development Cell, Parul University, Vadodara, Gujarat, India, vijay.upadhye82074@paruluniversity.ac.in

 | HTTPS://DOI.ORG/10.1515/9783112237083-204

Chapter 3

Dr. Satya Ranjan Das
Department of Computer Science and Engineering, Institute of Technical Education and Research, Siksha 'O' Anusandhan (Deemed to be University), Bhubaneswar, Odisha, India, Id-satyadas@soa.ac.in

Chapter 3

Shajahan Basheer
Department of Computer Science and Engineering, Faculty of Engineering and Technology, JAIN (Deemed-to-be University), Bengaluru, Karnataka, India, Id-shajahan.b@jainunievrsity.ac.in

Chapter 3

Bhupesh Goyal
Department of Allied Healthcare and Sciences, Vivekananda Global University, Jaipur, Rajasthan, India, bhupesh.goyal@vgu.ac.in

Chapter 4

Vishakha Khambhati
Department of Biomedical Engineering, Faculty of Engineering and Technology, Parul Institute of Technology, Parul University, Vadodara, Gujarat, India, vishakha-khambhati@gmail.com

Chapter 4

A. R. Chayapathi
Department of Information Science and Engineering, JAIN (Deemed-to-be University), Bengaluru, Karnataka, India, chayapathi.ar@jainuniversity.ac.in

Chapter 4

Anitha Josephine
Department of Information Science and Engineering, JAIN (Deemed-to-be University), Bengaluru, Karnataka, India, anitha.josephine@jainuniversity.ac.in

Chapters 5, 15

Abhishek Godara
Department of Pharmacy, Vivekananda Global University, Jaipur, Rajasthan, India, abhishek.godara@vgu.ac.in

Chapter 5

Vipul Vekariya
Department of Computer Science and Engineering, Faculty of Engineering and Technology, Parul Institute of Engineering and Technology, Parul University, Vadodara, Gujarat, India, vipul.vekariya18435@paruluniversity.ac.in

Chapter 5

Praveen Priyaranjan Nayak
Department of Electronics and Communication Engineering, Institute of Technical Education and Research, Siksha 'O' Anusandhan (Deemed to be University), Bhubaneswar, Odisha, India, praveennayak@soa.ac.in

Chapter 6

Ritu Das
Department of Allied Healthcare and Sciences, Vivekananda Global University, Jaipur, Rajasthan, India, ritu.das@vgu.ac.in

Chapter 6

Mamatha V.
Department of Computer Science and IT, Arka Jain University, Jamshedpur, Jharkhand, India, mamatha.v@arkajainuniversity.ac.in

Chapter 7

Tapasmini Sahoo
Department of Electronics and Communication Engineering, Institute of Technical Education and Research, Siksha 'O' Anusandhan (Deemed to be University), Bhubaneswar, Odisha, India, tapasminisahoo@soa.ac.in

Chapter 7

Sharon M.
Department of Computer Science and Engineering, Presidency University, Bengaluru, Karnataka, India, sharon.m@presidencyuniversity.in

Chapter 8

Sweta Kumari Barnwal
Department of Computer Science and IT, Arka Jain University, Jamshedpur, Jharkhand, India, sweta.b@arkajainuniversity.ac.in

Chapter 8

Avinash Samantra
Department of Centre for Data Science, Institute of Technical Education and Research, Siksha 'O' Anusandhan (Deemed to be University), Bhubaneswar, Odisha, India, avinashsamantra@soa.ac.in

Chapter 8

Ajeetsingh
Department of Allied Healthcare and Sciences, Vivekananda Global University, Jaipur, Rajasthan, India, ajeet.singh@vgu.ac.in

Chapter 9

Praveena K N
Department of Computer Science and Engineering, Presidency University, Bengaluru, Karnataka, India, praveenakn@presidencyuniversity.in

Chapter 9

Amit Prakash Sen
Department of Electrical and Electronic Engineering, Arka Jain University, Jamshedpur, Jharkhand, India, dr.amit@arkajainuniversity.ac.in

Chapter 9

Mahesh Singh Danu
Department of Allied Healthcare and Sciences, Vivekananda Global University, Jaipur, Rajasthan, India, mahesh.danu@vgu.ac.in

Chapter 10

Geetha Arjunan
Department of Computer Science and Engineering, Presidency University, Bengaluru, Karnataka, India, geetha.arjunan@presidencyuniversity.in

Chapter 10

Ayesha Taranum
Department of Computer Science and Engineering, Presidency University, Bengaluru, Karnataka, India, ayesha.taranum@presidencyuniversity.in

Chapter 11

Shalinee Pareek
Department of Information Science and Engineering, JAIN (Deemed-to-be University), Bengaluru, Karnataka, India, shalinee.pareek@jainuniversity.ac.in

Chapter 11

Soumya Sharma
Department of Allied Healthcare and Sciences, Vivekananda Global University, Jaipur, Rajasthan, India, soumya.sharma@vgu.ac.in

Chapter 11

Swadhin Kumar Barisal
Department of Centre for Internet of Things, Institute of Technical Education and Research, Siksha 'O' Anusandhan (Deemed to be University), Bhubaneswar, Odisha, India, swadhinbarisal@soa.ac.in

Chapter 12

Sudharsan M.
Department of Information Science, Presidency University, Bengaluru, Karnataka, India, sudharsan.m@presidencyuniversity.in

Chapter 12

Jaimeel Shah
Department of Computer Science and Engineering, Faculty of Engineering and Technology, Parul Institute of Engineering and Technology, Parul University, Vadodara, Gujarat, India, jaimeel.shah@paruluniversity.ac.in

Chapter 13

Santosh Kumar Behera
Department of Centre for Artificial Intelligence and Machine Learning, Institute of Technical Education and Research, Siksha 'O' Anusandhan (Deemed to be University), Bhubaneswar, Odisha, India, santoshbehera@soa.ac.in

Chapter 13

Asif Mohamed H. B.
Department of Computer Science and Engineering, Presidency University, Bengaluru, Karnataka, India, asif.mohamed@presidencyuniversity.in

Chapter 13

A. R. Chayapathi
Department of Information Science and Engineering, JAIN (Deemed-to-be University), Bengaluru, Karnataka, India, chayapathi.ar@jainuniversity.ac.in

Chapter 14

Dimpal Khambhati
Department of Biomedical Engineering, Faculty of Engineering and Technology, Parul Institute of Technology, Parul University, Vadodara, Gujarat, India, dimpalkhambhati@gmail.com

Chapter 14

Sanjay Kumar Jena
Department of Computer Science and Engineering, Institute of Technical Education and Research, Siksha 'O' Anusandhan (Deemed to be University), Bhubaneswar, Odisha, India, sanjayjena@soa.ac.in

Chapter 14

Logashanmugam Edeswaran
Department of Electronics and Communication Engineering, Sathyabama Institute of Science and Technology, Chennai, Tamil Nadu, India, logashanmugam.ece@sathyabama.ac.in

Chapter 15

Vishakha Khambhati
Department of Biomedical Engineering, Faculty of Engineering and Technology, Parul Institute of Technology, Parul University, Vadodara, Gujarat, India, vishakhakhambhati@gmail.com

Chapter 15

Dimpal Khambhati
Department of Biomedical Engineering, Faculty of Engineering and Technology, Parul Institute of Technology, Parul University, Vadodara, Gujarat, India, dimpalkhambhati@gmail.com

Chapter 16

G SUNDARI
Department of Electronics and Communication Engineering, Sathyabama Institute of Science and Technology, Chennai, Tamil Nadu, India, sundari.ece@sathyabama.ac.in

Chapter 16

Ashutosh Roy
Department of Computer Science and IT, Arka Jain University, Jamshedpur, Jharkhand, India, ashutosh.r@arkajainuniversity.ac.in

Chapter 17

Swapnil M. Parikh
Department of Computer Science and Engineering, Faculty of Engineering and Technology, Parul Institute of Technology, Parul University, Vadodara, Gujarat, India, swapnil.parikh17761@paruluniversity.ac.in

Chapter 17

Joany R. M.
Department of Electronics and Communication Engineering, Sathyabama Institute of Science and Technology, Chennai, Tamil Nadu, India, joany.ece@sathyabama.ac.in

Chapter 17

S. Lalithakumari
Department of Electronics and Communication Engineering, Sathyabama Institute of Science and Technology, Chennai, Tamil Nadu, India, lalitha.eni@sathyabama.ac.in

Chapter 18

Divya Paikaray
Department of Computer Science and IT, Arka Jain University, Jamshedpur, Jharkhand, India, divya.p@arkajainuniversity.ac.in

Chapter 18

Suman Sau
Department of Computer Science and Information Technology, Institute of Technical Education and Research, Siksha 'O' Anusandhan (Deemed to be University), Bhubaneswar, Odisha, India, sumansau@soa.ac.in

Chapter 18

A. R. Chayapathi
Department of Information Science and Engineering, JAIN (Deemed-to-be University), Bengaluru, Karnataka, India, chayapathi.ar@jainuniversity.ac.in

Chapter 19

K. V. Karthikeyan
Department of Electronics and Communication Engineering, Sathyabama Institute of Science and Technology, Chennai, Tamil Nadu, India, karthikeyankv.ece@-sathyabama.ac.in

Chapter 19

Rishali
Department of Nursing, Arka Jain University, Jamshedpur, Jharkhand, India, rishali.@arkajainuniversity.ac.in

Chapter 19

Raj Kumar Dhakad
Department of Paramedical Sciences, Vivekananda Global University, Jaipur, Rajasthan, India, rajkumar.dhakad@vgu.ac.in

Chapter 20

Rekha Chakravarthi
Department of Electronics and Communication Engineering, Sathyabama Institute of Science and Technology, Chennai, Tamil Nadu, India, dean.artsandsciences@sathyabama.ac.in

Chapter 20

Krishnamoorthy N. R.
Department of Electronics and Communication Engineering, Sathyabama Institute of Science and Technology, Chennai, Tamil Nadu, India, krishnamoorthy.eni@sathyabama.ac.in

Chapter 21

Mahendra Kumar Verma
Department of Paramedical Sciences, Vivekananda Global University, Jaipur, Rajasthan, India, mahendra.verma@vgu.ac.in

Chapter 21

Ipsita Dash
Department of Centre for Internet of Things, Institute of Technical Education and Research, Siksha 'O' Anusandhan (Deemed to be University), Bhubaneswar, Odisha, India, ipsitadash@soa.ac.in

CHAPTER 1

`Demystifying the Generative AI Paradigm

Manivannan Karunakaran, Suhel Khan, Akash Kumar Bhagat

Abstract The intersection of generative artificial intelligence (AI) and the Internet of Medical Things (IoMT) is a paradigm shift in intelligent healthcare, especially when it comes to the constant supervision and control of chronic illnesses. Long-term supervision and clinical interventions are necessary due to such long-term conditions such as diabetes, cardiovascular disorders, or respiratory diseases, which pose serious challenge to scalable and data-driven healthcare solutions. The IoMT-enabled sensors and medical devices enable the acquisition of physiological and behavioral data (such as vital signs, biochemical indicators, and patterns of activity) in real-time, which leads to large-scale and heterogeneous datasets in healthcare. Generative AI models allow further interpretation of these data streams through learning intricate temporal and context relationships and generate synthetic but clinically relevant healthcare data. Moreover, generative AI can also help to solve the issue of data shortage, data privacy, and system-to-system interoperability in IoMT systems. In systematic way, important uses, ethical issues and research problems are described in order to get a full picture of this new paradigm. This chapter provides the initial background to further investigations of the advanced methodologies and practical applications and the directions of future research of Generative AI based systems of IoMT healthcare.

Keywords Generative artificial intelligence – internet of medical things (IoMT) – chronic disease monitoring – wearable medical sensors – predictive diagnostics

1.1 Introduction

The integration of excellent computational intelligence and healthcare technology has brought a paradigm shift in the management of chronic disease. Generative artificial intelligence (AI), a subdivision of AI that aims to generate new and significant patterns of data based on learnt representations has become a potential weapon in the same area. In contrast to the classical methods of forecasting, the generated AI models, including generative adversarial networks (GANs), variational autoencoders (VAEs), and diffusion-based models, generate synthetic but real-world informational content without losing the statistical and structural features of the original data. This attribute is particularly helpful in the field of healthcare due to the lack of data, privacy, and heterogeneity which pose signifi-

 | HTTPS://DOI.ORG/10.1515/9783112237083-001

cant challenges [1]. Internet of Medical Things (IoMT) devices receive many streams of information, such as vital signs, blood glucose, changes in heart rate, breathing, and physical activity parameters [2]. With the IoMT, the real-time data capture functionality enables continuous monitoring in addition to offering a longitudinal perspective of the health of a patient. Combined with generative AI, IoMT-generated statistics are utilized to create robust models of predictions, identify the signs of an illness onset early, and aid with timely medical care.

Chronic diseases (CDs) such as diabetes, heart diseases, and respiratory diseases are subjecting enormous burden to healthcare systems in every corner of the world. Such conditions require prolonged monitoring and preventive management strategies that will help to prevent complications and improve patient outcomes. The conventional methods of therapy are often based on standard checkups and self-report of patients and this approach may not be effective in preventing any delays [3]. Moreover, variations in the results of individual patients, compliance with treatment, and lifestyle elements make uniform treatments ineffective. These problems are combined through incorporating generative AI and IoMT, which enable ongoing, personalized, and data-driven healthcare tracking [4]. The increased need to implement scalable, efficient, and patient-centered healthcare solutions increases the interest of researchers in exploring generative AI and IoMT convergence. As more and more devices based on the IoMT are introduced, and more healthcare data becomes accessible, there is an unprecedented opportunity in utilizing advanced AI methods to enhance patient care [5]. Within the framework of inference, the integration of generative AI and the IoMT has game-changing possibilities of chronic disease management [6]. In the field of inference, there is the understanding of the basic concepts of generative AI, assessment of its application in the healthcare ecosystem based on IoMT, an evaluation of the technological, ethical, and regulatory challenges in generative AI usage. The combination of a continuous data capture, a strong data modeling, and synthetic data synthesis makes this paradigm of healthcare interventions more precise, predictive, and personalized. These technologies, their applications, and challenges and opportunities in the current advanced healthcare systems are discussed in the following section.

Nowadays, practically anything with an internet connection uses the Internet of Things (IoT) to transport and access data. IoT has also done well in the healthcare industry by streamlining diagnostic processes and guaranteeing more effective patient monitoring. IoT has therefore transformed and upended almost every sector of the modern economy, from logistics management to academia. Automation is made easier by object detection in a number of ways, including recognizing an object, categorized product tracking, and characteristic recognition such as color, characteristics, and fine details of a specific object [7].

1.2 IoMT-Enabled Data Acquisition and Management

The data of IoMT is heterogeneous, dynamic, and obtained with different medical sensors and technologies. These data are also heterogenous in that they are of different formats, scales, accuracy and time, and these factors make them difficult to integrate, analyze, and make real time decisions in advanced medical centers. The sensors, communication gadgets, and the communication back-end systems are expensive and cannot be used on a large-scale basis. In order to reduce such costs, the shared-node IoT design is adopted, whereby the costly sensors and multiple users share networking facilities. This plan predisposes healthcare monitoring technologies to be more easily available [8].

1.2.1 Reliability

Accuracy and consistency are the key requirements with the health care monitoring system as medical decisions are made based on the correct and uninterrupted data. The breakdown of devices, network errors, and loss of data is all potentially dangerous to health. Body area networks enhance reliability because they identify communication faults and automatically reconfigure data transmission paths. Moreover, fault tolerance and suppression are also used to ensure continuous monitoring in the event of failure of the system.

1.2.2 Interoperability

Curriculum healthcare monitoring systems are often devices and platforms designed by numerous vendors based on various standards and protocols. This inconsistency makes it difficult to attain a seamless communication and data-sharing. The interoperability also facilitates the consistency of all the connected devices, sensors, and systems to communicate with each other, which is essential in sharing data accurately, delivering coordinated care, and running the systems efficiently.

1.2.3 Energy Limitations

Examples of IoT-based healthcare devices powered by batteries include wearable sensors, wireless nodes, and smartphones, which have a limited operating life due to their battery power. The energy consumption caused by frequent data transfer and constant monitoring can cause failure in the device when it is desperately needed [9]. To solve this problem, unnecessary data transfer and processing must be minimized and technology that is energy efficient in communication needs to be employed so that continuous and reliable surveillance is made possible.

IoT Solves the Problem of CDs: IoT is an opportunity to monitor CDs and patients constantly and in distance: The smart devices are able to transmit the physiological information about patients to the healthcare systems and healthcare specialists on a regular basis. The ability helps in early diagnosis and effective management of chronic conditions such as diabetes, cardiovascular conditions, asthma, obesity and cancer.

IoT-Based Healthcare Monitoring of CD Inspiration: The fact that the IoT gets to know the location of patients, medical staffs, and medical equipment within healthcare facilities is what is inspiring its application in monitoring chronic illnesses. Patient identification prevents treatment delays, pharmaceutical mistakes, and wrong dosage. Verification of medical practitioners will increase system security and confidence. It is the wearable and implantable gadgets that are smart in capturing and transmitting patient health data in real-time, as they enable safer, more reliable means of healthcare, and better quality of care to chronic patients.

Innovations in Healthcare Monitoring Systems to Deliver High-Quality Services: The IoT has enhanced the level of healthcare surveillance technologies significantly because they enable one to observe patients remotely and in real-time using sensors, tags, and actuators. Through such technologies, healthcare providers are able to track patient's status, medical devices as well as treatment processes remotely. Monitoring through IoT improves healthcare efficiency, reduces medical risks, and allows responding to an emergency faster. Figure 1.1 presents IoT-based sensor deployment within the healthcare environments.

IoT Building the Healthcare Monitoring of the Future: The IoT is transforming future healthcare monitoring as it delivers continuous, intelligent, and connected care, especially chronic disease care. Smart devices, sensors, smartphones, actuators, and medical equipment are used in IoT to monitor patients, medical workforce, and medical resources in real time. These interwoven systems allow identification of patients, tracking them and constant monitoring of patients no matter their location. The communication technologies used by IoT-enabled medical devices to collect, store, and analyze huge volumes of health data enable home-based and clinical monitoring in the case of chronic patients [10].

1.3 Challenges, Open Issues, and Future Directions of Healthcare Monitoring

In this chapter, the outstanding problems in the healthcare monitoring systems are discussed even though there have been improvements in the IoT technology. IoMT-based devices hold energy efficiency by designing them so that they use energy as efficiently as possible at each device or node sensor as well as the energy use in the entire network during its life cycle. This is specifically important to

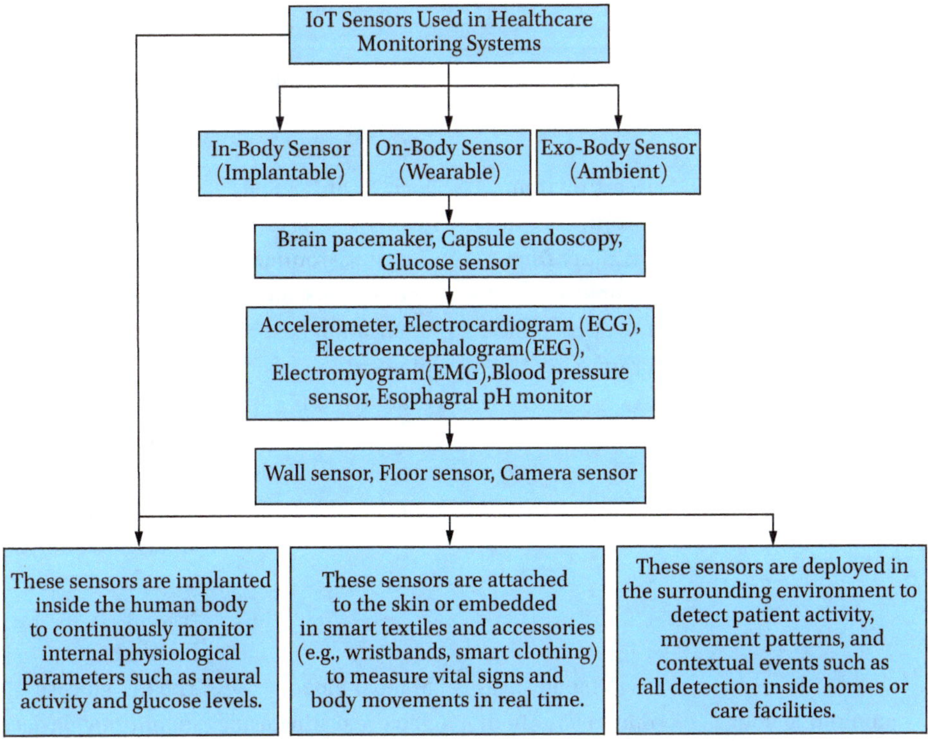

FIGURE 1.1 IoT sensors used in healthcare monitoring systems. (Source: Own elaboration)

surgical equipment in systems powered by IoMT, as power supply is the primary source of energy [11]. In order to overcome these limitations, future healthcare surveillance systems should be flexible, interoperable, energy-efficient, and context-sensitive so that they can be used to provide consistent, continuous, and patient-centric healthcare delivery in the field. All the issues, causes, traditional therapies, and how the IoT services enable continuous monitoring and timely therapy are brought to the fore in Table 1.1 which analyzes CDs.

1.4 Healthcare Data Characteristics in IoMT Systems

The biomedical measurement systems (BMS) play a crucial role in the contemporary healthcare as they constantly monitor vital signs and allow implementing required medical actions. Conventional BMS have often been problematic as they are intrusive, non-portables, or disruptive to normal operations. In addressing such limitations, the incorporation of the IoT has led to the development of the IoMT that enables wearable and implantable devices to access, process, and transmit health data with ease. IoMT-based BMS facilitates real-time monitoring, clinician

TABLE 1.1 Different types of chronic diseases, their challenges, and IoT solutions.

Types of CD	Associated challenges	Anomalies and reasons	Traditional solution	IoT solution
Cardiovascular	Monitoring cardiovascular conditions such as arrhythmia, heart rate anomalies, and blood pressure fluctuations is critical due to high mortality risk	Heart rhythm disorder is caused by injury from complications due to vessels of heart or muscle.	Electrocardiogram (ECG) testing in hospitals, routine doctor visits, and medication based on physical examination	The wearable ECG sensors have IoT capabilities to monitor the heart activity at any given moment. The information is sent to healthcare professionals so that it can be diagnosed and treated in real time to minimize risks of hospitalization.
Diabetes	Constantly monitoring your blood glucose levels can help prevent severe complications such as nephropathy (kidney problems) and neuropathy (nerve problems) as well as heart disease	High blood sugar is due to your body unable to produce or use enough insulin.	Manual blood glucose testing using glucometers and periodic clinical checkups	The IoT glucose monitoring devices help to track real-time sugar levels and alert patients and doctors, which will further allow managing insulin and controlling lifestyles.
Asthma	The abrupt onset or worsening of shortness of breath or the ability to control asthma attacks poses a serious risk to life.	Low oxygen levels and airway inflammation trigger acute asthma episodes.	Inhalers, nebulizers, and hospital emergency treatment	The sensors of wearable IoT can monitor the patterns of breathing and oxygen level and send an emergency notification to the healthcare workers during attacks.

Source: Own elaboration

remote access, and prompt emergency alarms, therefore, improving patient safety, independence, and efficiency of healthcare. Wearable devices capture circulatory data in the context of heart disease management, which is processed by clever algorithms to forecast future dangers and create preventative care warnings. Overall, IoMT-enabled BMS offers a dependable, user-friendly, and continuous monitoring framework that improves healthcare delivery, lowers hospitalization costs, and facilitates custom-made patient care in a variety of settings, including hospitals, residences, and remote locations [12]. Figure 1.2 shows the intelligent IoMT framework with generative AI for continuous health monitoring.

1.5 Generative Artificial Intelligence: Fundamentals

Generative AI models are rapidly altering healthcare by allowing the generation, exploration, and augmentation of medical data across many applications. The models such as (i) GANs, (ii) diffusion models, large language models (LLMs), and (iii) VAEs aid comprehending complex healthcare data by producing realistic medical images, synthetic patient records, and significant clinical insights. GANs are commonly employed to generate high-quality medical images and enhance restricted datasets, whereas diffusion models produce highly detailed and stable outputs via repeated noise refinement. Transformer-based LLMs provide clinical text analysis, decision-making, and expertise retrieval from huge medical data. VAEs provide efficient data organization and synthesis, which is particularly critical for dealing with data shortages and uncommon disease analysis. Together, these chapters enhance diagnostic accuracy, enable data-driven medical decisions, and increase the dependability of intelligent medical facilities [13].

GANs: The ability of GANs to produce excellent and realistic synthetic data has made them one of the significant generative AI models in healthcare. Recent advances demonstrate that GANs are widely used in healthcare image creation, segmentation, high-resolution, and multidimensional translation, specifically for MRI, CT, medical ultrasound, and histopathological imaging. They handle major concerns such as data scarcity, disparity in class, and privacy protection by producing synthetic images and patient records effectively. GAN-based frameworks have also shown excellent performance in illness diagnosis, by including brain tumors, cardiovascular problems, cancer, COVID-19, and neurological illnesses, as well as clinical decision support and therapy planning. Overall, GANs are particularly useful in improving the reliability of diagnosis, enriching the data, and performing useful medical analytics, thus being suitable in complex healthcare data analysis tasks.

LLMs: This chapter discusses the recent efforts to enhance medical practice with the help of generative AI and LLMs that also incorporates clinical administra-

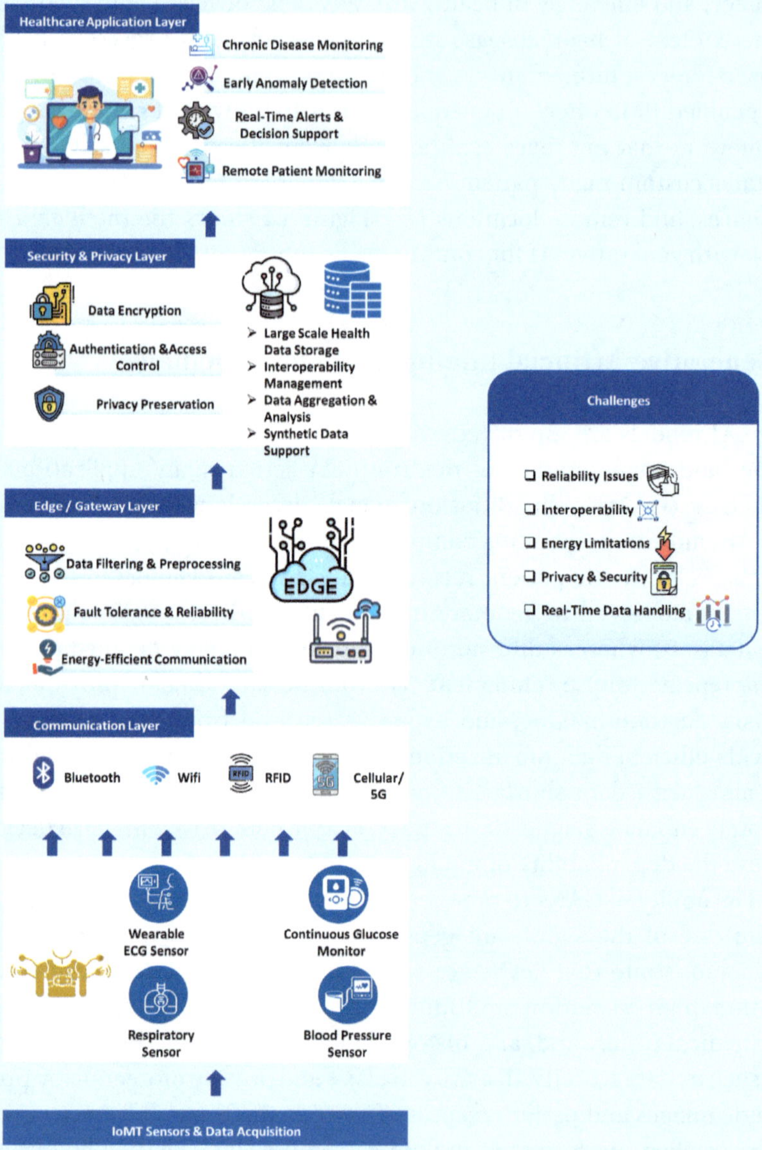

FIGURE 1.2 End-to-end IoMT architecture enabling secure, intelligent, and personalized healthcare services. (Source: Own elaboration)

tion, decision assistance, patient involvement, and data generation. They can optimize patient services, medical research, and reduce the workload of medical professionals by automating routine activities, making information more accessible, and obtaining more data. Generative AI is useful in the clinical administration by automating clinical documentation, recording, and coding. Virtual assistants, ChatGPT-based, Microsoft Co-pilot, Nuance, Suki, and Corti allow clinicians to make accurate notes, summarize meetings, transcribe conversations, and assign medical codes, reduce administrative stress and spend more time with patients.

Such clinical decision support systems as Glass AI, Regard, Redbrick AI, Paige FullFocus, and Kahun assist physicians in creating differential diagnoses, providing imaging analysis, and being interoperable with electronic health records. The tools enhance the accuracy of the diagnostic, speed up the workflow, and make the right decisions and complement the experience of clinicians. Hippocratic AI and Gridspace are empathy-based AI systems that enhance patient engagement. The tools are 24-h contact, automatized outreach, responding to patient inquiries, and administrative interaction, which enhances accessibility, confidence, and trust in medical care. Privacy-preserving exchange and upgrading can be done by synthetic data creation with models such as Syntegra and DALL · E 2. Artificial fabricated patient data and medical images assist in overcoming information gaps, reducing bias, researching, and training the AI model without violating patient confidentiality.

VAE: VAEs are helpful generative models that are used to learn an appropriate model of healthcare data and to generate realistic synthetic samples. They facilitate in capturing of complex data distribution and safeguard key clinical and architectural data. The use of VAEs has been typical in the production of medical images, data enhancement, and feature estimation to enhance the efficiency of the model in areas where limited data are available. They have successfully been applied in the analysis of brain tumors and the categorization of lung diseases. Diffusion models, VAEs, and LLMs, such as GANs, allow sensible, explainable, and trustworthy medical analytics [14].

1.6 System Architecture of Generative AI-Enabled IoMT Frameworks

Generative AI-enabled IoMT system architecture adopts a layered design to efficiently collect, transmit, and analyze healthcare data while supporting intelligent decision-making. This architecture enhances traditional IoMT by embedding generative AI capabilities across layers to enable predictive analytics, automation, and personalized healthcare services.

Perception layer: The perception layer to gather continuous health data from sufferers uses wearable, implanted, and ambient sensors. Cardiovascular temperature, and behaviors are merely some of the physiological, biochemical, and behavioral information that the gadgets record. Technologies for short-range communication allow for smooth data collection without interfering with normal tasks. In a generative AI-enabled IoMT system, this layer provides high-quality input data that serves as the foundation for intelligent learning, data synthesis, and early health risk detection.

Network Layer: The network layer ensures reliable and secure transmission of medical data from the perception layer to processing units. It integrates heterogeneous communication networks such as wireless, mobile, and internet-based systems to support continuous connectivity. This layer also supports interoperability and data integration, therebyenabling Generative AI models to access unified healthcare data streams for real-time analysis, prediction, and anomaly detection across distributed edge and cloud platforms.

Application layer: The application layer focuses on data analysis, visualization, and decision support for clinicians and patients. Generative AI models at this layer analyze medical data to generate predictive insights, personalized treatment recommendations, and timely alerts. This layer supports intelligent healthcare services such as disease prediction, remote monitoring, and clinical decision-making, thereby improving care quality, efficiency, and patient outcomes [15].

Radio-Frequency Identification (RFID): IoMT-based RFID tags are used to track supply prices and problems with pharmaceutical availability. The Food and Drug Administration (FDA) has made suggestions about RFID and medication logistics. For instance, adding labels to pharmaceutical packaging enables manufacturers to guarantee supply chain reliability.The wireless object recognition method known as RFID uses radio waves for transmission over incredibly tiny distances. RFID readers, which are reading devices, and radio frequency tags, which are radio transmission devices, are the two fundamental parts of RFID interactions [16]. Table 1.2 shows IoMT-enabled solutions for patient monitoring and care.

1.6.1 Cloud-Enabled IoMT Overview

Conventional IoMT systems have issues relating to linking various medical institutions and blending heterogeneous medical data resulting in isolated information silo, duplication of data, and wastage of resources. Another reason as well is that the massive volume and complexity of medical data augment storage, processing, and management requirements. Cloud computing provides high reliability, scalability, and sharing of data, through the capability of supporting privacy and informed decision-making in addition to providing secure and efficient as well as interconnected medical services [17].

TABLE 1.2 Major applications of the IoMT in smart healthcare.

Application area	Description	Key benefits
Patient identification and security	Discreetly identifies patients and handles sensitive health information using biometric systems (facial, fingerprint, and vein identification)	Increases information safety, decreases individual mistakes, and recovers faith
Vital signs monitoring	Makes use of IoMT devices that continuously monitor vital signs such as heart rate, temperature, and cholesterol levels	Allows initial analysis, actual well-being valuation
Personalized healthcare	Supports individualized assessment and therapy planning by analyzing patient-specific data	Recovers action correctness and enduring consequences
Telemedicine and mobile health	Allows for distant surveillance and discussions via laptops, tablets, and cellphones	Decreases infirmary appointments, advances admission to health care
Wearable healthcare devices	Includes bracelets and timepieces, smart bracelets, and embedded sensors for ongoing surveillance	Augments luxury, flexibility, and patient-medic organization
Hospital workflow optimization	Enhances hospital administration of resources and productivity via networked systems	Decreases prices, recovers competence, and facilitates excellence

Source: Own elaboration

1.6.2 Medical Service Layer

The medical service layer involves provision of computing, storage, and networking resources by means of virtualization. It has infrastructure, platform, and the software sub-layers. The infrastructure layer can provide scalable digital assets to hospitals, not requiring the construction of personal data centers. The platform layer helps developers to develop and scale medical applications without any problem. The software layer supports the end-users who get functional applications, and upheld the privacy of data and secure access by various users.

1.6.3 Service Management Layer

The application administrative layer is the basis of IoMT based on the internet. It ensures that the system is operating in a safe, stable, and efficient manner. This layer controls the software maintenance, resource distribution, and user interactions. It gathers, filters, and releases medical software and provides the communication among the users and developers.

1.6.4 User Layer

The end user layer consists of hospitals, clinics, disease control centers, and patients. The cloud is accessed by users (either smartphone or computer) and they share resources to provide collaborative care. Physicians will be able to access patient information, treatment schedules, and real time monitoring, and patients will be able to select appropriate medical services. Specialists in various facilities also work together on complicated cases resulting in better treatment success and the ability to share resources via the cloud network.

Edge-Cloud Integration: In IoMT systems, the use of cloud computing alone results in high latency, energy usage, and network congestion because of the huge medical data transmissions of wearable devices and hospital devices. Edge-cloud integration solves this through processing important data in the edge nodes, and storage as well as analytics in the cloud. The architecture consists of three layers: The terminal layer is equipped with IoMT devices that gather patient information, the edge layer is composed of immediate smart devices or gateways that enable the processing of the information to be fast, secure, and the cloud layer is a centralized storage, advanced analytics, and resource-intensive computations. Edge computing will facilitate the distributed processing of data at the source, lessening the latency, increasing the security, and enabling autonomous local operations even in the case of network interruptions. Variants such as multi-access edge computing, micro-cloud computing, and fog computing provide the flexibility of deployment according to computing power and application requirements. The edge cloud integration is the technique of making heavy computational loads on mobile or wearable devices and offloading them in edge servers or cloud servers to conserve energy and storage. Resource allocation and intelligent decision-making with the help of such techniques as game theory and optimization algorithms guarantee efficient work with the task, latence balance, and scalable, robust healthcare analytics.

1.7 Analytics: Predictive and Clinical Decision Support

Generative AI models are also revolutionizing the study of diseases by creating certain types of cells that would otherwise be hard or slow to access. The models make the process of researchers studying the biology of cells, comprehending how diseases occur, and trying possible treatments more efficient. Using the AI models, one can determine concealed trends and early signs of a health condition by comparing vast volumes of data on medical records, imaging reports, and wearable devices. Random forests, support vector machines, decision trees, and deep learning are some of the methods that can be used to predict the risk of diseases and assist in the early diagnosis. Besides this, natural language processing can be used to extract meaningful information in the notes of the doctors. On the whole, such systems allow making timely, effective, and data-oriented healthcare decisions, enhancing the outcome of patients and minimizing the cost of treatment.

Predictive Analytics and Early Intervention Strategies to Patient Outcomes: Early intervention strategies apply predictive analytics and generative AI to target health risks early and assist in making decisions on time [18]. These models are able to predict the onset and progression of diseases by examining the patient data related to health records, test outcomes, and physiological indicators.

1.8 Conclusion

The potential of generative AI-enabled IoMT systems in medical is outlined as a revolutionary one in this chapter. Developing an edge-cloud integrated IoMT architecture, applying predictive analytics to detect diseases early, providing individualized therapy, and managing a lack of data and preserving privacy are some of the key achievements. The chapter also focuses on ethical, privacy, and security of patient data. In general, the combination of AI, IoMT, and higher computing provides a way to a smarter, more efficient, and patient-centered healthcare, and the future of AI research should be explainable AI and the complete connection of smart healthcare systems.

References

[1] Radanliev P. Artificial intelligence: reflecting on the past and looking towards the next paradigm shift. Journal of Experimental & Theoretical Artificial Intelligence. 2025;37(7):1045–62. https://doi.org/10.1080/0952813X.2024.2323042.

[2] Sahoo S. Demystifying AI-enabled information systems for smarter business decisions. Paradigm. 2024;28(1):119–24. https://doi.org/10.1177/09718907241243112.

[3] Stoumpos AI, Kitsios F, Talias MA. Digital transformation in healthcare: technology acceptance and its applications. International Journal of Environmental Research and Public Health. 2023;20(4):3407. https://doi.org/10.3390/ijerph20043407.
[4] Garcia-Perez A, et al. Resilience in healthcare systems: cyber security and digital transformation. Technovation. 2023;121:102583. https://doi.org/10.1016/j.technovation.2022.102583.
[5] Talpur MSH, et al. Illuminating healthcare management: a comprehensive review of IoT-enabled chronic disease monitoring. IEEE Access. 2024;12:48189–209. https://doi.org/10.1109/ACCESS.2024.3382011.
[6] Mathkor DM, et al. Multirole of the internet of medical things (IoMT) in biomedical systems for managing smart healthcare systems: an overview of current and future innovative trends. Journal of Infection and Public Health. 2024;17(4):559–72. https://doi.org/10.1016/j.jiph.2024.01.013.
[7] Rani P, et al. Simulation of the lightweight blockchain technique based on privacy and security for healthcare data for the cloud system. International Journal of E-Health and Medical Communications (IJEHMC). 2022;13(4):1–15. https://doi.org/10.4018/IJEHMC.309436.
[8] Yadav SP, et al. An improved deep learning-based optimal object detection system from images. Multimedia Tools and Applications. 2024;83(10):30045–72. https://doi.org/10.1007/s11042-023-16736-5.
[9] Yu P, et al. Leveraging generative AI and large language models: a comprehensive roadmap for healthcare integration. Healthcare. 2023;11(20):MDPI. https://doi.org/10.3390/healthcare11202776.
[10] Abbasian M, et al. Foundation metrics for evaluating effectiveness of healthcare conversations powered by generative AI. NPJ Digital Medicine. 2024;7(1):82. https://doi.org/10.1038/s41746-024-01074-z.
[11] Bhushan B, et al. Towards a secure and sustainable internet of medical things (IoMT): requirements, design challenges, security techniques, and future trends. Sustainability. 2023;15(7):6177. https://doi.org/10.3390/su15076177.
[12] Zhang P, Boulos MNK. Generative AI in medicine and healthcare: promises, opportunities and challenges. Future Internet. 2023;15(9286.). https://doi.org/10.3390/fi15090286.
[13] Godwin RC, Melvin RL. The role of quality metrics in the evolution of AI in health care and implications for generative AI. Physiological Reviews. 2023;103(4):2893–95.
[14] Reddy S. Generative AI in healthcare: an implementation science informed translational path on application, integration and governance. Implementation Science. 2024;19(1):27. https://doi.org/10.1186/s13012-024-01357-9.
[15] Mazhar T, et al. Generative AI, IoT, and blockchain in healthcare: application, issues, and solutions. Discover Internet of Things. 2025;5(1):5. https://doi.org/10.1007/s43926-025-00095-8.

[16] Askar NA, et al. Architecture, protocols, and applications of the internet of medical things (IoMT). Journal of Communications. 2022. https://doi.org/10.12720/jcm.17.11.900-918.

[17] Tasnim F, Nilima SI. Harnessing predictive analytics: The role of machine learning in early disease detection and healthcare optimization. Journal of Ecohumanism. 2025;4(3):312–21. https://doi.org/10.62754/joe.v4i3.6642.

[18] Recharla M, et al. Revolutionizing Healthcare with Generative AI: Enhancing Patient Care, Disease Research, and Early Intervention Strategies. American Journal of Psychiatric Rehabilitation. 2025;28(1):98–111. https://doi.org/10.69980/ajpr.v28i1.70.

CHAPTER 2

Sensor Fusion Methods

Kalpesh Jadav, Manivannan Karunakaran, Shet Reshma Prakash

Abstract The rising global prevalence of chronic diseases, including cardiovascular disorders, diabetes, and respiratory illnesses, presents significant challenges to healthcare systems and emphasizes the need for technologies that enable continuous monitoring and timely clinical decision-making. This chapter synthesizes recent research and innovative approaches that integrate artificial intelligence (AI) with Internet of Medical Things (IoMT) to support real-time disease monitoring, diagnosis, and management. The IoMT ecosystems incorporate diverse sensors and devices, such as physiological sensors, wearable instruments, medical scanners, and intelligent monitoring systems, embedded in clothing, accessories, or clinical environments. These devices continuously acquire vital health parameters, including blood pressure, blood glucose levels, heart rate, respiratory signals, and related indicators. Robust analytical frameworks are required to manage and interpret these heterogeneous data streams, where sensor fusion techniques play a critical role in combining multimodal information and enhancing reliability. AI-driven analytics and generative intelligence models further facilitate complex data interpretation, address uncertainty, and extract actionable clinical insights. Moreover, AI-based data generation and augmentation techniques improve model robustness, generalization, and diagnostic accuracy. By addressing challenges associated with real-time data processing, scalability, and heterogeneity, the integration of AI and sensor fusion within IoMT infrastructures enables personalized, proactive, and preventive healthcare delivery for modern healthcare.

Keywords Chronic disease monitoring – sensor fusion – wearable health sensors – real-time healthcare analytics – smart healthcare systems

2.1 Introduction

Sensor fusion enhances the accuracy, robustness, and reliability of different intelligent systems by combining the information of multiple sensors to reduce uncertainty. Intelligent and connected patient care is being incorporated with sensor fusion with artificial intelligence (AI) and Internet of Medical Things (IoMT) due to the rising number of chronic diseases (CDs) in healthcare that needs constant monitoring and real-time decision support.

2.1.1 Global Burden of Chronic Diseases

Notably, the rates of CDs have risen during the past decades worldwide and it causes tremendous burden on the systems of healthcare. Noncommunicable diseases, including cardiovascular diseases, diabetes, chronic respiratory diseases, and hypertension, constitute a large proportion of the morbidity, mortality as well as long-term health expenditure of the entire global population [1]. Such conditions make it imperative to employ continuous and longitudinal examination and tracking, and not the utilization of infrequent clinical visits, since health conditions of the patients and views about the disease are dynamic and reliant on behaviors of the lifestyle, environmental status, and treatment compliance. This increase in the healthcare burden explains the need to urgently adopt scalable, cost-effective, and patient-centered monitoring systems [2].

2.1.2 Role of Continuous Monitoring and Real-Time Decision Support

The stress of managing CDs demands real time decision support and incessant surveillance. Heart rate (HR), blood glucose levels (BGLs), blood pressure (BP), oxygen saturation (SpO_2), and physical activity are some of the physiological parameters that are continuously tracked due to wearable devices, implantable devices, and home monitoring systems. These data streams when used with real-time analytics enable early identification of abnormalities, clinical action, and treatment modification at an individual level. The decision support systems can even be more helpful by providing physicians and patients with real-time data on the decision, as this will turn raw sensor data into useful information that contributes to improved clinical outcomes and fewer unnecessary hospitalizations [3].

2.1.3 Motivation for Integrating Sensor Fusion with AI and IoMT

The necessity to derive meaningful data and real information out of various and often noisy sources of health data is the motivation behind sensor fusion integration with AI and the IoMT. Sensor fusion techniques are what enable the aggregation of multi-modes sensor data and enhance patient health state contextual comprehension, accuracy, and robustness. IoMT provides the connectivity model of the data exchange between devices, healthcare providers, and cloud-based platforms without disruption, and AI-based models utilize this integrated data to conduct predictive analytics, risk stratification, and adaptive decision-making. The three concepts, AI, IoMT, and sensor fusion collaborate to form a synergistic system that enables intelligent, real-time, and customized delivery of healthcare to handle CDs [4]. Table 2.1 captures the most common reasons why AI-enabled sensor fusion and IoMT should be incorporated in CD management and also identifies the various technological aspects that facilitate the ongoing surveillance and real-time decision-making. It demonstrates how IoMT, sensor fusion and, in general, AI can be used to overcome the obstacles to the treatment of CD to provide personalized and proactive healthcare results.

TABLE 2.1 Key motivations and enabling technologies for AI-enabled sensor fusion in chronic disease management.

Aspect	Description
Global burden of CDs	The rising prevalence of diabetes, respiratory disorders, cardiovascular diseases, and hyper-tension requires long-term and continuous healthcare management.
Need for continuous monitoring	Chronic conditions demand ongoing assessment of physiological parameters rather than episodic clinical visits due to dynamic health variations.
IoMT-based data acquisition	Wearable, implantable, and home-based IoMT devices enable continuous collection of vital health signals such as HR, BP, BGL, and SpO_2.
Role of sensor fusion	Integrates heterogeneous and multimodal sensor data to reduce uncertainty, improve robustness, and provide a comprehensive view of patient health status.
Role of AI	AI models analyze fused sensor data to perform predictive analytics, risk stratification, and real-time decision support.
Real-time decision support	Enables early detection of abnormalities, timely clinical intervention, and personalized treatment adjustments.
Healthcare outcomes	Improves monitoring accuracy, reduces avoidable hospitalizations, and supports personalized, proactive, and preventive chronic disease care.

Source: Own elaboration

2.1.4 Aim of the Research

The proposed study will address the research question of how it is possible to integrate AI-based sensor fusion algorithms into IoMT systems to track chronic illnesses. It tries to analyze the means of incorporating heterogeneous physiological data to facilitate real-time and precise clinical decision-making. Moreover, the chapter will attempt to address the issue of data heterogeneity, uncertainty, and system scalability in IoMT environment. Lastly, it is committed to the advancement of personal, proactive, and preventive healthcare services.

2.1.5 Reset of the Study

There are eight sections of this chapter. Section 2.1 provides the rationale behind AI-based sensor fusion in IoMT to manage CDs. The sensors and data sources of the IoMT are discussed in Section 2.2 and the data, feature, and decision-level sensor fusion in Section 2.3. Section 2.4 discusses about sensor fusion architecture of the IoMT systems, while Section 2.5 describes the multimodal fusion AI techniques. Section 2.6 presents some of its uses in CD monitoring, Section 2.7 concerns itself with privacy and security issues, and Section 2.8 considers predictive and preventive healthcare systems. Section 2.9 recaps the main findings, explains the current limitations, and identifies future research opportunities relating to AI-assisted sensor fusion in IoMT-based healthcare.

2.2 IoMT: Sensors and Data Sources

In this section, the authors of the chapter describe the IoMT sensors and data sources that facilitate continuous health monitoring within AI-enabled healthcare systems. It concentrates on wearable and physiological medical sensors that yield multimodal and real-time health data that is vital in monitoring of CDs. The diversity, large volume, and time-sensitive nature of IoMT data are highlighted, due to the issues of integration, latency, and devices. Security, regulatory, and privacy concerns of sensitive medical information are also covered. On the whole, the section provides the background on sensor fusion by stating the need to adopt ingenious methods to ensure that the IoMT health data is implemented in the most accurate, secure, and efficient way.

2.2.1 Physiological and Wearable Medical Sensors

The objective and continuous measurement of the internal states of individuals by means of physiological sensors makes the real-time emotional and cognitive responses available, which can determine the choice of the travel mode. Physiological measures such as Galvanic Skin Response (GSR) unlike self-reported measures consider the instant alterations in stress and emotional arousal by variation in skin conductance which is controlled by the sympathetic nervous system. The time-resolved measurements are quite consistent with real travel experiences, and therefore, stress and arousal could be modeled as latent variables to mode choice models. Travel behavior analysis allows the inclusion of physiological data, providing a step beyond traditional observable travel behavior variables such as cost and travel time, to incorporate experiential and psychological factors, which raises an explanatory depth and realism in forecasting behavior of modal choice models [5].

2.2.2 Characteristics and Challenges of IoMT Health Data

The IoMT health data are highly heterogeneous and multisource because they originate due to a very large variety of medical sensors, wearable devices, and clinical systems working with different data formats, sampling rates, and communication protocols [6]. With such heterogeneity, data integration and data consistency is not easy, especially when a combination of several data streams needs to be combined to realize a harmonious picture of the health condition of a patient. Moreover, IoMT system is a continuous source of large amounts of real-time information, including vital signs and physiological measurements, which pose substantial requirements to transmission bandwidth, storage, and processing speed. It is important to make sure that there is a low latency and consistent delivery of data because it is important in providing timely clinical decisions, particularly in remote and continuous monitoring cases.

Medical information is the most sensitive and this poses serious issues associated with privacy and security as well as regulatory concerns. Multimode communication environments, which combine industrial IoT and medical IoT technologies, simultaneously complicate the process of making secure transmission and data loss or data tampering prevention. Network variability and device constraints may also have an effect on data synchronization and quality. Such challenges underscore the importance of ingenious integration approaches and dependable data transfer strategies toward the accurate, secure, and efficient utilization of IoMT health data. Figure 2.1 demonstrates a multimodal sensor fusion system powered by AI, which combines the heterogeneous data of IoMT through machine learning (ML) and deep learning (DL)-based feature- and decision-level fusion. It assists stress and health inference, disease monitoring, clinical decision support, and alerting of CD in real time.

2.3 Levels of Sensor Fusion (Data, Feature, and Decision)

This section describes how AI-enabled IoMT healthcare systems combine heterogeneous physiological data using three hierarchical levels of sensor fusion: feature-level, data-level, and decision-level. To manage sensor availability and reliability, adaptive weighting techniques are used to directly merge raw sensor values in data-level fusion. To increase robustness and applicability for ML models, the

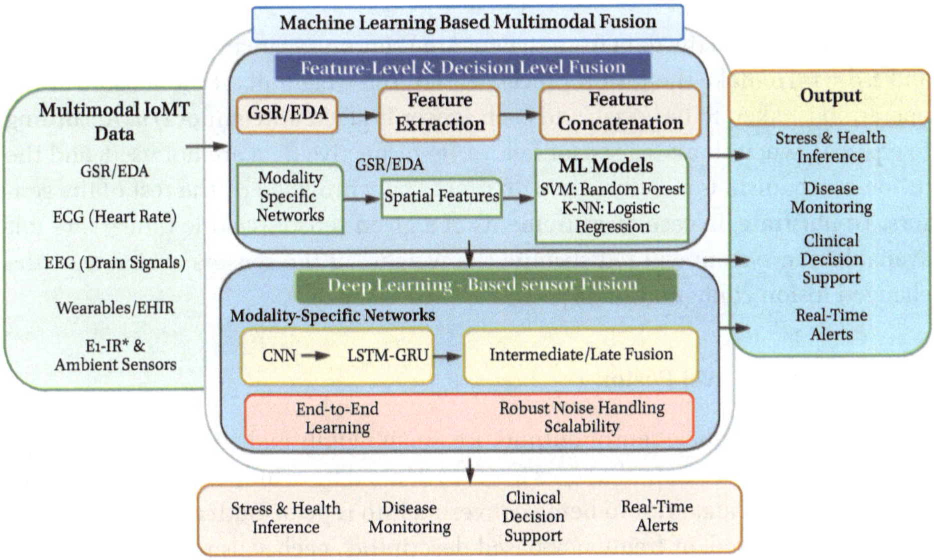

FIGURE 2.1 AI-enabled multimodal sensor fusion framework for IoMT-based chronic disease monitoring and decision support. (Source: Own elaboration)

next section goes over feature-level fusion, which focuses on the extraction and integration of informative features from various sensor data. For the purpose of accomplishing scalable and dependable health monitoring, finally decision-level fusion is introduced. This involves aggregating individual sensor decisions using voting-based or probabilistic algorithms. Collectively, this section demonstrates how various fusion levels manage issues with fault tolerance, data heterogeneity, uncertainty, and real-time clinical decision support in IoMT contexts.

2.3.1 Data-Level Fusion

Let $A = \{\alpha_1, \ldots, \alpha_m\}$ denote a set of heterogeneous IoMT sensors measuring physiological parameters (e.g., HR, BP, BGL, or respiratory signals), and let $\Gamma = \{\gamma_1, \ldots, \gamma_m\}$ represent the corresponding fusion weights, eq. (2.1) shows $\Sigma\,\gamma = 1$. Under normal operating conditions, when all sensors are functioning correctly, the fused system output is computed as:

$$o = \{\alpha_1\gamma_1, \ldots, \alpha_m\gamma_m\} \tag{2.1}$$

If one or more sensors become unavailable during operation due to communication loss, noise, or device failure, the fusion process is adaptively reconfigured. In such cases, the system output is updated as:

$$o' = \{\beta_1\gamma_{n1}, ..., \beta_n\gamma_{ni}\} \quad (2.2)$$

where β_i represents the data from the working sensors, γ_i are the updated weights, and $\Sigma\beta = 1$. To make the fusion process easier, the weight of all the sensors in the operation is taken to be equal, and each sensor is given a weight of $1/n$. According to eq. (2.2), when there is a sensor failure, its respective data are not used, and the fusion mechanism is recalibrated with inputs only provided by the rest of the sensors. To illustrate, in case measurements of a given sensor (calorie intake) are not available, the system will redistribute the weights of the sensors available to the changed fusion configuration [7].

2.3.2 Feature-Level Fusion

Heterogeneous medical sensor outputs are meaningfully and effectively combined due to feature-level fusion that uses features derived from individual sensor signals rather than raw data. Prior to being converted into representative features such as statistical, temporal, or frequency-based descriptors, each sensor's data is preprocessed. A comprehensive feature vector that captures complementary physiological data from various sensors is then created by combining these features [7]. This fusion level is ideal for ML and ensemble models because it lowers noise and data dimensionality, while maintaining key patterns. In AI-enabled IoMT environments, feature-level fusion enhances diagnostic robustness and accuracy by taking advantage of inter-sensor correlations, enabling dependable health monitoring and decision-making.

2.3.3 Decision-Level Fusion

The highest level of sensor fusion is known as "decision-level fusion," in which every IoMT sensor performs local classification on its own, and only the decisions that are produced are combined to produce a global result [8]. Because it lowers communication overhead, improves privacy, and allows scalable distributed monitoring, this strategy is ideal for AI-enabled IoMT healthcare systems. Each sensor assigns the observed physiological data to one of N_c health states and transmits decision information in one of the three forms: soft-level (probability vectors), ranked-level (ordered class labels), or label-level (single class label). Fusion is performed using simple, non-trainable rules to ensure robustness. For soft-level decisions, the mean rule is commonly used:

$$\hat{c} = \underset{c \in C}{\text{argmax}} \frac{1}{N_n} \sum_{n=1}^{N_n} d_{n,c}. \quad (2.3)$$

In eq. (2.3), $\hat{c}$ is the final predicted class after decision-level fusion, N_n indicates the total number of sensor nodes participating in the fusion process, C represents the set of all possible classes or health states, n is the index of a sensor node, and $d_{n,c}$ shows the confidence or probability assigned by the sensor node n to class c. Voting-based methods, like majority voting or weighted ranking, are used to determine the final class for ranked and label-level decisions. By combining several independent sensor decisions, decision-level fusion enhances overall reliability and fault tolerance, thereby making it useful for accurate and continuous healthcare monitoring.

2.4 Sensor Fusion Architectures for IoMT Systems

Sensor fusion in IoMT systems is implemented with the help of the architectural frameworks, and the focus is put on the efficient data processing and scalability as well as real-time provision of healthcare services. It studies how various architecture fusions structure sensing, computation, and communication between the different elements of the IoMT to handle heterogeneous and large volume medical data. Special consideration is paid to edge-cloud collaborative fusion models that allocate both tasks of fusion and analytics to in-edge and cloud solutions to strike a balance between the low-latency response and the large-scale capabilities. On the whole, the section underscores how the proper sensor fusion architectures can support strong, scalable, and smart IoMT systems to support continuous monitoring of CD and support clinical decision-making processes.

2.4.1 Edge-Cloud Collaborative Fusion Models

To enable effective and scalable sensor fusion in IoMT systems, edge–cloud collaborative fusion models combine local processing at edge devices with centralized analytics in the cloud. At the edge layer, before data is passed through the network, preliminary fusion, feature extraction, or inference is done, reducing the size of data transferred over the network and allowing low-latency response. The cloud combines the data of multiple edge nodes to create a single cloud and do global fusion, long-term analysis, and refine the models with more powerful computational capabilities. This scalable architecture provides the tradeoff between the real time responsiveness and computational efficiency and resolves the issues of bandwidth constraint, scalability and heterogeneity of the data. The edge and cloud collaborative models have also exhibited top-tier of robustness and scalability as they have decentralized the responsibility to consolidate data to edge and cloud layers thus have become very usable to AI-driven IoMT systems that require constant monitoring, and real-time decision-making abilities and healthcare analytics that can scale [9].

2.5 AI Techniques for Sensor Fusion

The practice of sensor fusion in the healthcare industry using AI methods will also facilitate the combination of heterogeneous medical data with intelligent learning models. Multimodal fusion is helped by ML techniques that learns combined representations of more than one sensor, thus contributing to robustness, tolerance to noise, and prediction. Fusion architectures based on deep learning (DL) enhance performance by automatically integrating spatial and temporal patterns in physiological signals, enabling accurate and scalable real-time healthcare monitoring and decision-making.

2.5.1 Machine Learning Methods for Multimodal Fusion

ML-based multimodal fusion enables the integration of heterogeneous sensor data by learning joint representations across multiple modalities [10]. ML models integrate characteristics from many sensor streams to capture complementary information and cross-modal correlations that are not visible from only single sensors, as shown in multimodal sensor fusion with ML for data-driven process monitoring. This method uses fused sensor characteristics to train supervised learning models to recognize intricate patterns, increase prediction accuracy, and improve robustness to noise. For AI-enabled IoMT systems, where, ML enables the efficient integration of various health sensor data for dependable monitoring and decision support, such data-driven fusion solutions are ideal.

2.5.2 Deep Learning (DL)-Based Fusion Models

DL-based fusion architectures can be utilized to provide a great framework that combines various data gathered by the IoMT devices [11]. These architectures automatically extract useful representations of multiple sensor modalities using deep neural networks and combine them into one model. DL models can capture complex spatial and temporal patterns of physiological signals by using layers, such as convolutional and recurrent networks. In a conventional architecture, data of different sensors is processed by feature extractors that are specific to that modality. The model then makes cross-modal correlations and complementary information by combining the learnt features at lower or higher levels in the network. End-to-end training helps the inference and fusion processes to be simultaneously enhanced and in turn enhance the ability to withstand noise and data variability. All in all, DL-based fusion architectures provide the scalable, exciting, and precise analysis of multimodal health-related data, and therefore, they are suitable and efficient in intelligent healthcare monitoring, diagnosis, and decision support in contemporary IoMT systems. Figure 2.2 depicts an end-to-end IoMT framework integrating wearable physiological sensors, edge devices, wireless communication, secure data management, and AI-driven sensor fusion. It highlights how multimodal health dataedge-cloud computing are processed and analyzed to enable continuous monitorin, clinical decision support, and CD management.

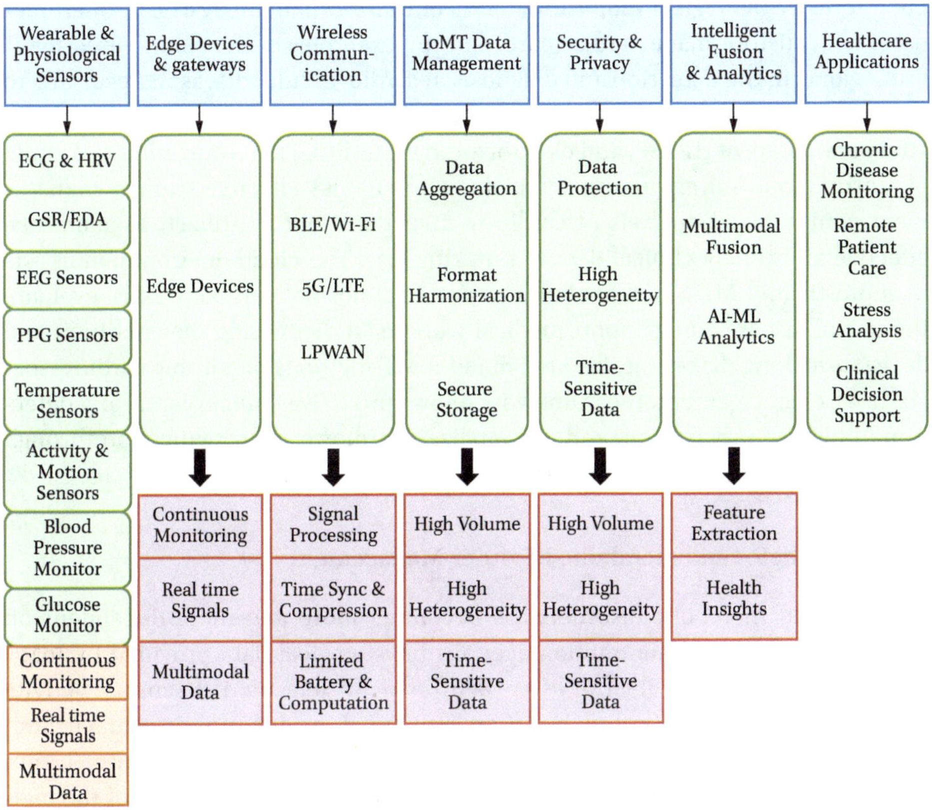

FIGURE 2.2 End-to-end IoMT architecture with edge–cloud sensor fusion for chronic disease monitoring. (Source: Own elaboration)

2.6 Applications of Sensor Fusion in CD Monitoring

This section explains how the sensor fusion method is practically implemented in monitoring CD and its cardiovascular diseases and diabetes-related metabolic disorders are given emphasis. It describes the process of fusing multimodal data provided by the IoMT devices such as wearable and physiological sensors so that continuous, precise, and personalized health monitoring can be available. Through sensor fusion and AI-based analytics, the section presents enhancements in the early detection of diseases, risk management, and chronic conditions in the long term. In general, it shows that sensor fusion aids in proactive, scalable, and data-driven medical care delivery in practice-based IoMT-based clinical applications.

2.6.1 Cardiovascular Disease Monitoring

Long-term physiological monitoring and centralized data analysis can offer end-users with custom-made cardiac care, and the examples of the cloud-based IoMT frameworks in the detection and diagnosis of cardiovascular diseases. According to the present model, wearable devices and IoMT gadgets constantly measure all the vital parameters of HR, BP, and electrocardiogram (ECG) and transmit it to a cloud platform. Cloud infrastructure is scalable and enables effective storage, management, and colossal analysis of cardiovascular data that is utilized to guarantee effective and timely clinical decision-making [12]. The cloud environment assists in sophisticated ML-based analytics to detect abnormal cardiac trends, evaluate the risk of disease, and support medical workers in diagnosing the condition and decision-making. Based on the AI-enabled IoMT methods of chronic cardiovascular disease management, the framework allows proactive clinical care, early detection of disease progression, and personalized cardiovascular medication through continuous monitoring and long-term follow-up.

2.6.2 Diabetes and Metabolic Disorder Management

Diabetes and metabolic disorders are becoming more and more dependent on sensor fusion to combine nonhomogenous physiological data provided by IoMT devices, such as blood glucose level (BGL) and BP sensors, HR sensors, activity trackers, etc. The analysis of each of these streams alone does not have much clinical use; however, the interpretation of both streams together with sensor fusion dramatically improves the precision, dependability, and completeness of the metabolic health measurement [13]. IoMT systems can help monitor glucose changes, metabolism, and predictive signs of complications such as heart and kidney diseases by means of data-, feature-, and decision-level fusion and AI-based analytics. This combination will provide the opportunity to detect abnormal trends in a timely manner, make proactive clinical decisions, and manage metabolic diseases in a personalized manner, in accordance with the current preventive and scalable chronic care strategies.

2.7 Privacy, Security, and Regulatory Constraints

The privacy and security are a great concern in AI-enabled sensor fusion in IoMT because, sensitive physiological data is continuously exchanged between connected medical devices and networks. Secure data transmission systems detect security gaps such as information interception, data manipulation, and hacking during wireless communication [14]. To mitigate these weaknesses, lightweight cryptographic tools, and secure generation of keys minimize these weaknesses to guarantee data integrity and confidentiality due to the low processing capabilities of IoMT sensors. To preserve integrated multimodal health information which remains useful in clinical decision-making where security measures take precedence.

Besides the technical issues of security, there is the additional limitation of sensor fusion systems in healthcare due to regulatory compliance. Such laws as the Health Insurance Portability and Accountability Act (HIPAA) and the General Data Protection Regulation (GDPR) demand very strict regulations on data access, storage, and transmission with a focus on patient privacy and consent. These needs drive the architectural design of AI-based fusion architectures by requiring secure authentication, encryption, and access-control without affecting real time performance. Therefore, the factors of privacy, security, and regulations are crucial in the development of trustworthy and adherent sensor fusion applications to the monitoring of CD based on IoMT.

2.8 Predictive and Preventive Healthcare Systems

Predictive and preventive healthcare systems aim to transform healthcare delivery by prioritizing early risk assessment, continuous monitoring, and timely intervention rather than relying solely on reactive treatment [15]. Prevention medicine plans are aimed at determining the behavioral, physiological, and environmental risks that lead to CDs including diabetes, respiratory, and cardiovascular problems. The frequent check-ups and early identification of the abnormal health trends, and preventive strategies can serve to retard the disease progression, minimize complications, and enhance the long-term patient outcomes and cut healthcare costs. By combining multimodal sensor data of wearable and clinical devices, a comprehensive picture of the health status of a person is obtained that reflects not only short-term changes but also long-term trends. This is because AI-based predictive models analyze the pool of combined data streams at an early clinical stage and aid in continuous care management to identify possible risks and facilitate the use of more specific preventive actions. These systems have superior population health outcomes, CD management, and sustainable health care delivery through prediction and prevention using data.

2.9 Conclusion

The chapter has presented sensor fusion with AI in the context of IoMT systems CD surveillance, and the advantages of combining multimodal physiological data to enhance accuracy, resilience, and timely clinical decision-making. IoMT systems provide scalable, responsive, and personalized healthcare provision by means of data, feature, and decision level fusion and edge and cloud collaborative architectures, and intelligent learning models. The applications that have been presented in the cardiovascular and metabolic disease management sector show that fusion-driven monitoring systems are useful in clinical practice. Overall, the sensor fusion related to AI is identified as an inevitable enabler of the next generation healthcare systems regarding proactive, preventative, and intelligent CD management. Despite these advantages, energy limitations of wearable devices, information privacy, model interpretability, and the complexity of wearable devices deployment in the clinical environment are some of the limitations. The future of AI-supported IoMT systems must address explainable fusion models, energy efficient, privacy saving learning frameworks, and adaptable architecture capable of enhancing the scalability, trust, and clinical adoption of the system.

References

[1] Airhihenbuwa CO, et al. Global perspectives on improving chronic disease prevention and management in diverse settings. Preventing Chronic Disease. 2021;18:E33. https://doi.org/10.5888/pcd18.210055.

[2] Thompson S, et al. Chronic disease management: how IT and analytics create healthcare value through the temporal displacement of care. Mis Quarterly. 2020;44(1):227–56. https://doi.org/10.25300/MISQ/2020/15085.

[3] Salama R, et al. (Benefits of Internet of Things (IoT) Applications in Health care-An Overview). In 2023 International Conference on Computational Intelligence, Communication Technology and Networking (CICTN) 2023 (Vol. 2.3, pp. 778–784). IEEE. https://doi.org/10.1109/CICTN57981.2023.10141452.

[4] Yadav SP, Yadav S. Image fusion using hybrid methods in multimodality medical images. Medical & Biological Engineering & Computing. 2020;58(4):669–87. https://doi.org/10.1007/s11517-020-02136-6.

[5] Henríquez-Jara B, et al. Modelling the effect of travel experiences in modal choice using virtual reality and physiological sensor data. Transportation Research Part C: Emerging Technologies. 2025;178:105178.

[6] Ding X, et al. A feasibility study of multi-mode intelligent fusion medical data transmission technology of industrial internet of things combined with medical Internet of Things. Internet of Things. 2023;21:100689. https://doi.org/10.1016/j.iot.2023.100689.

[7] Muzammal M, et al. A multi-sensor data fusion enabled ensemble approach for medical data from body sensor networks. Information Fusion. 2020;53:155–64. https://doi.org/10.1016/j.inffus.2019.06.021.

[8] Dekkers G, et al. Dynamic sensor activation and decision-level fusion in wireless acoustic sensor networks for classification of domestic activities. Information Fusion. 2022;77:196–210. https://doi.org/10.1016/j.inffus.2021.07.022.

[9] Gao Y, et al. Federated sensing: Edge-cloud elastic collaborative learning for intelligent sensing. IEEE Internet of Things Journal. 2021;8(14):11100–11. https://doi.org/10.1109/JIOT.2021.3053055.

[10] Petrich J, et al. Multi-modal sensor fusion with machine learning for data-driven process monitoring for additive manufacturing. Additive Manufacturing. 2021;48:102364. https://doi.org/10.1016/j.addma.2021.102364.

[11] Zhang T, et al. A joint deep learning and internet of medical things driven framework for elderly patients. IEEE Access. 2020;8:75822–32. https://doi.org/10.1109/ACCESS.2020.2989143.

[12] Adewole KS, et al. Cloud-Based IoMT Framework for Cardiovascular Disease Prediction and Diagnosis in Personalized E-Health Care. In Intelligent IoT Systems in Personalized Health Care 2021 (Vol. 4.2, pp. 105–45). Academic Press. https://doi.org/10.1016/B978-0-12-821187-8.00005-8.

[13] Alkaff FF, et al. The impact of the Indonesian chronic disease management program (PROLANIS) on metabolic control and renal function of type 2 diabetes mellitus patients in primary care setting. Journal of Primary Care & Community Health. 2021;12:2150132720984409. https://doi.org/10.1177/2150132720984409.

[14] Saif S, et al. A secure data transmission framework for IoT enabled healthcare. Heliyon. 2024;10(16). https://doi.org/10.1016/j.heliyon.2024.e36269.

[15] Mbata AO, et al. Preventative medicine and chronic disease management: reducing healthcare costs and improving long-term public health. International Journal of Multidisciplinary Research and Growth Evaluation. 2024;5(06):1584–600. https://doi.org/10.54660/.IJMRGE.2024.5.6.1584-1600.

CHAPTER 3

Multispectral Image Feature Extraction

Vijay Jagdish Upadhye, Satya Ranjan Das, Shajahan Basheer

Abstract Multispectral imaging records data across multiple spectral bands, providing detailed information that enhances medical images with Internet of Medical Things (IoMT) ecosystem. Feature extraction from multispectral images plays a critical role in identifying tissue types, detecting pathological regions, and supporting accurate diagnostic decision-making. This chapter discusses spectral, spatial, spectral-spatial feature extraction techniques, where spectral features exploit variations in reflectance across different wavelengths to differentiate tissues and abnormalities, while spatial features analyze texture, shape, and structural patterns to capture morphological details. It used multispectral medical image datasets, including fundus, brain (computed tomography/magnetic resonance imaging, chest X-radiation imaging (X-ray), kidney, and breast cancer images, to ensure diverse diagnostic representation. Data preprocessing step includes Gaussian filtering and median filtering. Dimensionality reduction is a technique that includes principal component analysis and wavelet transforms. Deep learning techniques such as autoencoders and convolutional neural networks, facilitate efficient, automated extraction of complex and informative features. Applications of multispectral image feature extraction in artificial intelligence-driven IoMT systems enable precise, real-time medical diagnostics, continuous patient monitoring, predictive healthcare, and intelligent clinical decision-making, advancing precision medicine and proactive intervention strategies.

Keywords Multispectral imaging – spectral-spatial analysis – internet of medical things (IoMT) – medical diagnostics – dimensionality reduction

3.1 Introduction

Multispectral imaging (MSI) has evolved as a significant instrument in the analysis of medical images since it records data in a wide range of spectral bands, which enables visualization of tissue features in a more detailed way compared to the conventional imaging methodology [1]. MSI is a process of collecting and examining images in a variety of spectral bands or wavelengths. It employs very sensitive sensors which are capable of sensing radiation that lies outside the spectrum of visible light that human eyes can perceive (400–700 nm) [2].

MSI is typically applied in many diverse fields such as agriculture, remote sensing, medical imaging, and environmental monitoring. It is also useful in the context

 | HTTPS://DOI.ORG/10.1515/9783112237083-003

of medical imaging, as it can help clinicians to make more successful decisions and diagnoses concerning the diseases and functions of tissues and organs due to the availability of valuable information about such entities [3]. Technology in the medical system plays a decisive role in new methods of preventing illnesses and treating them, as well as increasing resilience in the case of an epidemic in society [4]. Internet of Medical Things (IoMT) establishes a platform of information that integrates a wide range of smart devices such as hospital and medical examination equipment and wearables which are the most important elements of the IoMT system. The devices constantly collect and relay health information to central systems where they are further processed [5, 6].

The contemporary medical centers have also embraced the concept of a smart-bed whereby the bed would change shape and position depending on the sleeping position of the patient. Also, related devices that rely on the IoMT are changing conventional home healthcare services [7]. The smart system of medicines dispensing at home is intelligent to automatically transfer the medical history of the patient to the cloud where it serves as a reminder of the medication to be taken [8]. It uses artificial intelligence (AI) in multidisease classification, which determines patients with conditions such as COVID-19 and pneumonia and defines the severity of lung infections to treat them accordingly [9].

AI is capable of enhancing patient care and reducing the costs of healthcare as the need to expand the services offered has increased with the growth of the populations but facilitated through the breakthrough in AI and robotics [10]. The sphere of healthcare is also rapidly developing using AI and robotics, especially in the early diagnoses and implementation of diagnostic methods, which can help to increase efficiency and effectiveness in work traditionally carried out by a human [11]. Figure 3.1 illustrates the integration of MSI, IoMT devices, and AI systems for intelligent healthcare monitoring and diagnosis.

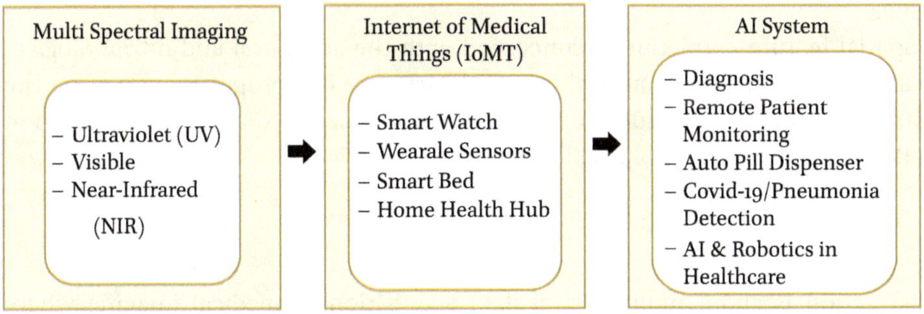

FIGURE 3.1 AI-driven multispectral and IoMT architecture for intelligent medical applications. (Source: Own elaboration)

3.1.1 Problem Statement

Multispectral medical imaging in the IoMT systems generate high-dimensional and complex data, which are difficult to analyze. Conventional feature extraction tools do not make the most of the combined spectral and spatial information and thus lead to redundancy, higher computational cost, and lower diagnostic accuracy. To facilitate sound, real-time, and intelligent clinical diagnosis, monitoring, and intelligent clinical decision-making, effective spectral-spatial feature extraction methods can be supported by dimensionality reduction algorithm and AI-based models.

3.2 Methodology

The deep learning (DL)-based methods of feature extraction are applied to enhance the accuracy, reliability, and efficiency of AI-enhanced multispectral medical image analysis in IoMT systems. Various Kaggle datasets provide multispectral medical-based data, which can be processed through the Gaussian and median filter. Principal component analysis (PCA), wavelet transforms, and convolutional neural networks (CNNs) feature extraction and autoencoders facilitate DL-based feature extraction, which allows feature extraction with high dimensionality reduction, the learning of spatial and spectral representations, and high-quality classification of medical images with the help of IoMT-based medical image analysis.

The rationale of spectral feature extraction is the use of differences in reflectance at various wavelengths as biologically different tissues have distinct spectral characteristics based on biological and biochemical attributes. The variations that are wavelength-dependent permit efficient differentiation of normal and diseased tissues.

Spatial feature extraction is concerned with the structural and morphological details that are available in multispectral pictures. The properties can store the texture patterns, tissue structure, and geometric structure that indicates variation on the anatomy and pathological alterations.

3.2.1 Data collection

This section is about multispectral data acquisition in medical imaging where there is a discussion of the significance of various spectral data. Kaggle datasets provided in brain, chest, kidney, breast cancer, and field image modalities presented multispectral and multimodal medical image data to have a variety of spectral and diagnostic representation.

The data information comprises stereo multispectral brain images obtained by magnetic resonance imaging and computed tomogrpahy (CT) scans of both healthy and tumor-affected patients. It offers spectral image pairs at high resolutions of various anatomical parts and clinical conditions. The data can be used to perform a detailed analysis of spatial, spectral, and depth features and analyze medical images.

Source:https://www.kaggle.com/datasets/zoya77/brain-multi-spectral-image-dataset

The data set includes multispectral field images which are obtained using Sentinel-2 L2A satellite. It has two data tables with a structure and a folder that contains multispectral images in JSON format.

Source:https://www.kaggle.com/datasets/honeybadger128bit/multespectral-field-images

The data set comprises multispectral images of the chest that is categorized into pneumonia and normal. It consists of total 2,465 pictures, 1,341 normal, and 1,124 pneumonia cases. MSI gives better diagnostic data in a variety of spectral bands. The dataset facilitates the use of medical images to classify them, identify pneumonia, and design AI-based diagnostic models.

Source:http://kaggle.com/datasets/programmer3/chest-multispectral-image-dataset

The database is composed of 10,000 breast cancer images which were obtained under different imaging modalities such as visible, infrared, and thermal spectrophotometers. The images are ranked into two groups, malignant and benign, and these support multimodal medical image analysis and breast cancer diagnosis.

Source:https://www.kaggle.com/datasets/programmer3/breast-cancer-multispectral-image-dataset

3.2.2 Data Preprocessing

The present section discusses the most utilized data preprocessing techniques, including Gaussian and median filtering, for multispectral medical image analysis. Technical preprocessing is the process of cleaning, normalizing, and improving the raw data to make it better and easier to understand. It eliminates noise, distortions, and main features, which guarantee quality and reliability of data. In particular, the images are smoothed with the help of Gaussian filter to minimize random noise but retain the overall structure of the image, which is vital to effectively represent tissues. Meanwhile, median filtering is used to eliminate impulse noise, e.g., salt-and-pepper artifacts, without blurring edges and fine textures.

Gaussian Filtering: Gaussian filtering is a linear image averaging that grouping technique to minimize the noise and detail to give a blurred yet structurally consistent image. To reduce high-frequency noise in multispectral images, Gaussian filtering was used through the convolving procedure of each spectral band with a Gaussian kernel to enhance signal smoothness, structural consistency, and feature's reliability of the further feature extraction and subsequent operations [12]:

$$\hat{x}_j = \left[\hat{x}_j^{\,1}, \ldots, \hat{x}_j^{\,P}\right] \tag{3.1}$$

In eq. (3.1), $\hat{x}_j$ denotes the original feature vector, where $\hat{x}_j^1$ is the feature value of the first spectral band, and $\hat{x}_j^p$ represents the normalized or transformed intensity value of jth sample in the pth spectral band:

$$\hat{y}_j = \left[\hat{y}_j^{\,1}, \ldots, \hat{y}_j^{\,P}\right] \tag{3.2}$$

In eq. (3.2), $\hat{y}_j$ denotes the preprocessed multispectral feature vector of the jth sample, $\hat{y}_j^1$ represents the normalized or transformed feature value of the first spectral band, and $\hat{y}_j^P$ represents the corresponding feature value of pth spectral band, where j is the sample index and P denotes the total number of spectral bands used for feature extraction:

$$\Delta_j = \frac{1}{P}\sum_{K=1}^{P}\left(w_j\hat{x}_j^{\,k} - w_j\hat{y}_j^{\,k}\right)^2 = w_j^T A_j w_j \tag{3.3}$$

In eq. (3.3), Δ_j measures class separability, P denotes the spectral bands, k indicates the spectral band index, w_j represents the discriminative weight vector, $\hat{x}_j^k$ and $\hat{y}_j^k$ are the preprocessed feature values of the kth spectral band for two classes or conditions, A_j is the symmetric scatter matrix constructed from spectral differences, and the quadratic form $w_j^T A_j w_j$ represents the same distance in matrix form for efficient optimization in medical imaging:

$$\frac{1}{2P}\left[\sum_{K=1}^{P} w_j\hat{x}_j^{\,k} + \sum_{K=1}^{P} w_j\hat{y}_j^{\,k}\right] = 0 \tag{3.4}$$

Here, the number of samples is indicated by P, w_j indicates the weight associated with the jth feature or component; $\hat{x}_j^k$ and $\hat{y}_j^k$ are the transformed components of the kth sample. The summation $\sum_{k=1}^{P}$ aggregates contributions of all P samples for each dataset, and the factor $1/2P$ normalizes the sum over datasets. The equation being equal to zero indicates a balance or centering condition across the weighted transformed components:

$$\frac{1}{2P}\left[\sum_{K=1}^{P}\left(w_j\hat{x}_j^{\,k}\right)^2+\sum_{K=1}^{P}\left(w_j\hat{y}_j^{\,k}\right)^2\right]=w_j^TB_jw_j=1 \tag{3.5}$$

In eq. (3.5), k denotes the sample index ranging from 1 to P, j represents the index of the selected feature, B_j indicates the scatter or covariance matrix corresponding to the jth feature, $w_j^TB_jw_j$ denotes the quadratic form expressing the weighted variance of the jth feature. The equality to 1 enforces a normalization condition that ensures comparable feature scales in multispectral images.

Median Filtering: Median filtering refers to a nonlinear image processing algorithm used to eliminate impulse noise by replacing each pixel with the median value of its local neighborhood, which effectively preserves edges while minimizing noise. Median filter was applied to remove impulse noise in multispectral images and to preserve edges and structural information with improving image quality to effectively extract and classify features [13]:

$$h(a,b)=\text{Med}\{f(a,b)\}\ \ (a,b)\in Y \tag{3.6}$$

Equation (3.6) represents a filtering operation or a median filter, where $h(a,b)$ is the filtered output pixel at spatial position, $f(a,b)$ refers to original input image intensity value at a spatial position, $\text{Med}\{f(a,b)\}$ denotes the median operator acting on a local neighborhood Y centered at (a,b), and (a,b) refers to pixel position.

3.2.3 Feature Extraction

The section discusses feature extraction in multispectral medical imaging. Feature extraction is a process that converts multispectral image data into useful representations that underscore important patterns, structures, or properties to be analyzed. Wavelet transforms and PCA are popular methods used to decrease redundancy, increase information of significance, and emphasize aspects of significance, making medical image analysis accuracy and efficiency enhancement strategies.

PCA: The feature extraction techniques used in multispectral medical imaging, with particular emphasis on PCA. PCA refers to one of the most common linear dimensionality reduction methods which forms a set of uncorrelated components ranked by their variance based on multispectral data. It gets rid of redundancy and retains the most significant data by transferring spectral data of high dimensionality in a low-dimensional space. In medical imaging, PCA is useful in improving tissue discrimination, minimizing noise, and improving the classification accuracy, especially in multispectral pipelines of feature extraction.

In which every feature gets added proportionately to its respective loading coefficient, allowing proper dimensionality reduction through transforming high-dimensional data into a lower-dimensional space, without loss of the most meaningful variation in the medical data analysis possible in the IoMT systems [14]:

$$PC_j = l_{1j}Y_1 + l_{2j}Y_2 + \cdots + l_{nj}Y_n \tag{3.7}$$

In eq. (3.7), PC_j is the transformed feature capturing maximum variance, Y_1, $Y_2, \ldots, Y_n$ are the original standardized input variables or spectral bands, $l_{1j}, l_{2j}, \ldots, l_{nj}$ are the corresponding loading coefficients that define the contribution of each variable to the jth component, and n denotes the number of input variables.

Wavelet Transforms: Wavelet transforms provide a multi-resolution representation of multispectral images by decomposing data into frequency sub-bands. The approach captures both spatial and spectral variations at different scales, enabling effective feature compression while preserving diagnostically relevant information. Wavelet-based dimensionality reduction is particularly useful in medical imaging for noise suppression, texture preservation, and efficient feature optimization, supporting accurate and real-time IoMT-enabled medical analysis.

Wavelet-based dimensionality reduction employs discrete wavelet transform for efficient multiresolution decomposition, continuous wavelet transform for detailed time frequency analysis, and stationary wavelet transform for shift-shift-resistant feature preservation, enabling compact yet informative spectral-spatial representation of multispectral medical images in IoMT-based diagnostic systems [15]:

$$W(FV, t) = \frac{1}{\sqrt{x}} \int_{-\infty}^{\infty} FV(t)^{*} \varphi\left(\frac{t-y}{a}\right) dt \tag{3.8}$$

In eq. (3.8), $W(FV, t)$ denotes the wavelet coefficient describing localized frequency content at time or position t, $FV(t)$ denotes the input signal or feature vector, $\varphi\left(\frac{t-y}{a}\right) dt$ indicates the mother wavelet function, a represents the scale parameter controlling the resolution of analysis, y denotes the translation parameter that shifts the wavelet along the signal, and the integral computes the correlation among the signal and shifted wavelet:

$$W(FV, t) = \frac{1}{\sqrt{x}} \int_{0}^{P} FV(t)^{*} \varphi\left(\frac{t-y}{a}\right) dt \tag{3.9}$$

This equation defines a wavelet-based a feature extraction process, where P indicates the signal length or total number of samples over which the transformation is performed, enabling multiresolution of multispectral features:

$$W(FV, t) = \frac{1}{\sqrt{x}} FV(t), \varphi\left(\frac{t-y}{a}\right) \tag{3.10}$$

In eq. (3.10), $W(FV, t)$ denotes the wavelet coefficient at position t, $FV(t)$ indicates the input feature vector or signal, a represents the scale parameter controlling the analysis resolution, y evaluates the translation parameter defining the wavelet shift, and $FV(t)$, $\varphi\left(\frac{t-y}{a}\right)$ denotes the inner product operation measuring the similarity among the signal and shifted wavelet for multiresolution feature extraction.

3.2.4 Deep Learning-Based Feature Extraction

This section presents DL-based feature extraction for multispectral image analysis. DL models enable automatic and hierarchical feature extraction directly from multispectral images, reducing dependence on handcrafted features. CNNs are DL methods that automatically extract spatial features from images using convolution operations for accurate classification. CNNs are particularly effective for learning spatial-spectral representations by capturing local textures, structural patterns, and spectral correlations simultaneously [16, 17]. Local textures are learned by fine-grained intensity and edge patterns, preserving spatial continuity for accurate tissue heterogeneity. structural patterns by learning hierarchical representations of edges, shapes, and spatial arrangements, enabling accurate anatomical and pathological structure identification. Spectral correlations by modeling cross-band dependencies and capturing wavelength relationships improve tissue property and abnormality discrimination in multispectral images. CNN models have shown superior performance in tasks like lesion detection, tissue segmentation, and disease classification. Figure 3.2 shows the overview of the CNN architecture used for classification:

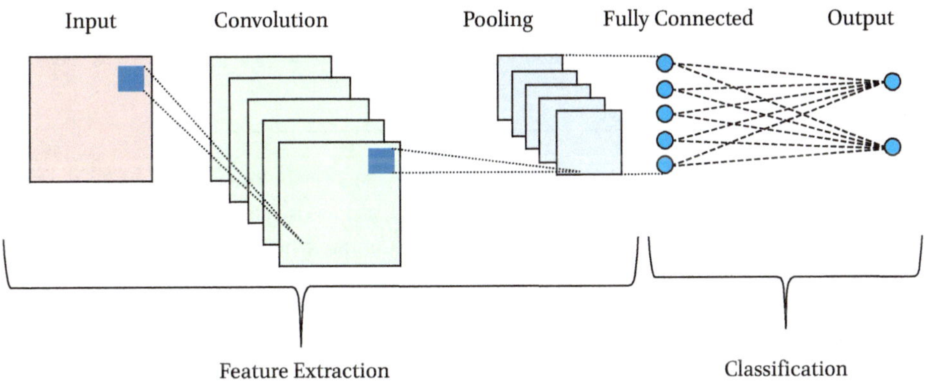

FIGURE 3.2 Workflow of CNN architecture. (Source: Own elaboration)

$$w_{ij}^{xy} = f\left(\sum_{m=1}^{M_i-1}\sum_{p=0}^{P_i-1}\sum_{q=0}^{Q_i-1} W_{ijm}^{pq} V_{(i-1)m}^{(x+p)(y+q)} + b_{ij}\right) \tag{3.11}$$

In eq. (3.11), w_{ij}^{xy} represents the activation value for the spatial location (x,y) of the jth feature map in the ith convolutional layer. The nonlinear activation function is represented by f, i indicates the current convolutional layer, j represents the output feature map, x, y denote the spatial coordinates, m indicates the input feature map, M_i represents the feature maps in $(i1)$th layer, and p, q denote the spatial positions within the convolution kernel. P_i, Q_i denote the height and width, respectively, W_{ijm}^{pq} denotes the convolution weight, $V_{(i-1)m}^{(x+p)(y+q)}$ represents the activation value from the mth feature map of the previous layer at location $(x+p, y+q)$, and b_{ij} indicates the bias term:

$$w_{ij}^{xyz} = f\left(\sum_{m=1}^{M_i-1}\sum_{p=0}^{P_i-1}\sum_{q=0}^{Q_i-1}\sum_{r=0}^{R_i-1} W_{ijm}^{pq} V_{(i-1)m}^{(x+p)(y+q)(z+r)} + b_{ij}\right) \tag{3.12}$$

In eq. (3.12), w_{ij}^{xyz} represents the output activation, z denotes the depth coordinate of the 3D output feature map, r represents the kernel index along the depth dimension, R_i indicates theconvolution kernel, and $V_{(i-1)m}^{(x+p)(y+q)(z+r)}$ represents the input activation from the previous layer:

$$f(y) = \max(0, y) \tag{3.13}$$

In eq. (3.13), $f(y)$ represents the output of the activation function. y denotes the input value of the Rectified Linear Unit (ReLU) function. DL-based feature extraction utilizing CNNs and autoencoders is shown in Figure 3.3.

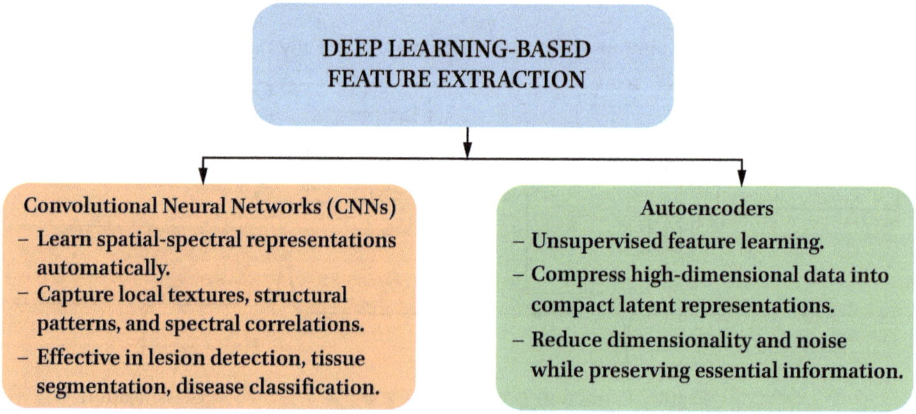

FIGURE 3.3 Deep learning-based feature extraction using CNNs and autoencoders. (Source: Own elaboration)

3.2.5 Autoencoders

The another method is autoencoders, a neural network that develops efficient data representations through encoding inputs into latent features and reconstructing them with minimal loss [18]. Autoencoders provide an unsupervised approach to feature learning by compressing high-dimensional multispectral data into compact latent representations.

The feature extraction process consists of several sequential stages. Given an unlabeled input image X, initial preprocessing is performed using basic data augmentation techniques, including cropping, translation, scaling, and horizontal or vertical flipping, to enhance data diversity and robustness. Subsequently, a set of N distinct rotation transformation operations are described as follows:

$$g_1(X|x_i) = \text{Rot}\,(X, (x_i - 1) \times 90) \tag{3.14}$$

Here, $g_1(X|x_i)$ denotes the first rotation-based transformation function applied in the process of feature extraction, where X represents the original input image, x_i is the index of transformation, $\text{Rot}(X, \theta)$ denotes the rotation operator, which rotates image X by an angle theta degrees and $(x_i - 1) \times 90$ denotes the rotation angle that allows rotation-invariant and discriminative image features toeffectively learned.

3.3 Results

The results provide CNN-based hybrid feature extraction assessment, reliability-efficiency trade-offs by checking measures such as precision, accuracy, recall, F_1-score, Kappa coefficient, dimensionality of features, and computation time, which illustrate the efficacy of the methods and efficiency of computing in multispectral medical image examination.

Recall: It tests the capability of a model to detect all the pertinent occurrences and is done by testing the proportion of true positives. Recall determines the ability of the model to identify all the pathological areas of interest to have minimum false detections in medical multispectral images.

True positives (TP) in eq. (3.15) denotes the number of correctly identified pathological or target regions whereas false negatives (FN) is actually the pathological regions that are missed by the model:

$$\text{Recall} = \frac{\text{TP}}{\text{TP} + \text{FN}} \tag{3.15}$$

Precision: Precision compares the TP predictions by computing the percentage of correct identifications of occurrences among the predicted positive. Precision measures the ability of the system to detect true pathological locations with minimal false occurrences in the multispectral medical imaging using eq. (3.16):

$$\text{Precision} = \frac{\text{TP}}{\text{TP} + \text{FP}} \tag{3.16}$$

Accuracy: It is defined as the rate of correctly classified medical images of all samples, which means the successful extraction of multispectral features and stable AI-based diagnosis in the IoMT systems of healthcare and is presented in the form of eq. (3.17):

$$\text{Accuracy} = \frac{\text{TP} + \text{TN}}{\text{TP} + \text{TN} + \text{FP} + \text{FN}} \tag{3.17}$$

F1-Score: It represents harmonic mean of recall and precision, which shows balanced classification in terms of minimizing FP and FN in multispectral medical image analysis in eq. (3.18):

$$F_1 - \text{score} = 2 \times \frac{\text{Precision} \times \text{Recall}}{\text{Precision} + \text{Recall}} \tag{3.18}$$

Kappa Coefficient: The Kappa coefficient ($\varkappa$) denotes a measure of inter-rater or classifier agreement applied in categorical data, which eliminates the possibility of chance agreement. It measures the range of values between –1 (disagreement) and 1 (perfect agreement) as shown in eq. (3.19):

$$k = \frac{P_o - P_e}{1 - P_e} \tag{3.19}$$

TABLE 3.1 Performance evaluation of CNN-based hybrid feature extraction methods.

Method	**Accuracy (%)**	**Precision (%)**	**Recall (%)**	**F_1-score (%)**
CNN + PCA	89.42	88.75	87.96	88.35
CNN + autoencoder	92.15	91.84	91.32	91.58
CNN + wavelet transform	93.87	94.15	92.68	93.41

Source: Own elaboration.

Table 3.1 and Figure 3.4 illustrate the performance of various feature extraction techniques integrated with CNNs for multispectral medical image analysis. Approaches, such as PCA, autoencoder, and wavelet transform are tested by measuring their performance in terms of precision, accuracy, recall, and F_1-score. PCA is used in conjunction with CNNs to decrease the large-dimensional spectral data and still retain the key features, enhancing the computational efficiency and showing the most informative patterns to be used in classification. CNNs are combined with autoencoders with the purpose of performing deep, unsupervised feature extraction, and learn compact and meaningful representation of multispectral images in an automatic way. Wavelet transform is combined with CNNs to obtain the spatial and frequency content to increase the depth of feature representation based on the analysis of both textural and spectral features on various scales.

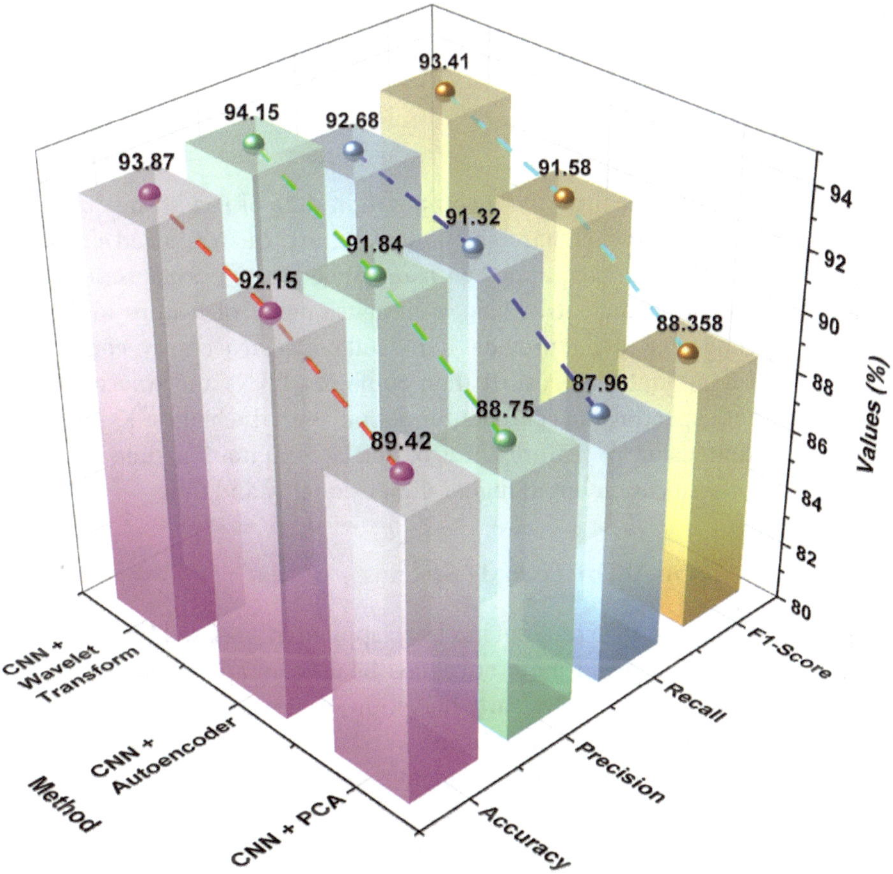

FIGURE 3.4 CNN-based feature extraction techniques. (Source: Own Elaboration)

Table 3.2 indicates the Kappa coefficient that indicates the agreement between the predicted and true classification with high values implying higher reliability and consistency. The concept of feature dimensionality refers to the count of all features or variables that have been obtained out of an image to be analyzed. The time spent computing a task represents the method efficiency, as shown in the computation time. It takes a combination of CNNs and PCA with a Kappa coefficient of

TABLE 3.2 Performance of CNN-based feature extraction methods.

Method	Kappa coefficient	Feature dimensionality	Computation time (s)
CNN + PCA	0.86	128	1.42
CNN + autoencoder	0.89	192	1.63
CNN + wavelet transform	0.91	256	1.78

Source: Own elaboration

0.86, 128 feature dimensionality, and a computation time of 1.42 s. The approach offers rapid computation, which has a small set of features but a reduced classification reliability. It trains CNNs equipped with autoencoders to learn automatically, resulting in a Kappa coefficient of 0.89, a dimensionality of feature 192 and a computational time of 1.63 s that provides a tradeoff between accuracy, complexity of features, and computational cost. It then combines CNNs and wavelet transforms to learn both spectral and spatial information, with the highest Kappa coefficient of 0.91, dimension of features of 256, and a computation time of 1.78 s, requiring better classification but at higher dimensionality and time.

3.3.1 Applications in AI-Driven IoMT Systems

Multispectral image feature extraction has been applied in AI-based IoMT systems to use multispectral images in more advanced healthcare applications. Spectral and spatial pattern analysis allows early detection of diseases and provides real-time diagnostics and predictive healthcare. Wearable MSI sensors are used to ensure continuous monitoring of patients and provide physiological data of high quality to be constantly evaluated [19]. Moreover, the concept of multispectral characteristics benefits clinical decision support systems by making specific, data-driven recommendations to guide clinicians to make correct diagnosis, treatment planning, and informed decision-making which, in turn, improves patient outcomes and facilitates proactive and intelligent healthcare delivery [20].

3.4 Discussion

Explore spectral, spatial, spectral-spatial and DL based feature extractions to improve accuracy, efficiency, and reliability of AI-based multispectral medical image analysis in IoMT systems. Though, hybrid feature extraction using transfer learning enhances the segmentation accuracy, the model is very dependent on large well-labeled datasets, which restricts its flexibility in a variety of medical imaging settings [21, 22]. Multilevel deep feature extraction network causes an increase in computational complexity, limiting real-time deployment and scalability in resource-limited IoMT-based healthcare systems [23, 24]. To address these drawbacks, this chapter uses hybrid spectral-spatial DL with dimensionality reduction, low-weight architectures, noise-resistant fusion of features, and better generalization methods of scalable, efficient IoMT medical imaging.

3.5 Conclusion

The chapter provided a detailed system of multispectral image feature extraction in IoMT systems based on AI. The chapter has shown the application of multispectral data in capturing tissue characteristics, pathological differences, and structural patterns that are not readily available by other imaging modalities through the integration of spectral, spatial, and spectral-spatial features extraction methods. As key preprocessing measures, Gaussian filtering and median filtering were used. Dimensionality reduction techniques such as PCA and wavelet transform demonstrated that the two techniques reduce redundancy and computational load whilst retaining diagnostically useful information. Additionally, DL based methods, such as CNNs and autoencoders, made it possible to learn hierarchical features automatically, which advanced the accuracy of classification, the strength of these methods, and real-time processing to a much better level.In general, the chapter makes multispectral feature extraction an essential facilitator of precise diagnosis, dynamic monitoring, and intelligent clinical decision support of IoMT systems to enable scalable, efficient, and precision-oriented healthcare applications in real-time medical settings. Further research will focus on transformer-based attention models, multimodal fusion, and real time implementation to make multispectral image analysis scalable.

References

[1] Hoyt CC. Multiplex immunofluorescence and multispectral imaging: forming the basis of a clinical test platform for immuno-oncology. Frontiers in Molecular Biosciences. 2021;674747:8. https://doi.org/10.3389/fmolb.2021.674747.

[2] Ilișanu M-A, Moldoveanu F, Moldoveanu A. Multispectral imaging for skin diseases assessment – state of the art and perspectives. Sensors. 2023;23(8):3888. https://doi.org/10.3390/s23083888.

[3] Mengu D, et al. Snapshot multispectral imaging using a diffractive optical network. Light: Science & Applications. 2023;12(1):86. https://doi.org/10.1038/s41377-023-01135-0.

[4] Al Khatib I, Shamayleh A, Ndiaye M. Healthcare and the internet of medical things: applications, trends, key challenges, and proposed resolutions. Informatics. 2024 11(3):MDPI. https://doi.org/10.3390/informatics11030047.

[5] Kim B, et al. Application of an Internet of Medical Things (IoMT) to communications in a hospital environment. Applied Sciences. 2022;12(23):12042. https://doi.org/10.3390/app122312042.

[6] Manickam P, et al. Artificial intelligence (AI) and internet of medical things (IoMT) assisted biomedical systems for intelligent healthcare. Biosensors. 2022;12(8):562. https://doi.org/10.3390/bios12080562.

[7] Mahiddin N, et al. An interrelated decision-making model for an intelligent decision support system in healthcare. IEEE Access. 2022;10:31660–31676. https://doi.org/10.1109/ACCESS.2022.3160725.

[8] Thandapani S, et al. IoMT with deep CNN: AI-based intelligent support system for pandemic diseases. Electronics. 2023;12(2):424. https://doi.org/10.3390/electronics12020424.

[9] Sunarti S, et al. Artificial intelligence in healthcare: opportunities and risk for future. Gaceta Sanitaria. 2021;35:S67–S70. https://doi.org/10.1016/j.gaceta.2020.12.019.

[10] Shafiq M, et al. Advances in IoMT for healthcare systems. Sensors. 2023;24(1):10. https://doi.org/10.3390/s24010010.

[11] Dave M, Patel N. Artificial intelligence in healthcare and education. British Dental Journal. 2023;234(10):761–764. https://doi.org/10.1038/s41415-023-5845-2.

[12] He Y, et al. Multispectral image change detection based on single-band slow feature analysis. Remote Sensing. 2021;13(15):2969. https://doi.org/10.3390/rs13152969.

[13] Shen P, et al. A Fourier transform-based calculation method of wilting index for soybean canopy using multispectral image. Agronomy. 2022;12(7):1650. https://doi.org/10.3390/agronomy12071650.

[14] Zhang X, et al. Combining spectral and texture features of UAS-based multispectral images for maize leaf area index estimation. Remote Sensing. 2022;14(2):331. https://doi.org/10.3390/rs14020331.

[15] Hong K. Classification of emotional stress and physical stress using a multispectral based deep feature extraction model. Scientific Reports. 2023;13(1):2693. https://doi.org/10.1038/s41598-023-29903-3.

[16] Kong F, et al. A spectral-spatial feature extraction method with polydirectional CNN for multispectral image compression. IEEE Journal of Selected Topics in Applied Earth Observations and Remote Sensing. 2022;15:2745–2758. https://doi.org/10.1109/JSTARS.2022.3158281.

[17] Sharifi O, Mokhtarzadeh M, AsghariBeirami B. A new deep learning approach for classification of hyperspectral images: feature and decision level fusion of spectral

and spatial features in multiscale CNN. Geocarto International. 2022;37(14):4208–4233. https://doi.org/10.1080/10106049.2021.1882006.

[18] Chamola V. Evolutionary computation-based self-supervised learning for image processing: a big data-driven approach to feature extraction and fusion for multispectral object detection. 2024;11:130. https://link.springer.com/content/pdf/10.1186/s40537-024-00988-5.pdf.

[19] Aggarwal 1st Lt Pushkar, Papay FA. Applications of multispectral and hyperspectral imaging in dermatology. Experimental Dermatology. 2022;31(8):1128–1135. https://doi.org/10.1111/exd.14624.

[20] Yadav SP, Yadav S. Fusion of medical images in wavelet domain: a hybrid implementation. Computer Modeling in Engineering & Sciences. 2020;122(1):303–321. https://doi.org/10.32604/cmes.2020.08459.

[21] Çevik N, et al. Advancing patient care with AI: a unified framework for medical image segmentation using transfer learning and hybrid feature extraction. Frontiers in Medicine. 2025;12:1589587. https://doi.org/10.3389/fmed.2025.1589587.

[22] Munawwar S, Rao PVGK. An efficient deep learning based multi-level feature extraction network for multi-modal medical image fusion. International Arab Journal of Information Technology (IAJIT). 2025;22(3). https://doi.org/10.34028/iajit/22/3/2.

[23] Rani P, et al. Simulation of the lightweight blockchain technique based on privacy and security for healthcare data for the cloud system. International Journal of E-Health and Medical Communications (IJEHMC). 2022;13(4):1–15. https://doi.org/10.4018/IJEHMC.309436.

[24] Aboneh T, Rorissa A, Srinivasagan R. Stacking-based ensemble learning method for multi-spectral image classification. Technologies. 2022;10(1):17. https://doi.org/10.3390/technologies10010017.

CHAPTER 4

Multifaceted Wearable Sensor Technologies for Chronic Disease Management

Bhupesh Goyal, Vishakha Khambhati, A.R. Chayapathi

Abstract Chronic conditions, including diabetes, respiratory disorders, and cardiovascular issues, are becoming more common globally and need ongoing, long-term therapy to avoid complications and enhance patient outcomes. Traditional monitoring techniques are often intermittent, reactive, and resource-intensive, highlighting a critical gap for continuous, real-time, and personalized healthcare solutions. This chapter explores multifaceted wearable sensor technologies integrated within the chronic disease management (CDM), which involves real-time observation using the Internet of Medical Things (IoMT). Wearable sensors embedded in clothing, accessories, or medical devices collect physiological and biochemical data, including blood pressure, glucose levels, and heart rate. Advanced data analytics and artificial intelligence (AI) transform these data into actionable insights, enabling early interventions, predictive care, and personalized disease management. AI also enhances analytics accuracy by generating supplementary healthcare data. The chapter analyzes the design, applications, and challenges of wearable IoMT devices, including data security, interoperability, and device reliability. By addressing the existing gaps in continuous monitoring, this chapter highlights how the integration of wearable sensors, IoMT, and AI is reshaping CDM toward preventive, personalized, and precision healthcare.

Keywords Wearable sensors – chronic disease management – Internet of Medical Things – artificial intelligence – physiological monitoring

4.1 Introduction

The chronic conditions represent a major problem in global population health. Depression is a widespread condition, some patients have chronic ailments such as hypertension, coronary heart disease, and diabetes. Depression is thoroughly associated with cardiovascular risk. Individuals with multimorbid diseases are also more likely to experience depression by more than twice that of individuals without multimorbidity [1]. The diseases, such as diabetes and heart disease, were identified as major risk factors for severe morbidity and mortality from COVID-19. Diabetes mellitus occurs with an increase in blood glucose levels. Heart diseases are considered as another main reason for death in middle and elderly people. The

 | HTTPS://DOI.ORG/10.1515/9783112237083-004

World Health Organization's (WHO) cardiovascular data show that heart stroke accounts for almost 17.9 million fatalities [2].

Smartwatches and fitness trackers are some of the commercial wearables that have been most prevalent in recent years. The devices are commonly used in the monitoring of biological signals, especiallythe absence of conformal skin touch while doing regular tasks, and are still unable to offer reliable physiological data, such as blood oxygen levels, heart rates, and blood pressure [3]. In chronic disease management (CDM), continuous health status monitoring enables the performance of essential clinical assessments to evaluate a patient's overall condition while automatically sharing the collected data with all members of the healthcare team, thereby supporting timely and effective care coordination [4, 5].

Today, healthcare has increased in demand to shift their services to provide patient-based and smart healthcare, including biomedical diagnostics, health management, and patients with chronic conditions being monitored remotely [6, 7]. In this regard, Internet of Medical Things (IoMT) has huge potential and implies that a number of disciplines (bioengineering, computer science, informatics, etc.) are involved in the efficient healthcare systems [8].

The advancement of artificial intelligence (AI) and related digital health devices, such as wearable technology and mobile health (mHealth) is likely to overcome these obstacles and reconsider chronic illness self-management [9]. AI can analyse large amounts of patient-generated data across a large number of sources, such as wearable devices, mHealth applications, and electronic health records (EHRs) to offer interactive assistance, proactive insights, and predictive warnings, making healthcare approach closer to a more paradigm of predictive, preventative, and personalized care [10].

The IoMT also enables patients to have vital signs, medication adherence, and the development of the disease constantly and remotely monitored. IoMT results in more effective management of illnesses, and better quality of life of patients [11–12]. Figure 4.1. illustrates chronic disease challenges, wearable limitations, and the role of IoMT and AI in predictive healthcare.

4.1.1 Problem Statement

Cardiovascular, diabetes, and respiratory disorders are chronic diseases that are increasing worldwide and necessitate long-term management. Conventional methods of monitoring are fragmented, reactive, and consumptive, and usually do not identify the impediment in time or assist individual care. IoMT technologies, such as wearable sensors, provide real-time data acquisition and remote monitoring, but the quality of data, security, and compatibility and integration with AI-based analytics are problematic. The AI-enabled wearable systems are needed to facilitate continuous, predictive, and personalized CDM. The chapter refers to the design,

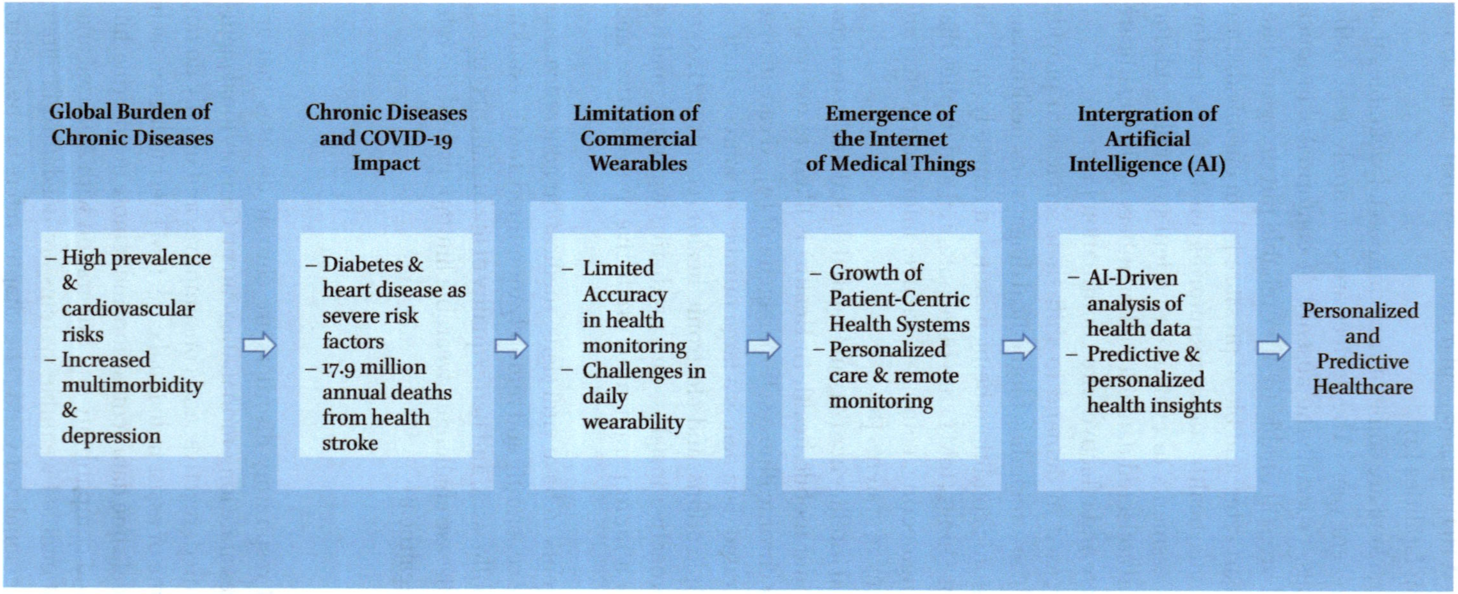

FIGURE 4.1 Evolution of CDM from global burden to AI-enabled personalized healthcare through IoMT integration. (Source: Own elaboration)

usage, and issues of incorporating wearable sensors, IoMT, and AI to overcome these constraints and improve the care of chronic diseases.

4.2 Characteristics of Chronic Diseases

Chronic diseases are long-term diseases which often require constant medical care, lifestyle change, and long-term management measures. Diabetes mellitus, cardiovascular diseases (CVDs) and chronic respiratory diseases such as the chronic obstructive pulmonary disease (COPD) are the most common chronic diseases in the world. The diseases constitute major causes of disability, compromised quality of life and death among people in the world, and contribute to a huge part of the world disease burden [13].

4.2.1 Diabetes Mellitus

Diabetes is a long-term illness which needs continuous management with the help of proper medical attention, diet planning, and physical exercise. Wearable technologies, such as insulin pumps and continuous glucose monitoring (CGM) devices, have made diabetes management much more effective as they allow tracking the level of blood glucose in real-time. CGMs are small handheld devices that constantly monitor glucose by a sensor on the skin and relay the information wirelessly to a special device or smartphone application to aid in prompt and knowledgeable decision-making.

4.2.2 Cardiovascular Diseases (CVDs)

CVDs are the major chronic health issues across the globe, and the morbidity cause of death. According to recent estimates made globally, it was estimated that approximately 17.9 million deaths could be attributed to cardiovascular conditions as at 2022. The increasing rate of the heart disease is closely connected with poor lifestyle, lack of physical activity, and bad habits such as smoking.

In this regard, wearable technologies fulfil an important task in favor of cardiovascular health by making it possible to monitor and take early preventive measures. The use of smart wearable devices in monitoring, diagnosing, and assisting in the management of heart disease is becoming more popular with positive clinical results. These devices are available in many different varieties such as smartwatches, patches, rings, and finger bands, and enable users to monitor the most important physiological attributes such as, physical activity, blood pressure, heart rate, thoracic fluid levels, and electrocardiographic indicators:

$$Y = \{y_j\}(-\infty < j < +\infty) \tag{4.1}$$

In eq. (4.1), Y denotes the entire collection of elements. $\{y_j\}$ represents individual elements of the sequence, where each element is indexed by j. j denotes an integer-valued index that identifies the position of each element in the sequence. $-\infty < j < +\infty$ indicates that the index j ranges over all integers, meaning the sequence is bi-infinite, extending infinitely in both negative and positive directions [14]:

$$y_{i-M}, \ldots, y_j, \ldots, y_{i+M}, \tag{4.2}$$

In eq. (4.2), y_{i-M} denotes the element located M indices before the central element. y_{i+M} indicates the element located M indices after the central element. y_j represents a generic element within the window:

$$x_i = \text{Med}[y_{i-N}, \ldots, y_i, \ldots, y_{i+N}] \tag{4.3}$$

Equation (4.3) defines a median filtering operation applied to an electrocardiogram (ECG) signal, where x_i is the filtered ECG sample at index i, N denotes a positive integer specifying the half-window length of the filter, and the set $\{y_{i-N}, \ldots, y_{i+N}\}$ represents a local window of $2N+1$ consecutive ECG samples centered at i.

The main advantage of wearable cardiovascular devices is their ability to provide real-time, accurate information that can be safely transmitted to healthcare professionals to undergo clinical assessment and offer individualized advice. State-of-the-art wearables, such as fitness trackers, smartwatches, and so on, incorporate such functions as viewable daily activity tracker and an inbuilt ECG, thus enabling proactive and data-infused management of heart-related diseases.

4.2.3 Chronic Obstructive Pulmonary Disease (COPD)

COPD denotes a chronic respiratory disease that impacts millions of people across the entire globe, leading to a high rate of morbidity and mortality. Wearable gadgets assist in real-time control and monitoring of COPD, which can be diagnosed and self-managed. Blood oxygen saturation and heart rate are the most important indicators that should be monitored. In the above case, regular exercise and body movement could help improve the outcomes of COPD patients.

4.3 Overview of Wearable Sensor Technologies

The digital healthcare systems comprises wearable sensor technologies that allow real-time, noninvasive physiological monitoring, biochemical, and behavioral parameters. With the integration of these sensors and the IoMT platforms, it assists in long-term monitoring and individualized management of the chronic diseases [15].

4.3.1 Biochemical Wearable Sensor

Wearable electrochemical sensors that do not involve invasive procedures have already been developed to identify target analytes in biological fluids which are easily accessible, including perspiration, saliva, and tears. Nevertheless, the constant availability of blood through extracorporeal wearable systems is technically difficult because of challenges of invasiveness, biocompatibility, and user comfort. Wearable biochemical sensors are normally placed near the local physiological stimuli or source of the biological fluid, thus allowing real-time measurements of the relevant biomarkers. The sensors have a high-resolution of time by recording fluctuations in both the concentration levels and dynamic changes in target analytes on long durations of monitoring based on their designed placement.

4.3.2 Wearable Physiological Sensor

Wearable physiological sensors can monitor the most important physiological parameters including body temperature, respiratory rate, pulse rate, and other noninvasive bioelectrical energy that is given out through the skin. Physiological wearable sensors are gaining flexibility, stretchability, and portability, such that the skin can be firmly attached without being detached when during physical activity or exercise. The frequencies of the biopotential signal recorded with noninvasive wearable sensors generally lie between 1 Hz and a few kilohertz with voltage ranges between microvolts (1 μV) and volt (1 V). To effectively record these feebly electrophysiological signals but without using conventional sensing technology, wearable electrophysiological sensors should be characterized by high electrical conductivity, increased signal-to-noise ratio, rapid response time, and low repeatability error, hence, enabling correct and quality signal processing and recording.

4.3.3 Wearable Sensors Based on Textiles

Textile-based wearable sensors attract considerable research attention because of their light weight, pliability, and mechanical deformation resistance. Their high softness, abrasion resistance, and smoothness in being integrated in daily apparel, make them viable in wearable health monitoring and human motion detection, smart home care, and medical diagnostics. Additional considerations like on-body requirements demand issues of wearing comfort, flexibility, durability, washability, and biocompatibility besides sensitivity and strain range. Nevertheless, there is a big challenge to get high-performance electronic textiles which have long-term mechanical robustness and environmental stable.

4.3.4 Continuous Glucose Monitoring Sensor

CGM sensors are wearable biochemical devices which can monitor glucose level in interstitial fluid continuously, and provide critical information in the management of diabetes. The essential data on the biomarker dynamics detected in biological fluids would indicate important information on the tissue physiology and could be applied in the context of providing individual medical care. Wearable microdialysis based sensors permit continuous in situ sampling and chemical analysis of the target biomolecules with micro peristaltic pumps and droplet microfluidics. Although quite precise, these machines are supposed to be placed by professionals, are sensitive to constant electricity, and need emptying on a regular basis, which is why they are currently only applicable in hospitals. Future developments can enable them to be incorporated into home medical systems.

4.4 IOMT Architecture for Smart Wearable Systems

The IoMT architecture offers a structured framework for connecting wearable devices within healthcare systems, supporting scalable, secure, and interoperable monitoring solutions. In the context of smart wearables, a typical IoMT architecture is organized into multiple layers, as depicted in Figure 4.2.

4.4.1 Sensor Layer

The layer includes miniaturized wearable/implantable sensors that obtain patient-specific biochemical and physiological data, such as measures of activity, blood pressure, glucose levels, and heart rate. The signals gathered are then sent wirelessly to the next layer with a low power communication protocol such as MedRadio spectrum, Bluetooth, and Wi-Fi such that they can be tracked in real time and offers a high degree of reliability.

4.4.2 Gateway Layer

The small processing and storage of wearable sensors, raw data are not processed significantly locally before being sent to the gateway layer. The layer is usually composed of the smartphone of the patient or a special access point that has superior processing and storage facilities. Gateway devices are used to conduct initial processes, such as validation of data, short-term buffering, and preprocessing with

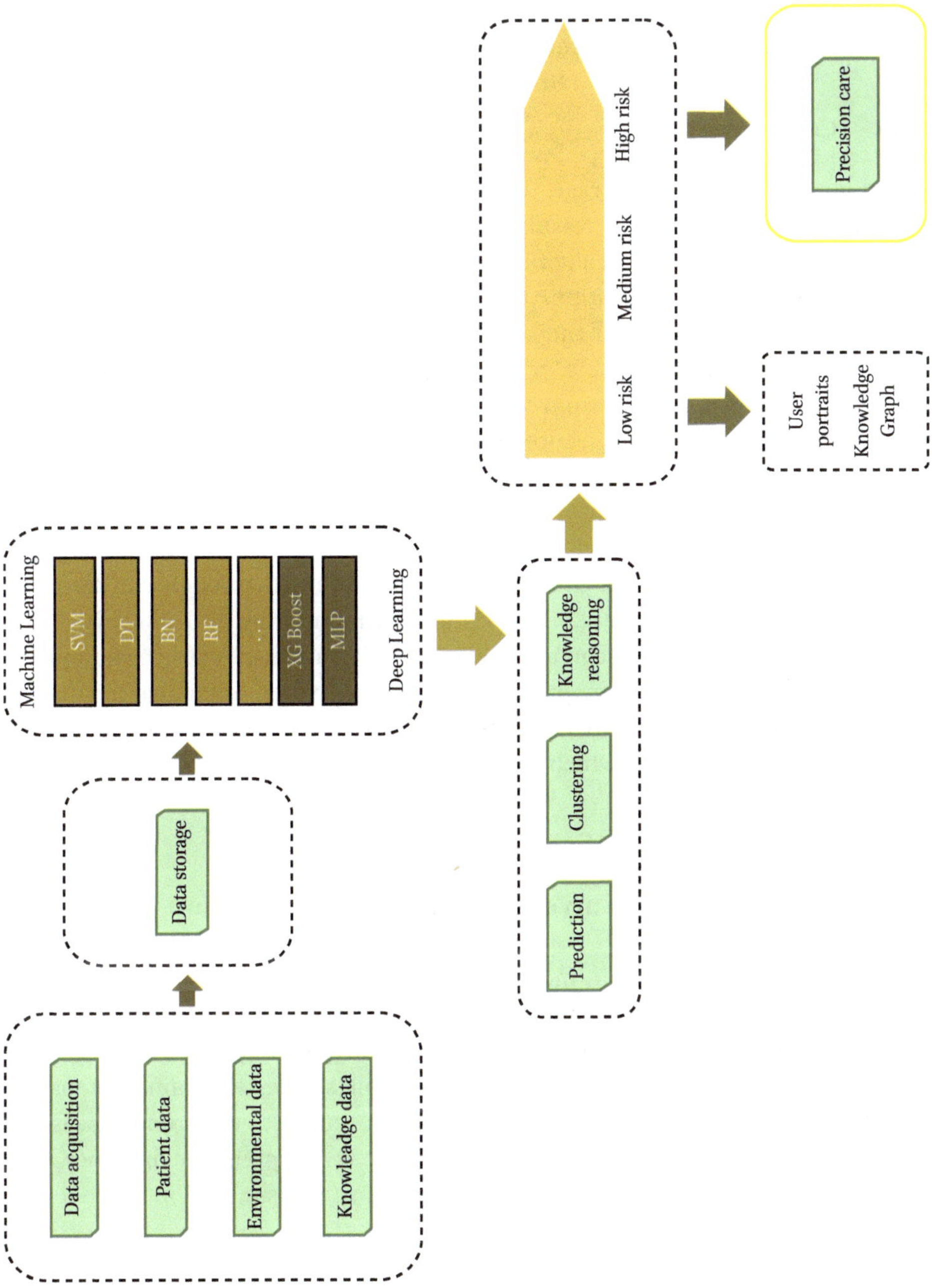

FIGURE 4.2 Architecture of IoMT systems. (Source: Own Elaboration)

the help of AI. The resulting data is then sent safely to cloud solutions where sophisticated analytics, long term storage, and clinical integration are done.

4.4.3 Cloud Layer

Cloud layer offers secure storage, data volume management, and password controlled access to information relating to patients. The incoming data is processed by advanced analytics and AI algorithms, which identify the trends, abnormalities, and possible disease exacerbations and provide clinicians and patients with actionable insights. The security of the system is also addressed by this layer such as generation of unique identifiers, cryptography key management, and access control. It also enables remote-controlling and managing wearable devices, which ensure scalable, reliable, and secure monitoring of chronicdiseases.

4.4.4 Visualization Layer

The processed data and analytical insights are presented through clinician and patient-friendly interfaces, enabling continuous monitoring and informed decision-making. These insights can assist clinicians to customize intervention plans, change the treatment regimen, prescribe physical activities, or change the dose of medication. The layer enables customized care, distance monitoring, and increased interaction with patients, which will assist in proactive and accuracy-focused CDM [16].

4.5 Chronic Disease Surveillance and Early Prevention

AI is moving away to a more model-focused paradigm into data value-focused, contextually integrated, and adaptive-focused approaches. In this regard, the intersection of medical and engineering practices with AI is a key point in developing CDM. In this section, the use of AI-based wearable sensors to maintain constant monitoring, early detection, and timely alert to reoccurrence of chronic diseases is outlined. Combining multimodal physiological data with intelligent analytics, they assist in the proactive interventions, prediction of the risks, and individual care delivery.

Combining artificial intelligence (AI) with medical engineering technologies enables effective chronic disease management, accurate monitoring, and prediction (Figure 4.3). Wearable sensors and noninvasive devices continuously collect data on environmental factors, behavior, and physiology of patients and send them to the cloud-based platforms via the wireless communication networks. The analytics based on AI is then implemented to analyze and read the data and determine

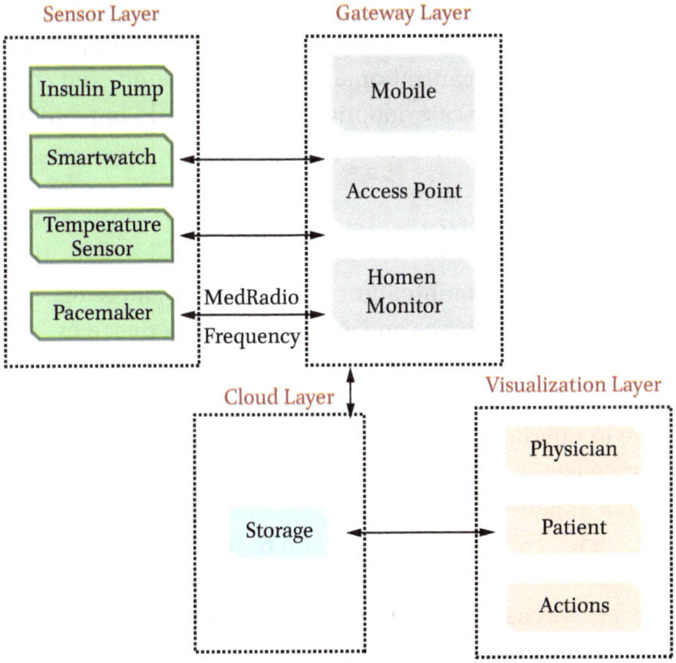

FIGURE 4.3 Architecture of an intelligent healthcare system using ML/DL models for user profiling and precision care. (Source: Own elaboration)

the progression of diseases and predict risks in time. The approaches of machine learning (ML) and deep learning (DL) can be used to combine heterogeneous sources of data in order to detect health risks, detect underlying patterns in pathophysiology, and aid in personalized care planning. Moreover, such intelligent systems are dynamic and optimizing treatment and monitoring strategies according to changes in the health condition of patients, which enhances clinical performance and general quality of care.

4.5.1 Chronic Disease Precision Care

Engineering and medical science have been integrated, which has accelerated the rate of AI integration in CDM. Regarding the utilization of the human-computer interaction (HCI) concept, this part examines how AI may facilitate individual chronic disease care to promote patient engagement, usability, and acceptance.

4.5.2 AI-Driven User-Centered Care

With the advancement of "Internet plus" nursing, AI enables personalized chronic disease care through dietary guidance, exercise recommendations, and intelligent health coaching systems. User-centered design improves patient engagement and acceptance. However, limitations such as restricted monitoring capabilities, uncertain clinical effectiveness, insufficient organizational and policy support, and constrained funding resources hinder large-scale adoption and real-world implementation of AI-driven chronic disease interventions.

4.5.3 Human-Computer Interaction (HCI)

The consideration of the user's needs, a combination of medicine, engineering, and AI, should result in increased user acceptance and engagement experience by balancing HCI. Indicatively, it, which incorporated wearable devices, open environmental data, AI-enhanced telecare systems, and environmental intelligence technologies, has been effective in offering patients with chronic disease personalized health services. Other lifestyle factors that require additional investigation of data collection technology are not exhaustively covered by the lifestyle data including sleep, heart rate, and steps. Digital twin models would also be implemented to offer automated personalized recommendations to patients with chronic diseases on health promotion. Moreover, AI-based chatbots use multimodal engagements, and natural language processing, to improve self-management when chronic disease patients are involved [17].

Physiological and biomedical data collected from wearable sensors are transmitted to the m-Health platform, where they are aggregated and processed using Big Data frameworks such as Apache Spark. These platforms enable efficient handling of high-volume, high-velocity, and heterogeneous health data generated by wearable and IoMT devices.

The processed data is the input of the AI models to infer valuable patterns, forecast health-related risks, and create actionable clinical insights. On the basis of these understandings, the m-Health application triggers alerts, ideas or clinical interventions, which preserves a patient under constant supervision and provides healthcare in a timely manner. The real-time decision-support service is particularly relevant to treat chronic conditions that require constant monitoring and interventions at an early stage.

The design uses a methodical approach that consists of information gathering, preprocessing, AI model development and validation, and the combination of AI and Big Data analytics with ease. The system will be scalable and flexible in architecture that will support the changing healthcare requirements and the increasing amount of data. The m-Health framework enhances the clinical decision-making process, patient outcomes, and individualized CDM with the help of the instruments of highly-developed analytics and intelligent automation [18].

4.6 Applications of Wearable IoMT in CDM

The wearable technologies complement the telemedicine by enhancing personalized treatment, continuous monitoring, and early diagnosis. The adoption needs patient trust, data security, and digital literacy to increase engagement, outcomes, and equitable CDM [19].

4.6.1 Wearable Technology in Telemedicine: Improving Chronic Care

Telemedicine has gained a significant role in the contemporary health services particularly in the long-term care of chronic illnesses. It enhances the access to healthcare services, decreases healthcare cost, and enhances patient outcomes. Telemedicine systems can also be used to augment CDM by providing qualitative remote, continuous, and real-time patient monitoring through the use of wearable devices.

Wearable technology collects data that is physiological and behavioral in nature, such as vital signs, frequency of heart rate, exercise, and sleep patterns, in a continuous manner. The data can be helpful in discovering the possible problems early and the progression of the disease, which will enable medical practitioners to take preventative measures. In addition, wearable technologies would be useful in promoting self-management in patients, including compliance to an intervention regimen and healthy lifestyle change, and improving patient engagement.

Similarly, in the treatment of chronic heart failure, wearable technology is used to monitor the intensity of movement and heart rate, so as to detect early signs of exacerbation. The approach, along with long distance, and remote assessment and treatment provided by telemedicine, can help prevent early intervention, reduce hospital stays, and healthcare costs.

The wearable sensors and other performing devices including smartwatches, patches, rings, and sensors in clothing enable the monitoring of a wide range of physiological measurements, including oxygen saturation (SpO_2), heart rate, activity, blood pressure, and sleep quality. IoMT-enabled CDM systems are based on the multimodal data streams.

Wearable sensors, mobile apps, and AI have proven to have great potential in diabetes care. CGM and lifestyle-based AI-driven analytics are effective in terms of glycaemic control, decreasing the rates of hypoglycemia and improving patient satisfaction. These technologies respond to the issues of heterogeneous and high-dimensional healthcare data since they allow making personalized and predictive diabetes management possible.

4.6.2 Patient Trust and Compliance in Digital Health

Wearable health technologies in CDM requires patient trust. The issues of data privacy, security, and AI utilization can act as obstacles to engagement. Providers need to enforce effective security protocols, convey them effectively, and provide their patients with the control of their data.

Education to address digital literacy in both the patients and clinicians is an important topic, and the affordability of access and a steady internet connection is vital to prevent expanding health-related inequalities. Although digital health provides an opportunity to offer personalized care, the number of administrative burdens, including EHR tasks, can decrease the productivity of clinicians and influence patient care. Wearable IoMT systems should be streamlined and easy to use hence critical to successful adoption.

4.7 Challenges

The IoMT systems have major challenges such as data privacy and security, interoperability, big data management, energy consumption, data reliability, and complexities of AI network design. The management of the issues is vital to monitor them accurately, securely and continuously to have effective, scalable, and reliable CDM.

4.7.1 Data Privacy and Security

The continuity of generating and sharing sensitive health data makes IoMT systems susceptible to cybersecurity attacks. Hackers, breaches of information, and malicious cyberattacks may compromise patient safety and trust. To ensure protection of patient data, it is necessary to use strong encryption, authentication, secure communication protocol, and access control.

4.7.2 Interoperability and System Integration

IoMT ecosystems comprise heterogeneous devices, platforms, and standards of communication. Assuring a clear interoperability and integration of various healthcare information systems is still difficult, and it needs standard format of the data, open architecture, and compatible communication protocols.

4.7.3 Big Data Management and Analytics Challenges

IoMT devices generate huge amounts of real-time healthcare data. Raw sensor data needs to be converted into actionable clinical knowledge but the mechanisms of data storage, processing, analysis, and sharing must be efficient, scalable, and reliable.

4.7.4 Energy Efficiency and Power Management

Majority of the IoMT devices have limited power requirements because they have batteries. Constant sensing and wireless communication are highly energy consuming and therefore power-efficient hardware design, energy conscious routing protocol, and adaptive duty-cycling are needed.

4.7.5 Data Reliability and Quality Assurance

The accuracy of data produced by IoMT is critical to clinical decision-making. Noise in sensors, interference in the environment, and device failures may compromise the quality of data. To attain reliable healthcare analytics, it is important to implement data validation, error correction, and quality control mechanisms [20].

4.8 Conclusion

The chapter has explored the transformative nature of multifaceted wearable sensor technologies that are incorporated in the IoMT and AI in CDM. Such chronic conditions as diabetes, CVDs, and breathing problems demand constant and individual care, which is not always afforded by traditional healthcare organizations. Wearable sensors include physiological, biochemical, textile, and CGM sensors, which enable noninvasive health data gathering in real time and prevention of disease management and early detection of the disease. Such sensors are incorporated in the layered IoMT architectures to aid in the provision of controlled transmission of data, cloud analytics, and data visualization that is easily accessible to the clinicians and the affected patients. The other aspects of AI and Big Data analytics are enhanced decision-making that benefits in terms of predictive insights, system of early warning, and precision care based on individual needs. However, interoperability, data security, energy efficiency, and data reliability still exist as problems. Altogether, wearable sensor convergence, IoMT, and AI are a paradigm shift into preventive, predictive, and personalized healthcare, and it has a great potential to assist in improving the clinical outcome and quality of life of long-term illnesses. The future focus should aim at making ultra-low-power, high-precision, and biocompatible wearable sensors to enable sustained and continuous monitoring. The focus on interoperability standards and privacy-preserving data sharing between the platforms of the IoMT should be even greater. Real-time personalized decision support at lower latency can be made a possibility through the integration of advanced AI models, digital twins, and edge intelligence. Also, it requires user-centric design, clinical validation, and enabling regulatory environments to guarantee scalability, confidence, and pervasive user of wearable IoMT to CDM solutions.

References

[1] Ma Y, et al. Relationship between chronic diseases and depression: the mediating effect of pain. BMC Psychiatry. 2021;21(1):436. https://doi.org/10.1186/s12888-021-03428-3.

[2] Rashid J, et al. An augmented artificial intelligence approach for chronic diseases prediction. Frontiers in Public Health. 2022;10:860396. https://doi.org/10.3389/fpubh.2022.860396.

[3] Lee Y, et al. Standalone real-time health monitoring patch based on a stretchable organic optoelectronic system. Science Advances. 2021;7(23):eabg9180. https://doi.org/10.1126/sciadv.abg9180.

[4] Mucchi L, et al. Tele-monitoring system for chronic diseases management: requirements and architecture. International Journal of Environmental Research and Public Health. 2021;18(14):7459. https://doi.org/10.3390/ijerph18147459.

[5] Hacker K. The burden of chronic disease. Mayo Clinic Proceedings: Innovations, Quality & Outcomes. 2024;8(1):112–19. https://doi.org/10.1016/j.mayocpiqo.2023.08.005.

[6] Yadav SP, Yadav S. Image fusion using hybrid methods in multimodality medical images. Medical & Biological Engineering & Computing. 2020;58(4):669–87. https://doi.org/10.1007/s11517-020-02136-6.

[7] Rani P, et al. Simulation of the lightweight blockchain technique based on privacy and security for healthcare data for the cloud system. International Journal of E-Health and Medical Communications (IJEHMC). 2022;13(4):1–15. https://doi.org/10.4018/IJEHMC.309436.

[8] Shafiq M, et al. Advances in IoMT for healthcare systems. Sensors. 2023;24(1):10. https://doi.org/10.3390/s24010010.

[9] Du Y, et al. Artificial intelligence in chronic disease self-management: current applications and future directions. Frontiers in Public Health. 2025;13:1689911. https://doi.org/10.3389/fpubh.2025.1689911.

[10] Xie Y, et al. Integration of artificial intelligence, blockchain, and wearable technology for chronic disease management: a new paradigm in smart healthcare. Current Medical Science. 2021;41(6):1123–33. https://doi.org/10.1007/s11596-021-2485-0.

[11] Yadav E, et al. Internet of Medical Things (IOMT) for chronic disease management. 2024;13:2278–0181. https://doi.org/10.17577/IJERTV13IS060134.
[12] Alamri M, et al. Edge-driven disability detection and outcome measurement in IoMT healthcare for assistive technology. Bioengineering. 2025;12(10):1013. https://doi.org/10.3390/bioengineering12101013.
[13] Ali U. The impact of wearable devices on the management and treatment of chronic medical conditions. Archives of Clinical Psychiatry. 2022;49(2). https://doi.org/10.15761/0101-60830000000415.
[14] Li H, et al. Wearable wireless physiological monitoring system based on multi-sensor. Electronics. 2021;10(9):986. https://doi.org/10.3390/electronics10090986.
[15] Luo H, Gao B. Development of smart wearable sensors for life healthcare. Engineered Regeneration. 2021;163–70:2. https://doi.org/10.1016/j.engreg.2021.10.001.
[16] Ghubaish A, et al. Recent advances in the internet-of-medical-things (IoMT) systems security. IEEE Internet of Things Journal. 2020;8(11):8707–18. https://doi.org/10.1109/JIOT.2020.3045653.
[17] Dong C, et al. Precision management in chronic disease: an AI empowered perspective on medicine-engineering crossover. iScience. 2025;28(3). https://doi.org/10.1016/j.isci.2025.112044.
[18] Sitaraman SR. Optimizing healthcare data streams using real-time big data analytics and AI techniques. International Journal of Engineering Research and Science & Technology. 2020;16(3):9–22. https://ijerst.org/index.php/ijerst/article/view/58.
[19] Adepoju VA, et al. Wearable technology in the management of chronic diseases: a growing concern. Chronic Diseases and Translational Medicine. 2025;11(02):117–21. https://doi.org/10.1002/cdt3.156.
[20] Mathkor DM, et al. Multirole of the internet of medical things (IoMT) in biomedical systems for managing smart healthcare systems: an overview of current and future innovative trends. Journal of Infection and Public Health. 2024;17(4):559–72. https://doi.org/10.1016/j.jiph.2024.01.013.

CHAPTER 5

Smart Healthcare Services and Applications

Anitha Josephine, Abhishek Godara, Vipul Vekariya

Abstract Smart healthcare has become a revolution that is being led by the development of digital technologies in the form of the Internet of Medical Things (IoMT), wearables, artificial intelligence (AI), and cloud computing (CC). These technologies provide the chance to observe constantly, predict disease in its initial stages, and provide individualized care, which the traditional healthcare model fails to accomplish. Challenges include the disintegration of services, low interoperability, the security of data, and the absence of integrated smart health systems. This overview offers broad knowledge of smart healthcare services and applications by analyzing their most significant technological enablers, principles of their functional architecture and real-life applications of smart healthcare services in current healthcare systems. This chapter analyzes the fundamental services and applications in smart healthcare, such as data acquisition, processing, analytics, and service delivery, and examines how these technologies support patient care and clinical processes. Smart healthcare services are designed to be used in real-time health monitoring, remote clinical support, intelligent decision-making, and patient engagement. Applications such as telemedicine, wearable health monitors, and AI-assisted diagnostics improve the level of and access to healthcare. They endorse patient-focused and scalable delivery coupled with defining issues of interoperability, data security, and ethical deployment to enable sustainable innovation.

Keywords Smart healthcare – Internet of Medical Things (IoMT) – healthcare services – wearable sensors – healthcare applications

5.1 Introduction to Smart Healthcare

The development of smart healthcare services and programs is one of the most considerable turns in the healthcare system of contemporary times as it accepts using the latest digital resources along with the traditional healthcare measures. One such technology in focus is the Internet of Medical Things (IoMT) that enables the simplified connection between medical equipment, wearable and implantable sensors, patients, and medical staff. Smart healthcare is an IoMT application, which is integrated with artificial intelligence (AI), big data analytics, cloud computing (CC), and mobile health (mHealth) technologies to deliver effective, precise, and patient-centric healthcare services. This digital revolution makes healthcare systems more responsive and allows care to be provided beyond the hospital setting [1, 2].

The IoMT ecosystem includes the wearable devices, implantable devices, smart diagnostic devices, and connected hospital systems that collect and transmit physiological and clinical data in real time. With the help of those data streams, one can receive helpful information concerning the health of a patient and identify the diseases at the earliest stage, manage chronic diseases effectively, monitor patients at a distance, and plan the individual treatment. The medical information available in real time that has the option of being shared over medical networks will enhance clinical decision-making and reduce the time gap in the diagnosis and treatment process hence enhancing patient safety and treatment outcome [3].

IoMT supports a broad portfolio of smart healthcare applications, which include telemedicine systems, smart emergency response systems, electronic health records (EHRs), and automated hospital management systems. These applications enhance access to healthcare, especially for rural and underserved groups, and also facilitate the effective use of healthcare resources and the decrease in operations costs. Moreover, the IoMT-based systems would help in preventive and predictive healthcare through the analysis of vast amounts of patient data to detect patterns of risk, predict complications, and prescribe early intervention [4].

In general, IoMT implementation of smart healthcare services improves the scalability of the systems, quality of care, and patient engagement. Given the current issues of aging population, increased costs, and lack of medical professionals, which are affecting healthcare systems all over the world, smart healthcare methods, in this case, provide a viable, interconnected and intelligent method of delivering healthcare. Figure 5.1 depicts the integration of wearable, implantable sensors, telemedicine systems, and hospital systems that can support the operation of continuous data collections and intelligent healthcare delivery to support big data analytics, EHRs, automated hospital management, and preventive healthcare to achieve better patient outcomes.

5.1.1 Focus of the Research

The objective is to critically observe and report on the role of smart healthcare services and applications in changing the modern healthcare delivery model by means of incorporating the IoMT, AI, wearable technology, and cloud-based solutions. These innovations improve patient-centered care delivery, clinical efficiency, and data-driven decision-making, while overcoming the challenges associated with interoperability, data security, and sustainable deployment.

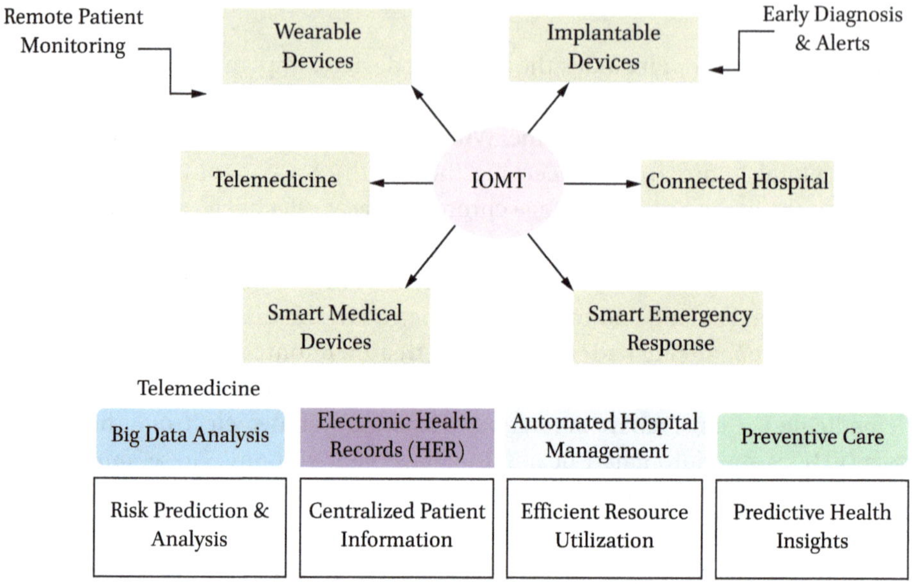

FIGURE 5.1 IoMT-based smart healthcare ecosystem. (Source: Own elaboration)

5.1.2 Research Organization

This chapter is structured as follows: In Section 5.1, the importance of smart healthcare is discussed. Section 5.2 describes the most important enabling technologies, including AI and IoMT. Section 5.3 describes the smart healthcare structure and design. Section 5.4 includes the main applications of smart healthcare. Section 5.5 brings out the advantages and implications to the healthcare setting. Section 5.6 is the closing part of the chapter that summarizes the main results and recommendations on the further work.

5.2 Technological Enablers of Smart Healthcare

This section outlines the core technologies that enable smart healthcare, focusing on the IoMT and AI/machine learning (ML). It discusses how these technologies support real-time health monitoring, intelligent data analysis, early disease detection, and patient-centric healthcare services.

5.2.1 IoMT

IoMT is one of the key technology enablers of smart healthcare, making it possible to track and monitor health problems remotely and in real time by connecting wearable sensors, medical devices, and intelligent data processing systems [5]. IoMT makes it possible to purchase and deliver physiological readings in real time, which are very helpful for keeping monitoring on senior citizens' health. ML techniques are used to process the sensed health data in IoMT-enabled smart healthcare systems. An artificial neural network is fed input data collected by IoMT devices, and the hidden-layer computation is expressed as follows:

$$\psi_j = b_1 + \sum_{i=1}^{m} \left(w_{ij}{}^{*} r_i\right), \ \varphi_j = \frac{1}{1+e^{-w_j}} \tag{5.1}$$

where r_i represents the IoMT sensor inputs, the term ψ_j denotes the net input to the hidden neuron, and φ_j is the sigmoid-activated output. The weight between the input and hidden neurons is denoted by w_{ij}, and b_1 is the bias. The index $m$indicates the total number of IoMT sensor inputs, and e is Euler's number (≈2.718). This enables effective feature extraction and learning from continuous health data. The performance of the IoMT-enabled model is evaluated using the error function:

$$E = \frac{1}{2}\sum_{k} \left(\tau_k - \varphi_k\right)^2 \tag{5.2}$$

It guides the optimization of network parameters for accurate health condition detection. τ_k is the target health condition, φ_k is the predicted output, E represents the training error used to optimize the model, and k is the index representing the output neuron or health class given in eq. (5.2). Overall, IoMT enables scalable, data-driven, and patient-centric smart healthcare services by integrating continuous monitoring with intelligent analytics.

5.2.2 AI and ML

ML and AI are considered to be key technological facilitators of smart healthcare that can help convert vast amounts of healthcare data into valuable clinical knowledge. ML algorithms enable early disease detection, accurate diagnosis, and individual treatment planning by analyzing patterns in a range of data, including medical images, EHRs, and sensor data generated by IoMT. Clinical decision support systems are AI-based programs that support healthcare professionals in their work by analyzing specific information about patients and giving recommendations in a timely manner to enhance the accuracy of the decision and decrease the clinical workload. Besides that, predictive analytics can aid in preventive healthcare because it predicts the likelihood of developing a health issue [6]. Despite the challenges, including the quality of data, the ability to interpret their models, and privacy concerns, AI and ML are at the core of smart healthcare services based on intelligence, using a patient-centered approach, and on a large scale.

5.3 Smart Healthcare Architecture and Framework

In this section, a layered model of smart healthcare systems is described in detail, as it explains the relations between IoMT devices, data processing platforms, communication networks, and healthcare applications to support long-term monitoring, intelligent analysis, and efficient service delivery. It identifies the significance of the combined functioning of data acquisition, analytics, communication as well as application layers to guarantee interoperability, scalability, security, and real-time clinical decision support to patient-centricity and data-driven healthcare.

5.3.1 Smart Healthcare Systems' Layered Architecture

This section discusses the layered architecture of smart healthcare systems and how all the levels collaborate to create effective and smart healthcare. It addresses the data acquisition layer functions, data processing and analytics layer functions, communication layer functions, and application and service layer functions to enhance the continuous data collection, safe data transfer, smart data interpretation, and service provision. This level is outlined in the section by providing how the tiered framework that is the basis of patient-centric and data-driven smart healthcare systems can be used to guarantee interoperability, scalability, and real-time decision support.

5.3.2 Data Acquisition Layer

Data was gathered from Puli Christian Hospital's outpatient clinic database and contained both structured and unstructured data from patient consultation records. Data preparation must be done initially due to the variety of data types and data structures [7]. Using the Mobius platform, one of the oneM2M standard frameworks constructed a server for collecting data. This platform server was set up in a setting with sufficient computational resources to enable efficient data collection from several users [8].

5.3.3 Data Processing and Analytics Layer

The data processing and analytics layer forms a strong part of the smart healthcare system architecture because it transforms raw and irregular health data into clinically meaningful and useful information. This layer receives the majority of the data generated by the IoMT devices, wearable sensors, and EHRs. It implements relevant processes and procedures such as data aggregation, data cleaning, data filtration, data normalization, and data preprocessing to verify the accuracy, uniformity, and dependability of the information [9]. The system eliminates noise and discrepancies to make good datasets to be further analyzed.

Additionally, state-of-the-art analytics and ML models are implemented to detect aberrant physiological changes, predict disease progression, and assess the risk of patients. They are smart algorithms and assist clinicians with preliminary alerts, diagnostic assistance, and prescriptions. Furthermore, it will be supported by cloud and distributed computing platforms that shall be employed to facilitate real-time and large-scale batch processing in such a way that, the patients will be monitored on a 24/7 basis and reacting speedily to severe health complications. Overall, this layer allows delivering the healthcare on a scale, where the lens of patient-centric and data-driven and transforming the intricate medical data to useful clinical knowledge enhancing decision-making and patient outcome.

5.3.4 Communication Layer

The smart healthcare architecture contains a communication layer that facilitates the exchange of medical information between the IoMT devices, gateways, cloud platforms, and clinical applications in a trusted and secure manner. It also promotes heterogeneity of the networking technology of Wi-Fi, Bluetooth low energy, cellular network, and standardized communication protocols to increase limitless connectivity between various healthcare facilities. The layer is significant in the integrity of data and low latency as well as interoperability in the transmission of real-time physiological signals, alerts, and clinical updates. However, IoT-based smart healthcare systems are problematic on this level, which includes congestion in the network, bandwidth, and inconsistency in latency and security threats during data exchange. Addressing these deficiencies through the standardized communication schemes, adaptive routing, and secure data transmission systems is a precondition to the provision of reliable, scalable, and patient-centered smart healthcare services according to your system structure and framework [10].

5.3.5 Application and Service Layer

The highest component of a smart healthcare system structure is called the application and service layer, and it provides healthcare services to end users. It converts analyzed health data into useful applications to support clinical decision-making, patient care, and real-time tracking. The ability of the healthcare professionals to constantly observe the status of patients and react to abnormal incidences in time makes telemedicine systems, remote patient monitoring platforms, and mHealth dashboards continuously relevant [11].

ML models and cognitive analytics are also included in this layer to aid in risk assessment, early disease detection, and personal care advice. Alerts and warning systems assist in alerting clinicians and caregivers to critical levels, when they are exceeded to take some action. Besides this, the application and service layer enhances the efficiency of healthcare by assisting in securing access to data, interoperability with EHR systems, and communication between patients and healthcare providers. Ultimately, it is crucial to allow scalable, data-driven, and patient-centric smart healthcare services.

5.3.6 Integration of Devices, Networks, and Platforms

To provide constant patient surveillance and high-quality information analytics a smart healthcare architecture based on an IoMT system requires a successful combination of sensing tools, communication systems, and healthcare applications [12]. Wearable and medical sensors measure physiological parameters that include temperature, heart rate, electrocardiogram, and oxygen saturation, and transmit them via wireless networks to a centralized e-healthcare platform where they are aggregated and analyzed. To minimize noise and ensure that the sensors have guaranteed monitoring, the sensor readings are often averaged across time using an averaging mechanism, which can be written as:

$$S_{\text{total}} = \sum_{i=1}^{n} S_{\text{Hvpp}} \tag{5.3}$$

where S_{total} denotes the total aggregated physiological data, S_{Hvpp} represents the individual sensor measurements of physiological parameters, and n is the number of sensors considered. This centralized data is then transmitted to cloud-based systems to be stored, visualized and accessed by clinicians. This type of integration guarantees intra- and interoperability of heterogeneous devices, transfer of data on networks, and integrated provision of healthcare services, which contribute to live tracking and remote forging of clinical choices. In Figure 5.2, the smart healthcare architecture is layered with patient data obtained through sensors and transmitted to secure networks to be analyzed by intelligent systems and offered as healthcare services. It allows real-time notifications, clinical intelligence, and patient care by combining IoMT, communication networks, analytics, and applications.

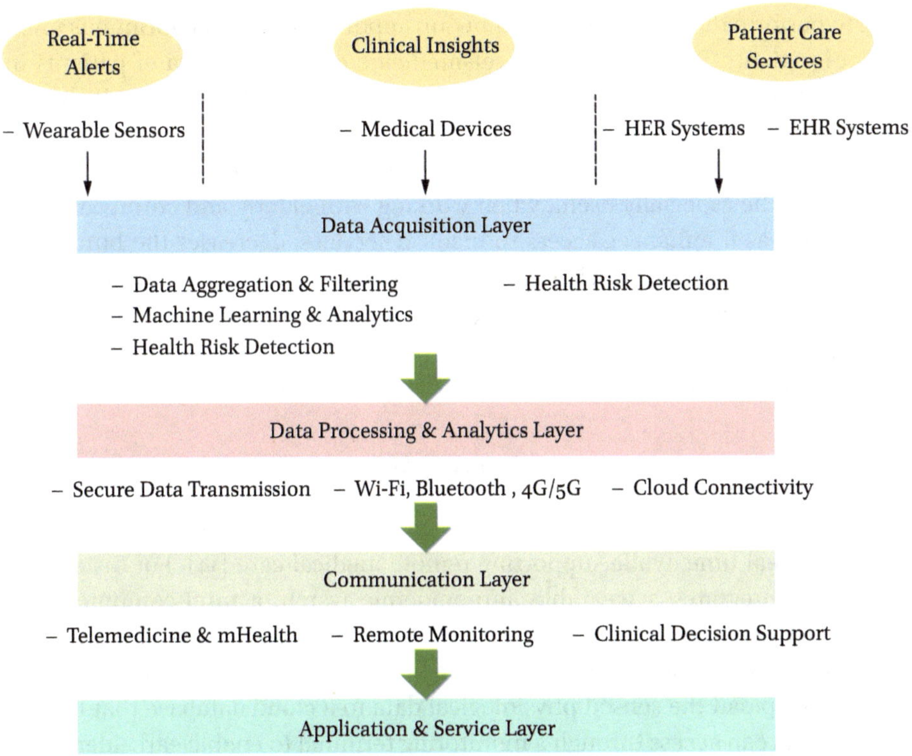

FIGURE 5.2 Layered architecture of a smart healthcare system. (Source: Own elaboration)

5.4 Smart Healthcare Applications

The section is about large-scale smart healthcare applications that use digital technologies to enhance the accessibility, efficiency, and quality of care in healthcare. It includes telemedicine and remote consultation systems, wearable-based health monitoring applications, mHealth applications, and AI-assisted diagnostic and predictive systems. The section emphasizes the fact that these applications facilitate remote care provision, persistent health, intelligent clinical decision-making, and patient interaction, thus, assist in patient-centered, data-driven, and scalable smart healthcare solutions.

5.4.1 Telemedicine and Remote Consultation Systems

Such intelligent applications of healthcare as telemedicine and remote consultation systems are essential to offer the patient a new chance to receive medical institutions and specialist consultation without physically coming to a hospital.

These systems are based on smart devices, communication system and digital health to perform distant diagnosis, follow-up appointments and continuous monitoring of patients. Smart healthcare telemedicine entails live data of patients in which wearable and IoMT-based devices gather patient's data and permit the virtual consultation software to enable the healthcare practitioners to determine the status of their patients and deliver healthcare services in a timely fashion [13]. Such a strategy can be especially useful when working with elderly and chronically disease patients, as it enhances access to medical services, decreases the burden on travelling, and allows for timely medical treatment. Telemedicine systems use of real-time communication, secure data transfer, and remote clinical decisions, have led to the improvement of the efficacy of healthcare approach by taking into account patient-centered and cost-effective smart healthcare delivery.

5.4.2 Wearable-Based Health Monitoring Applications

IoT-enabled wearable devices allow vital health parameters to be monitored continuously in real time, while supporting remote medical care [14]. For instance, a system that integrates a wearable anti-epidemic watch, a mini-computer, and cloud services can track body temperature, heart rate, and blood oxygen levels (SpO_2), and make this information available for medical evaluation. Such a system cannot only upload the sensed physiological data to a cloud database that healthcare providers can access through a monitoring terminal to enable early identification of critical deviations and timely medical intervention without necessarily having to be in physical contact with patients. With these wearable devices, the healthcare responsibilities of healthcare professionals are reduced as well since, they do not require manual monitoring to enable remote care to be scaled, particularly in critical and emergency situations, such as during the pandemic isolation or during the management of chronic conditions.

5.4.3 Mobile Health (mHealth) Applications

mHealth apps are critical factors in smart healthcare, as they allow providing healthcare services using smartphones and tablets as a part of IoMT and wearable devices. The applications can assist in constant tracking of physiological indicators, distant consultation, medication notifications, and customized health management. Through the collection and processing of patient information in real time, mHealth applications can enable the healthcare professional to monitor the conditions of their patients, give them timely feedback, and enhance clinical decision-making [15]. They promote patient involvement and self-management as well, through friendly user interfaces and real-time health data. In general, mHealth applications enhance access, efficiency, and continuity of care, which contribute to patient-centric and data-driven smart healthcare services.

5.4.4 AI-Assisted Diagnostic and Predictive Systems

The most prominent examples of smart healthcare applications are AI-assisted diagnostic and predictive systems that can be help to improving clinical decision-making by providing intelligent analysis of medical data. These systems use ML and deep learning algorithms with patient data obtained through EHRs, medical images, and IoMT devices to identify diseases and abnormal patterns, and predict the possibility of health risk at its earliest stage. The AI-based systems promote correct diagnosis, an individualized treatment plan, and predict the results by relying on past and present data. They also help healthcare professionals to reduce diagnostic workload and enhance efficiency, as well as provide preventive and patient-centric healthcare by enabling timely insights and automated clinical support [16].

5.5 Benefits and Impact on Healthcare Ecosystem

The section describes how the smart healthcare systems change the healthcare ecosystem to improve clinical decision-making, personalized and patient-centric care, and scalability and efficiency in healthcare delivery. It highlights the role of data-driven technologies, intelligent analytics, and interconnected healthcare systems in improving the quality of care, operational efficiency, and interaction with the patient, which leads to a more responsive, sustainable, and efficient healthcare system.

5.5.1 Enhanced Clinical Decision-Making

It is viewed as one of the main advantages of smart healthcare systems because the ability to transform clinical decision-making using data analytics, ML, and predictive models can be incorporated into clinical processes. The huge amounts of heterogeneous patient data received electronically, on the medical imaging and lab reports levels, and also the sensor data streams provided by the IoMT are analyzed by intelligent algorithms to identify hidden patterns and correlations that are hard to identify in the traditional methodology [17]. The given models allow predicting the progression of a disease, a degree of risk exposure of a patient, and potential complications in real time and can assist clinicians in making quicker and more precise diagnoses. Additionally, predictive analytics will allow responding early (by forecasting the unwanted events (disease worsening, hospital readmission, or medical emergency)) before they occur. Since smart healthcare systems can deliver timely and evidence-based information and personalized risk assessments it can enhance treatment planning, decrease diagnostic confusion, and increase the overall quality of care within a patient-centered, data-driven healthcare eco-system.

5.5.2 Patient-Centric and Personalized Healthcare

The focus on patients and personalized healthcare is one of the key advantages of smart healthcare ecosystems, as patients are directly involved in the management and control of their own health data and care procedures. The personal health records integrated with smart healthcare solutions, including data from wearable devices, clinical systems, and medical histories, can be securely accessed and shared at the will of patients [18]. This enables the medical practitioners to have a comprehensive and continuous view of the health condition of a patient, thus aiding in the improved diagnosis and treatment regimen. The process of care plan personalization based on individual analytics is also possible regarding a lifestyle, history, and active physiological measurements of a patient. Smart healthcare, therefore, leads to involvement, increased compliance with treatment and care guidelines, and improved the quality of care without compromising privacy, transparency, and trust in the healthcare ecosystem.

5.5.3 Scalability and Efficiency in Healthcare Delivery

Scalability and efficiency are the main benefits of smart healthcare systems that enable healthcare to provide its services to a huge and constantly growing number of individuals without deteriorating the performance and quality of services they provide. The latter can be used to adjust to vast volumes of health data generated by thousands of connected patients and sensors by connecting IoMT devices, cloud infrastructures, and AI-controlled resources on a smart healthcare system. The smart optimization methods contribute to the reduction of data flow on a limited basis, the decrease of the network saturation, and the reduction of the energy consumption to avoid the loss of the constant monitoring and the real-time analytics under the banner of the high cost [19]. There is the opportunity to improve access to healthcare services among the remote and rural groups and underserved populations and make the system stable and responsive. Scalable and efficient smart healthcare delivery improves the operational productivity, low healthcare costs, and more sustainable and quality patient care across the healthcare ecosystem.

5.6 Conclusion

The research proposed to aim at investigating how smart healthcare services and applications based on the IoMT, AI, wearable devices, and CC may transform healthcare nowadays and introduce it as a data-driven, patient-centered, and efficient system. The purpose of this chapter was to analyze the most influential technological enablers, layered system architecture, and the main purpose of telemedicine, mHealth, and AI-assisted diagnostics, as well as to evaluate the impact of these technologies and tools on clinical decision-making, patient engagement, and healthcare efficiency. Although the benefits of smart healthcare are enormous, its widespread adoption is not possible because of reasons such as the jeopardy to data security and privacy, the incompatibility of the different devices, the high implementation and maintenance cost, and the reliance on the accessibility of good networks and quality data. The future state of such restrictions can be reduced by enhancing the interoperability standards, data protection, and embedding new technologies such as advanced AI, 5G, edge computing, and blockchain that can enable the realization of safer, more scalable, and smarter healthcare.

References

[1] Vergütz A, et al. An architecture for the performance management of smart healthcare applications. Sensors. 2020;20(19):5566. https://doi.org/10.3390/s20195566.

[2] Paul S, et al. Industry 4.0 applications for medical/healthcare services. Journal of Sensor and Actuator Networks. 2021;10(3):43. https://doi.org/10.48550/arXiv.2110.04997.

[3] Rani P, et al. Simulation of the lightweight blockchain technique based on privacy and security for healthcare data for the cloud system. International Journal of E-Health and Medical Communications (IJEHMC). 2022;13(4):1–15. https://doi.org/10.4018/IJEHMC.309436.

[4] Pujahari RM, Khan R, Yadav SP. Federated Autonomous Deep Learning for Distributed Healthcare System. Federated Intelligent System for Healthcare: A Practical Guide. 2025;25–59. https://doi.org/10.1002/9781394271382.ch2.

[5] Khan MF, et al. An iomt-enabled smart healthcare model to monitor elderly people using machine learning technique. Computational Intelligence and Neuroscience. 2021;2021(1):2487759.

[6] Vuong QP. The potential for artificial intelligence and machine learning in healthcare: The future of healthcare through smart technologies. Future Healthcare Journal. 2024;6(2):94–98.

[7] Yang W-C, et al. Using medical data and clustering techniques for a smart healthcare system. Electronics. 2023;13(1):140. https://doi.org/10.3390/electronics13010140.

[8] Ji G, Woo J, Lee G, Msigwa C, Bernard D, Yun J. AIoT-based smart healthcare in everyday lives: Data collection and standardization from smartphones and smartwatches. IEEE Internet of Things Journal. 2024;11(16):27597–619. https://doi.org/10.1109/JIOT.2024.3400509.

[9] Kumar A, et al. A novel smart healthcare design, simulation, and implementation using healthcare 4.0 processes. IEEE Access. 2020;8:118433–71. https://doi.org/10.1109/ACCESS.2020.3004790.

[10] Upadhyay S, et al. Challenges and limitation analysis of an IoT-dependent system for deployment in smart healthcare using communication standards features. Sensors. 2023;23(11):5155. https://doi.org/10.3390/s23115155.

[11] Quy VK, et al. Smart healthcare IoT applications based on fog computing: Architecture, applications and challenges. Complex and Intelligent Systems. 2022;8(5):3805–15. https://doi.org/10.1007/s40747-021-00582-9.

[12] Igo EE, Agana MA. healthcare sensor-based physiological parameters data aggregation and analytics scheme for monitoring of patients in internet of medical things (iomt) enabled e-healthcare platform. Information Technology. 2025;8(2):1–14. https://doi.org/10.52589/BJCNIT-VRUHJJLN.

[13] Abugabah A, Nizamuddin N, Alzubi AA. Decentralized telemedicine framework for a smart healthcare ecosystem. IEEE Access. 2020;166575–88:8. https://doi.org/10.1109/ACCESS.2020.3021823.

[14] Wu J-Y, et al. IoT-based wearable health monitoring device and its validation for potential critical and emergency applications. Frontiers in Public Health. 2023;11:1188304. https://doi.org/10.3389/fpubh.2023.1188304.

[15] Albahri OS, et al. Smart Real-Time IoT mHealth-based Conceptual Framework for Healthcare Services Provision during Network Failures. Applied Data Science and Analysis. 2023;2023:110–17.

[16] Pradhan B, et al. An AI-assisted smart healthcare system using 5G communication. IEEE Access. 2023;11:108339–55. https://doi.org/10.1109/ACCESS.2023.3317174.

[17] Ghimire P, Adhikari S. Advancing Smarter Healthcare Through Data Analytics: A Study on the Integration of Machine Learning and Predictive Models in Clinical Decision-Making. Advances in Computational Systems, Algorithms, and Emerging Technologies. 2025;10(4):21–36.

[18] Aldamaeen O, Rashideh W, Obidallah WJ. Toward patient-centric healthcare systems: Key requirements and framework for personal health records based on blockchain technology. Applied Sciences. 2023;13(13):7697. https://doi.org/10.3390/app13137697.

[19] Rattal S, et al. AI-Driven Optimization of Low-Energy IoT Protocols for Scalable and Efficient Smart Healthcare Systems. IEEE Access. 2025;13. https://doi.org/10.1109/ACCESS.2025.3551224.

CHAPTER 6

Event-Driven and Microservices Architecture (EDA and MSA)

Praveen Priyaranjan Nayak, Manivannan Karunakaran, Ritu Das

Abstract Selecting the appropriate architecture for big data and IoT-enabled applications is a complex task, requiring knowledge of the problem domain, anticipated system requirements, and the evolving technology landscape. Traditional monolithic architectures often struggle to handle high-throughput data streams and dynamic workloads, making scalable and extensible solutions essential. This chapter presents an integrated system leveraging micro-services architecture (MSA) and event-driven architecture (EDA) for real-time data processing in big data and IoT environments, monitoring using wearable and contactless sensors. EDA concentrates on asynchronous communication via events, enabling organizations to respond to modifications in real time while keeping elements loosely coupled. MSA divides systems into discrete, separate services, each with a specialized business purpose, allowing for quicker creation, implementation, and adaptability. Combining EDA with MSA improves system durability, scalability, and reliability by allowing microservices to communicate via events streaming instead of direct calls, decreasing interconnections, and enhancing reliability. A conceptual flow model is presented, demonstrating how event-driven communication orchestrates microservices to execute complex business processes efficiently. The chapter underscores the practical benefits of combining EDA and MSA, including improved systems responsiveness, modularity, and support for real-time analytics, making them indispensable for designing contemporary cloud-native applications.

Keywords Event-driven architecture (EDA) – microservices architecture (MSA) – internet of medical things (IoMT) – wearable and contactless sensors – real-time systems

6.1 Introduction

Modern software design and implementation system has been transformed radically with the rapid advancement of big data technologies and Internet of Things (IoT) gadgets. Sensors such as wearable sensors and contactless sensors are now generating vast volumes of real-time information that is on-demand in healthcare, smart environments, industrial monitoring, and intelligent transport systems. Such fast flows of data present severe structural issues in processing and analysis in a reliable and prompt manner. Traditional monolithic software designs using tightly

 | HTTPS://DOI.ORG/10.1515/9783112237083-006

coupled subunits and synchronous versions of communication could often not scale well and responsiveness to dynamic workloads, bottlenecks to performance, lack of system resilience, and scalability [1]. To address the constraints, the new cloud-native systems are shifting toward the more architectural paradigms to support modularity, scalability, and responsiveness. Microservices architecture (MSA) and event-driven architecture (EDA) are one of them and have become the popular variants to create the distributed systems capable of supporting the applications requiring real time and data-intensive. MSA breaks down big applications into a group of small and independent services, each of which deals with a desired business capability [2]. This division enables autonomous deployment and scaling of services and increases the system maintainability and accelerates the innovation cycle. Within the medical field context, the architectural issues are additionally exacerbated by the development of the Internet of Medical Things (IoMT), in which connective medical equipment, wearable sensors, and contactless surveillance systems constantly produce sensitive physiological information [3]. Applications of IoMT require real-time processing, high reliability, and high-security processes to assist in clinical decision-making and patient safety. As a result, architectural designs that could withstand the burdens of the continuous stream of medical data are needed in a large scale. EDA overcomes these shortcomings by allowing asynchronous communication with events that can be interpreted as an important change in state [4]. The producers place an event in an event system and are unaware of the consumers; it therefore implies that multiple services can react to an event in an independent way. This loose coupling allows system flexibility, fault isolation as well as scalability, particularly those systems with continuous data streams and unpredictable workloads [5]. The purpose of this chapter is to consider the integration of the EDA and the MSA to obtain the effective processing of the information in real-time in large-scale data and IoT-based systems, in particular, when they are applied to wearable and contactless sensors. The chapter compares the main concepts, the structural features, and the interaction patterns of EDA and MSA to help illustrate how their combination lead to improved real-time monitoring, analytics, and decision-making.

6.2 IoMT Devices and Sensor-Based Data Collection

The IoMT is transforming the healthcare, as it enables lasting and remote monitoring of patients with the help of a connection of interconnected medical devices and sensors. Unlike the traditional episodic check-ups, the IoMT solutions access the dynamic physiological information in real-time, and this feature provides clinicians with a real-time view of the health of a patient. The information flow contributes to the prompt clinical decision-making process, allows detecting the abnormalities early, and permits designing tailored treatment regimes. The keys to the success of IoMT and the guarantee of the accurate, reliable, and secure healthcare monitoring are efficient sensing technologies and the powerful data management infrastructure. Wearable and contactless medical sensors are at the centre of these systems, and they enable noninvasive vital sign measurement without causing discomfort and immobility to the patient. Combined with the sophisticated transmission and cloud-based data infrastructure, IoMT systems are used to provide scalable, receptive, and patient-based solutions to healthcare.

6.2.1 Wearable and Contactless Medical Sensors

IoMT-enabled healthcare monitoring is based on wearable and contactless medical sensors. Smartwatches, fitness bands, and adhesive patches, as well as clothing, all contain wearable sensors. The devices contain important physiological measurements such as:

- Heart rate
- Blood oxygen saturation (SpO_2)
- Body temperature
- Blood pressure

With sensors integrated into everyday items, patients can be continuously monitored without disrupting their regular routines; as a result, compliance is high and discomfort is minimal. Noninvasive contactless sensors, such as optical, thermal, or radar sensors, offer an addition to noninvasive monitoring of patients unable to carry a device, or requiring an unobtrusive presence, such as in a hospital ward or nursing home.The main strengths of these sensors include their passivity and real-time characteristics of data gathering, which provide the ability to identify long-term trends and even minor physiological changes that could not be detected during regular clinical evaluation. This is a vital quality, particularly in chronic disease management in which all the conditions of cardiovascular diseases, diabetes, hypertension, and stress-related disorders are to be tracked down through out. Furthermore, the longitudinal data that has been collected can be used to identify anomalies at an early stage, predictive analytics, and personal healthcare interventions. IoMT systems could be viewed as a comprehensive view of the health of a patient because vital signs are integrated with behavioral and environmental data through several sensor modalities to help make more informed clinical decisions [6].

6.2.2 Data Transmission and Infrastructure

The IoMT systems can be successfully implemented not only due to the quality of the sensors but also because of the strength of the infrastructure used and the effectiveness of the data transmission framework. Upon capturing the physiological signals, data transfer occurs between the sensors and the IoMT gateways, e.g., smartphones, tablets, or even dedicated edge devices. These gateways serve as channels of transmitting sensor data to the cloud servers where first processing and storage are done. In order to ensure that the data flow between the patient and the healthcare platform is smooth and safe, wireless communication protocols such as Bluetooth, Wi-Fi, Zigbee, or cellular networks are employed.

The preprocessing by the edge/gateway level is required to enhance the quality of data received and save bandwidth. The methods that reduce the meaningless data transfer and make the system more efficient are noise filtering, outlier detection, and compression. Once transmitted, data gets stored in the cloud-based solutions that can scale, has high computing capacity to facilitate analytics, and real-time accessibility to assist healthcare workers. The cloud-based systems can process sophisticated functions, including trend process, anomaly, predictive, and integration of electronic health records (EHRs). This top-down architecture ensures the IoMT systems are flexible, resilient, and reactive. This results in the system being capable of a very high rate of data of possibly many patients simultaneously with low latency as sensing, transmission, and analysis layers are decoupled. Furthermore, the implementation of the real-time clinical alerts with this architecture enables healthcare providers to respond as promptly as feasible to extreme events in the shape of the abnormal heart rate, oxygen desaturation, or unforeseen alterations in blood pressure [7].

6.3 Event-Driven Architecture (EDA) for IoMT

EDA offers a software engineering strategy that supports applications in responding proactively to events produced by distant elements. In the context of the IoMT, any significant change in the physiological parameter of a patient such as abnormal heartbeat, abrupt reduction of oxygen saturation, or an abnormal blood pressure could be regarded as an event. The EDA is used to manage such events in real time wherein the healthcare systems are able to make the necessary timely interventions and decisions. EDA improves scalability, fault tolerance, and flexibility of the systems through decoupling event producers (medical sensors) and consumers (analytics modules, alerting services, or dashboards). This particularly qualifies it to be used in continuous patient monitoring, predictive healthcare analytics, and alert-based clinical decision support in IoMT ecosystems.

6.3.1 Overview and Characteristics of EDA

EDA is based on three fundamental concepts, namely, event generation, event detection, and event processing. Within the IoMT systems, medical sensors and wearable devices constantly produce data that are translated into discrete events.

An event is caused by a heart rate sensor detecting a sudden episode of tachycardia, which is then fed to analytics services or monitoring dashboards. Applications or microservices that receive the events, produce insights, and cause notifications or interventions are called event consumers.

Asynchronous communication is one of the characteristic features of EDA that enables events to be executed separately and not to interrupt the activity of other systems. This makes sure that a critical notification is not postponed by the daily routine data processing jobs. Loose coupling between event producers and consumers also has a positive influence on fault tolerance: the failure of one service does not stop the whole system, but lets other services keep watching and processing events. Besides, EDA facilitates real-time responsiveness, in that the system responds directly to the medical events, instead of using periodic polling or batch processing. This sensitivity is essential in care environments, where a couple of seconds' delay could influence patient outcomes [8].

6.3.2 Role of EDA in IoMT Systems

EDA is a key to enhance patient safety, the modularity of systems, and real-time decision-making in the IoMT healthcare applications. The constant monitoring with the EDA provides the ability to generate a notification of an event: an unusual reading, a sudden drop in SpO_2, or an irregular heart rhythm sends an immediate message to a healthcare provider, caregiver, or automatic intervention system. This shortens the reaction to emergency cases and gives the opportunity to preventively treat the case.

EDA is also used to coordinate several microservices into an IoMT platform. Depending on its role, every microservice should subscribe to certain events. A data analytics service is going to process incoming vital signs, a risk assessment module is going to measure possible complications, and a visualization service is going to refresh the clinician dashboard. This modular design means that the system is easily scalable, can easily manage large volumes of events simultaneously, and can easily accommodate new medical devices or services without affecting current workflows Such event coordination plays a vital role in the dynamic patient data, variety of medical devices, and multifaceted clinical processes [9]. Figure 6.1 depicts the event-driven IoMT architecture for real-time monitoring.

6.4 Microservices Architecture (MSA) for IoMT

The MSA is a contemporary software design framework in which applications are decomposed into small, independent, and loosely coupled services, which perform a particular business or technical task. MSA facilitates the creation of medical plat-

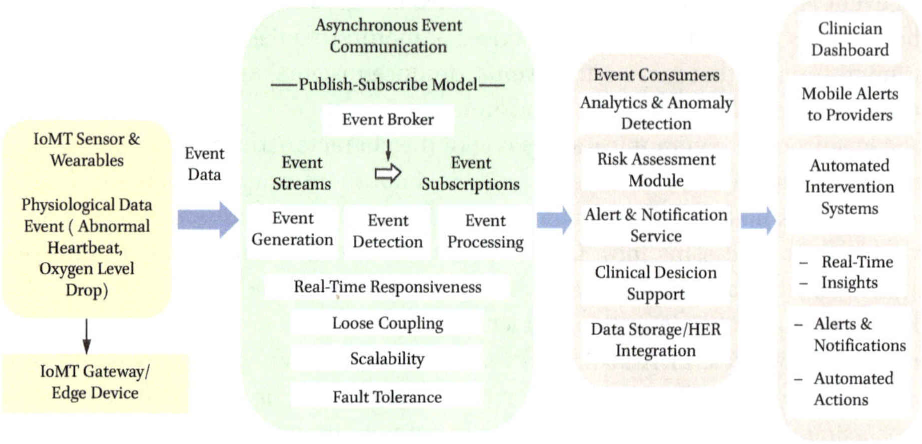

FIGURE 6.1 Event-driven IoMT architecture for real-time monitoring. (Source: Own elaboration)

forms in the context of IoMT to handle complex and heterogeneous medical services, including the following:

- Sensor data collection
- Analytics and alerts
- Patient monitoring and
- Clinical decision support

In contrast to monolithic systems, where all system parts are closely interconnected, MSA enables single services to be created, scaled, and distributed. This flexibility is essential to manage continuous streams of medical information from various sensors and devices, and to be highly available, fault-tolerant, and responsive. MSA facilitates quick innovation, simpler inclusion of new IoMT devices, and adjustable healthcare procedures through encouraging modularity and service autonomy. Therefore, IoMT systems that are developed based on microservices have the possibility to adapt to the new medical technologies, support diverse patient needs, and offer strong support to real-time, data-driven healthcare processes [10].

6.4.1 Core Principles and Components

Service decomposition is the main concept of MSA, according to which a complex IoMT system is broken into smaller, domain-specific services. Individual microservices serve to fulfill a single purpose, such as sensor data ingestion, real-time patient analytics, alert generation, or dashboard visualization, and are autonomous. Each service is accessible to others via a well-defined application programming interface (API), and it facilitates the exchange of data in a seamless manner without reliance on others. The API gateways serve as the major points of entry for client requests and handle authentication, routing, rate limiting, and load balancing, which make the interaction of services simpler. Lightweight, isolated environments for each service, with modern containerization technologies such as Docker or Kubernetes, make it possible to independently develop, deploy, and scale a service without interfering with other services. A modular design helps keep IoMT systems flexible and scalable, enabling them to adapt to evolving healthcare requirements by allowing the seamless addition of new sensors, services, or analytics modules. MSA generates resilience of systems, tolerance to faults, and operational efficiency in a dynamic healthcare setting as well because it decouples functionalities and standardizes communication.

6.4.2 Benefits and Challenges in IoMT

MSA provides a number of major benefits to healthcare platforms based on IoMT. Fault isolation can be used to make sure that the failure of one microservice, a sensor ingestion service, does not affect the whole system, and the reliability of the entire platform remains intact. This increases maintainability since the maintenance or patching of individual services is easier. Quicker functional releases enable programmers to launch new analytics models, warnings, or integrations into devices rapidly, to enable ongoing innovation in patient tracking and treatment. Nonetheless, the implementation of MSA in the IoMT creates technical issues. The data can be stored in several services, which could create a problem of data consistency and synchronization. Moreover, the presence of several microservices means that they have to be effectively monitored and managed, which necessitates powerful logging, tracing, and orchestration solutions. At that, the advantages of modularity, scaling, and adaptability make MSA a very successful architectural solution to the modern, real-time IoMT healthcare system [11]. Figure 6.2 shows microservices-based IoMT architecture for scalable healthcare.

6.5 Integration of EDA and MSA in IoMT

EDA combined with MSA is a potent model for creating the latest, scalable, and reactive IoMT systems. EDA provides real-time creation and response to medical sensor events, whereas MSA provides modularity, isolation of faultiness, and independent service control. A combination of these two architectures enables the IoMT platforms to effectively manage the consistent stream of medical data, the heterogeneous sensor devices, and the dynamic clinical processes that are typical in modern health care environments. With this integration, system flexibility is improved because event generation is decoupled from service execution, such that the services can handle medical events in an asynchronous manner. The inte-

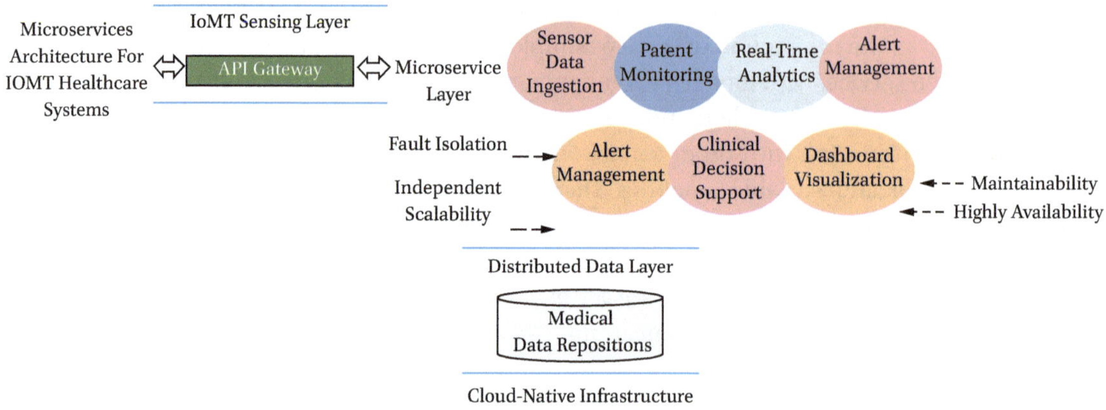

FIGURE 6.2 Microservices-based IoMT architecture for scalable healthcare. (Source: Own elaboration)

gration of EDA and MSA fosters active patient care, quick clinical response, dynamic follow-ups of complex health concerns which is particularly relevant in a high-demand setting such as a hospital ward, intensive care unit (ICU), and telemedicine platforms.

6.5.1 Motivation and Communication Patterns

The motivation behind integrating EDA and MSA in IoMT systems is to bring together asynchronous and event processing with scalability and modularity of microservices. The model assumes that medical sensors and wearable devices are continually generating data which is transformed into discrete events and transmitted to a central event broker. Microservices with task-specific functionality such as real-time analytics or anomaly detection or notification generation or dashboard visualization subscribe to the relevant events and handle them individually. With this type of communication, the IoMT systems become very resilient with the most critical alerts being relayed in cases where part of the services fails in the short run. This flow of events, from sensor to gateway, event brokers, and ultimately to microservices and end-user interfaces such as dashboards or alert systems, allows timely, reliable, and secure delivery of medical data, which is essential in a quick decision-making process in patient care. It is also compatible with the addition of new services that are made modularly (such as the addition of predictive analytics modules or AI-driven decision support) without having to design the entire system [12].

6.5.2 Conceptual System Flow Model

The conceptual system flow model shows the relationship between EDA and MSA to offer patient monitoring and clinical decision-making in real-time. Under this model, the IoMT devices continuously record the physiological readings of heart rate, blood pressure, oxygen saturation, and temperature. These readings are then converted to events, which are published to an event broker. Microservice subscribers to particular event types are involved in data validation, preprocessing, analysis, and risk assessment, and work with each other dynamically with no service-to-service dependencies. Orchestration is event-driven, meaning that services respond to events in real-time, which facilitates proactive interventions and minimizes the delay in decisions. The given model would improve the responsiveness, reliability, and adaptability of the systems and enable IoMT platforms to handle large-scale sensor networks and various clinical situations effectively. It further enables continuous improvement, since by adding new microservices or new types of events, functionality can be extended or patient care outcomes can be improved seamlessly [13].

6.6 Applications of EDA and MSA in IoMT Systems

EDA and MSA have been integrated into IoMT systems to turn them into highly flexible and smart healthcare systems. These architectures enable the continuous monitoring of patients, timely clinical decision-making, and efficient management of large-scale medical data streams by integrating the real-time responsiveness of EDA and the modularity and scalability of MSA. Sensors are considered the event producers in the IoMT systems, whereas the microservices are the event consumers that consume and act on the event autonomously. The resultant architecture supports active and continuous care delivery, enhances patient outcomes, and clinical resource optimization. It can be modified to suit a variety of healthcare uses such as chronic disease management, smart hospital environment, telemedicine, and geriatrics due to its versatility. With the support of EDA-MSA architectures, modern healthcare systems can be of great use to make sure that the critical health events are processed and responded as quickly as possible which allow to minimize the latency during the alerting process and provide real-time clinical interventions to be completed.

6.6.1 Chronic Disease and Patient Monitoring

EDA-MSA in IoMT systems play a major role in the management of chronic conditions, such as diabetes, cardiovascular disease, and high blood pressure. Medical sensors are contactless and wearable thus recording the most significant physiological information including blood glucose, heart rate, blood pressure, and blood oxygen. These readings are translated to discrete events and delivered to the event broker of the system. Subscribing microservices validate real-time data, perform data analysis, and anomaly detection. Similarly, there are the cardiac monitoring services that detect abnormal heart rates and send out emergency care alerts. The active and passive accumulation of patient data allows tracking the patient during a long time period, evaluating the trends, and providing personal medical services. Through EDA-MSA, healthcare professionals can offer real-time, information-based, and patient-centered health care and improve the management of the disease, reduce complications, and augment the patient adherence to care alternatives.

6.6.2 Smart Healthcare Environments

The EDA-MSA architectures support real-time predictive healthcare provision and monitoring of smart healthcare environments, such as hospital settings, ICUs, telemedicine environments, and elder care environments. Different IoMT devices generate streams of events that are continuous and point to vital signs of a patient, environmental conditions, and operational status of a device. Such events are sent to specialized microservices to be processed predictively and to be assessed in terms of risks and optimization of resources. Similarly, in facilities that provide elderly care, wearable gadgets are used to check the mobility, heart rate, and fall detection to provide timely responses. Telemedicine platforms require the use of EDA-MSA to integrate information provided by patients, automated notifications, and distance consultations to improve patient accessibility and continuity. The system promotes high efficiency of operation, patient safety, and quality of care through ensuring that microservices are coordinated according to event response. Overall, by integrating both EDA and MSA, the intelligent healthcare systems can be dynamic, responsive, and scaled to the needs and demands of the contemporary settings in the clinical environment [14]. Figure 6.3 shows EDA-MSA-based IoMT in smart healthcare.

6.7 Challenges and Future Directions

The IoMT systems, despite having apparent improvement in the field of scalability, real-time analytics, and patient-centered care, have multiple barriers to their implementation in clinical practices with the introduction of EDA and MSA. These problems span technical limitations, security and privacy risks, integration challenges in the systems, and human adoption. IoMT systems are expected to

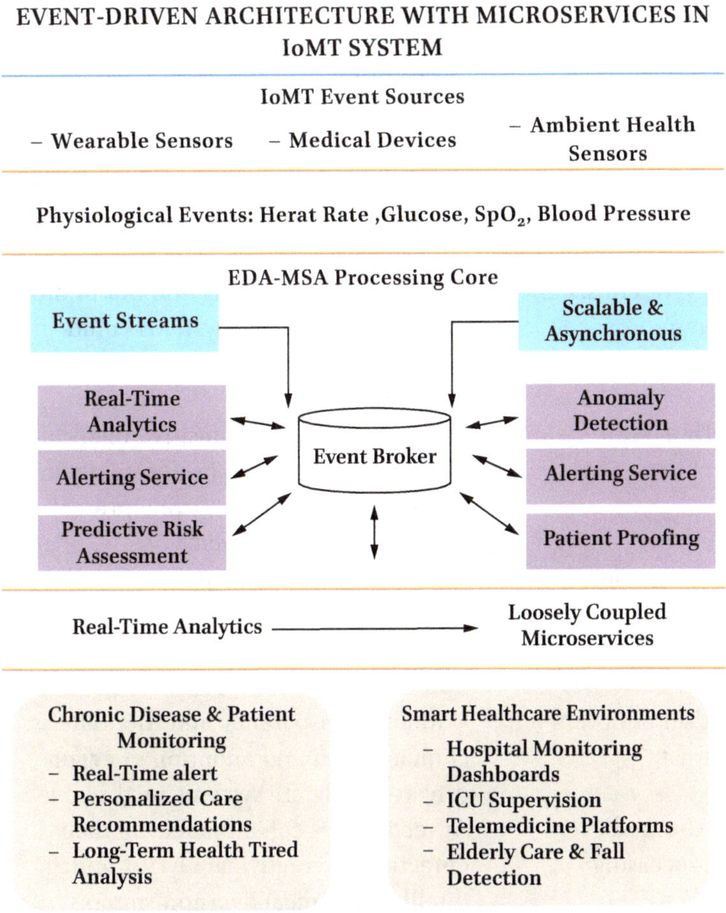

FIGURE 6.3 EDA-MSA-enabled IoMT applications in smart healthcare. (Source: Own elaboration)

manage high sensor data volatility and render dependability, steadiness and low lag processing. Furthermore, implementation of EDA-MSA architectures requires high-level infrastructure, powerful orchestrator, and continuous monitoring mechanism to maintain the performance and stability of the systems. To offer safe, reliable, and sufficient solutions to healthcare, which is IoMT-enabled, one needs to address these issues [15]. Table 6.1 offers the challenges and future research of EDA-MSA-based IoMT systems.

To allow implementing EDA-based IoMT systems using MSA in a real-life healthcare environment, the technical, security, and adoption concerns of these systems are essential issues that should be considered. The future research areas, intelligent event processing, edge enabled architecture work, and standardized frameworks will enhance future system scalability, reliability, and interoperability.

TABLE 6.1 Challenges and future research directions in EDA-MSA-based IoMT systems.

Category	Aspect	Description
Technical and adoption challenges	Sensor accuracy and reliability	Noisy or inaccurate sensor data could trigger false event and alert notifications, which could influence patient safety and clinical trust
	Event consistency and synchronization	Inconsistency in timing and out of order events in distributed systems could cripple real time clinical decision making.
	Data security and patient privacy	The transmission of sensitive medical data constantly poses both security and privacy issues and the strong protection mechanisms are needed.
	Integration with legacy systems	The incompatibility of protocols in legacy healthcare systems also makes it a challenge to integrate with modern EDA -MSA systems.
	User adoption and trust	Without transparency and friendliness, clinician trust and adoption of event-driven healthcare systems might be low
Future research directions	AI-based event processing	AI techniques can enhance anomaly detection, predictive analytics, and intelligent clinical decision support.
	Edge and fog computing integration	Processing data closer to sensors reduces latency and improves responsiveness for time-critical healthcare applications
	Security-aware microservices	Embedding security mechanisms within microservices strengthens privacy, authentication, and access control
	Standardization and interoperability	Standardized frameworks and protocols enable seamless integration of heterogeneous IoMT devices and platforms
	Adaptive orchestration and monitoring	Intelligent orchestration tools improve resource management, fault detection, and system availability

Source: Own elaboration.

These constraints can be overcome by developing an architecture of the EDA, which can subsequently liberate its potential in the early development of secure, real-time, and patient-centric healthcare systems.

6.8 Conclusion

The combination of EDA and MSA is an extremely effective platform that can assist to sustain modern IoMT based healthcare with flexibility, scalability, and sensitivity to time. EDA also allows the IoMT applications to react immediately to the significant changes in the indicators of patient health, such as irregular heartbeats, a reduction in oxygen saturation, or sudden glucose levels by applying asynchronous event processing to react to the changes in a short period of time. This ensures the timely notification and interventions which are essential to the active monitoring of the patient and improved clinical outcomes. Meanwhile, MSA decomposes rich healthcare applications to independent and modular services enabling planned systems to scale, isolate failures, and remain maintainable. Each of the microservices can develop, update, and scale automatically, i.e., the healthcare providers can add new monitoring or analytics capabilities without disrupting existing workflows. The synergy between EDA and MSA is also needed to promote real-time analytics and clinical decision support to allow healthcare systems to process large amounts of physiological data in a highly efficient manner and provide personalized recommendations. Additionally, the architecture paradigm results in enhancing the resiliency of the system, reducing downtimes, and improving reliability of dynamic clinical environments with heterogeneous IoMT devices. Overall, the EDA-MSA methodology can guarantee patient-centered care by properly managing the data, constantly monitoring the patient, and delivering smart care services. It is a key method of developing future ready, scalable, and reconfigurable IoMT health systems capable of fulfilling needs of modern day medicine, which are dynamic and changing.

References

[1] Rodríguez-Rodríguez I, et al. IoMT innovations in diabetes management: predictive models using wearable data. Expert Systems with Applications. 2024;238:121994. https://doi.org/10.1016/j.eswa.2023.121994.

[2] Pala N. Understanding event-driven architecture: a framework for scalable and resilient systems. International Journal on Science and Technology. 2025;1–12.

[3] Akerele JI, et al. Improving healthcare application scalability through microservices architecture in the cloud. International Journal of Scientific Research Updates. 2024;8(02):100–09. https://doi.org/10.53430/ijsru.2024.8.2.0064.

[4] Yadav SP, et al., eds. Advances in Fuzzy-Based Internet of Medical Things (IoMT). John Wiley & Sons, (2024).https://doi.org/10.1002/9781394242252.

[5] Pujahari RM, Khan R, Yadav SP. Applications of AI and Deep Learning in Biomedicine and Healthcare. In Applications of Parallel Data Processing for Biomedical Imaging 2024 (pp. 93–124). IGI Global Scientific Publishing. https://doi.org/10.4018/979-8-3693-2426-4.ch006.

[6] Shafiq M, et al. Advances in IoMT for healthcare systems. Sensors. 2023;24(1):10. https://doi.org/10.3390/s24010010.

[7] Singh G, et al. Hybrid deep learning model for wearable sensor-based stress recognition for internet of medical things (IoMT) system. International Journal of Communication Systems. 2024;37(3):e5657. https://doi.org/10.1002/dac.5657.

[8] Singh S, et al. A secure energy-efficient routing protocol for disease data transmission using IoMT. Computers and Electrical Engineering. 2022;101:108113. https://doi.org/10.1016/j.compeleceng.2022.108113.

[9] Khriji S, et al. Design and implementation of a cloud-based event-driven architecture for real-time data processing in wireless sensor networks. The Journal of Supercomputing. 2022;78(3):3374–401. https://doi.org/10.1007/s11227-021-03955-6.

[10] Herrero R. REST and EDA architectures in IoT actuation. Internet of Things and Cyber-Physical Systems. 2023;205–12:3. https://doi.org/10.1016/j.iotcps.2023.05.002.

[11] Calderón-Gómez H, et al. Evaluating service-oriented and microservice architecture patterns to deploy ehealth applications in cloud computing environment. Applied Sciences. 2021;11(10):4350. https://doi.org/10.3390/app11104350.

[12] Surantha N, et al. Intelligent sleep monitoring system based on microservices and event-driven architecture. IEEE Access. 2022;10:42069–80. https://doi.org/10.1109/ACCESS.2022.3167637.

[13] Muthusamy K. Event-driven data engineering in microservices architectures. International Journal of Emerging Trends in Computer Science and Information Technology. 2025;6(1):36–43. https://doi.org/10.63282/3050-9246.IJETCSIT-V6I1P104.

[14] Konkepudi B, Rao TKRK. Enhancing elderly care: microservices-enabled IoT models for addressing and monitoring needs. Frontiers in Health Informatics. 2024;13(3).

[15] Laigner R, et al. "Data management in microservices: state of the practice, challenges, and research directions." *arXiv preprint arXiv:2103.00170* (2021). https://doi.org/10.14778/3484224.3484232.

CHAPTER 7

Computer Vision (CV) and Natural Language Processing (NLP)

Mamatha V., Tapasmini Sahoo, Manivannan Karunakaran

Abstract Digital healthcare settings are becoming more and more dependent on artificial intelligence to process large amounts of heterogeneous medical data. Picture and video data, as well as unstructured written data in clinical records, have essential information that cannot be analyzed with conventional analysis tools. Computer vision (CV) and natural language processing (NLP) respond to these issues by creating the ability to understand visual and linguistic health data in real-time automatically. This chapter aims to explain the role of CV and NLP in enabling intelligent analysis, monitoring, and decision support within modern healthcare systems. The chapter describes CV techniques applied to medical imaging, video-based monitoring, and sensor-derived visual data for disease detection and functional assessment. NLP methods for extracting clinically relevant information from electronic health records, clinical narratives, and patient-reported text are presented. Multimodal frameworks that integrate visual and textual intelligence enhance diagnostic accuracy and contextual interpretation. Deployment within Internet of Medical Things (IoMT) architectures enables scalable, low-latency analytics. CV and NLP significantly advance digital healthcare by transforming visual and textual data into actionable clinical intelligence, supporting accurate monitoring, and efficient decision-making and patient-centered care.

Keywords Computer vision (CV) – natural language processing (NLP) – digital health – medical imaging – clinical text analysis – Internet of Medical Things (IoMT)

7.1 Introduction

The active growth of digital technologies has transformed the present healthcare systems radically. The recent evolution of data collection, data storage, and data computing technologies has enabled access to large volumes of clinical data that has been collected from various sources. However, translating this data into clinical knowledge remains one of the major concerns. The example of an artificial intelligence (AI) is computer vision (CV) and natural language processing (NLP); it has become a useful application in decoding complex medical information [1]. The reasons as to why CV and NLP have been introduced into the healthcare sector

 | HTTPS://DOI.ORG/10.1515/9783112237083-007

are explained where both of them reflect the prospects of enhancing the process of diagnosis, monitoring, and decision support and raising the degree of efficiency and patient-centered care [2]. When medical devices are linked via the Internet of Things (IoT) technology to monitor patient health, they form a subclass known as Internet of Medical Things (IoMT). Healthcare IoMT, also known as IoMT, involves integrating to incorporate a combination of machine learning (ML)-based AI, robotics, and interlay sensors to provide medical surveillance and surveillance, in the absence of human participation. The IoMT technology links physicians and patients through medical equipment, which provides a level of autonomy to the process of collecting, analyzing, and transferring medical data through a secure connection [3]. Visuals provide valuable data in detecting diseases, anatomy, and functional analysis, whereas the textual data are those clinical scenarios, physician arguments, and patient-reported outcomes. The sheer size and volume of such kinds of data, their variety, render them unwieldy to read through manually. It is, therefore important to introduce methods of computation that can utilize visual and textual information jointly in order to achieve complete and accurate healthcare analytics [4]. AI is one of the values that can enable the healthcare to become intelligent by automating the process of data interpretation and supporting the clinical decision-making process. CV enables machines to process medical images and videos to detect, segment, and monitor, whereas NLP extracts clinically important information in unstructured text. All these technologies together allow complete familiarity with patient information in forms. AI-based systems will enhance more accurate diagnoses, enable timely treatment and enable personalized care. The combination of CV and NLP can give an intelligent healthcare system scalable, efficient, and contextually aware clinical solutions [5]. This chapter seeks to examine the worth of CV and NLP in the interpretation of diverse healthcare data by defining the application of such tools in the medical imaging, clinical text interpretation, and multimodal decision support domains, and discussing the challenges of implementation, ethics, and future of intelligent and patient-centered digital healthcare system.

7.2 Digital Healthcare Data Landscape

7.2.1 Scalable Data Processing Frameworks

The modern medical analytics rely on distributed computing architectures to manage the volume, velocity, and diversity of the complex medical data. The key frameworks of big data that are significant in supporting scalable and efficient health care data processing are Apache Spark, Hadoop, and cloud-based data warehouses. Spark is tailored to enable high-performance, in-memory analytics, and repetitive

ML to be required in the predictive model and population-level studies. Hadoop also offers a low-cost way of storing and processing unstructured data like clinical text and medical images. Amazon Web Services (AWS) Redshift is another type of cloud-based tool that helps improve its analytical performance based on elastic, column-oriented data warehousing, which lowers the cost of infrastructure and supports large-scale healthcare decision-making.

7.2.2 Cloud-Based Healthcare Data Lakes

Cloud-based healthcare data lakes are now a scalable solution when managing the growing mass and complexity of medical data. Their use has been increasing with medical institutions having confidence in the use of cloud technologies in handling secure health data. Data lake architectures may solve major problems related to the data fragmentation; i.e., the heterogeneous data sources, clinical, genomic, imaging, and administrative can be united into one repository. The integration enables cross-domain analytics, which enables population health management and precision medicine initiatives. Compared to the traditional data warehouses, healthcare data lakes are naturally accommodative to both structured and unstructured data that constitute the majority of healthcare data.

7.2.3 ETL Pipelines for IoMT-Based Healthcare Data

IoMT-based healthcare analytics relies on efficient extract, transform, and load (ETL) pipelines since the data transmitted by connected health devices, wearable sensors, imaging systems, and clinical platforms are continuous and heterogeneous. The electronic health records can be integrated with the device data in real-time through automated ETL workflows, which in turn facilitate the timely delivery of clinical insights. Interoperability frameworks, standardized data extraction, and data assurance ensure that data is seamlessly ingested within distributed IoMT systems; terminology normalization using standards of Systematized Nomenclature of Medicine – Clinical Terms (SNOMED CT), Logical Observation Identifiers Names and Codes (LOINC), and RxNorm ensure that there is semantic consistency. The problems in healthcare-specific data quality validation include signal noise, lack of values, and synchronization errors. Data is made more useful by transformation processes, such as temporal alignment of sensor streams and context-sensitive feature extraction [6].

7.3 Computer Vision (CV) in Healthcare

7.3.1 Oncology Image Analysis, Disease Detection, and Diagnosis

Medical image analysis applies CV methods in order to extract clinically relevant information in the complex medical imaging data. DL has significantly improved the precision of the diagnostic process, speed, and consistency.

7.3.1.1 Clinical Applications

- Early Disease Detection: The implementation of subtle pathological changes prior to the manifestation of the symptoms.
- Severity of Disease Assessment: Measuring progression and response to treatment.
- Computer-Aided Diagnosis (CAD): Aiding radiologists with second-opinion systems.
- Treatment Planning: Surgery and radiotherapy guidance and targeting.

7.3.1.2 Clinical Practice Benefits

- Enhanced Accuracy of Diagnosis: Decreased human error and inter-observer error.
- Improved Efficiency: Image interpretation and workflow optimization are increased.
- Scalability: Capacity to handle great amounts of imaging data.
- Decision Support: Strengthening clinician knowledge with information.

7.3.2 Key Components of a CV System

Several crucial phases are usually involved in a conventional CV system:

Image Collection: Medical images such as X-rays, MRI, CT scans, and movies of endoscopic procedures are collected with the help of specialized imaging equipment.

Preprocessing: To enhance quality and reliability, the images gotten are enhanced or smoothed. The process of clinical decision making relies on the analysis that is reliable, and such a quality is guaranteed by such techniques as noise reduction, normalization, and contrast modification.

Characteristic Extraction: There are characteristic features, which are vascular patterns, tumor boundaries, and tissue edges. These properties are the basis of detecting abnormalities or the history of a disease.

Object Tracking: Real-time treatments and surgical guiding are supported by the system's ability to follow the progression of organs, instruments, or cells across images in dynamic imaging, such as an endoscope or ultrasound [7].

Interpretation: In the end, this could help medical professionals make well-informed judgments by emphasizing tumors, measuring the severity of the disease, or supporting robotic surgery.

7.3.3 Dermatology (Dermoscopic Imaging)

Explainable AI is beneficial in dermoscopic image analysis because it has a high lesion variability and clinical risk. Grad-CAM, SmoothGrad, and Score-CAM are attribution methods that offer visual explanations that are in accordance with dermatological guidelines and irregularity of edges, asymmetry, and color difference in skin lesions. The methods increase the transparency of melanoma and basal cell carcinoma classification, clinician trust, and multi-class lesion analysis. Transformer-based methods and explainable DL models extend the interpretability and diagnostic consistency of automated skin cancer detection systems.

7.3.4 Ophthalmology (OCT and Fundus Images)

OCT and fundus images have wide applications in ophthalmology in the early diagnosis of eye disorders like glaucoma and diabetic retinopathy. Explainable AI algorithms (e.g., CAM and Grad-CAM) emphasize clinically significant features, e.g., the optic disc and cup-to-disc ratio, which allow both glaucoma and cataracts to be detected with a high degree of sensitivity in an explainable way. Diabetic retinopathy has received applications of integrated gradients and LIME, which give lesion-level explanations and lesion severity. The visualization methods can synchronize the model focus with the ophthalmic biomarkers and improve the transparency, clinical confidence, and accuracy of the diagnosis in automated retinal disease screening systems [8].

7.3.5 Medical Imaging Analysis: Pathology and Histopathology Imaging in IoMT-Based Healthcare

The analysis of histopathological images of the tissue is of great importance in the diagnosis of the diseases, especially in oncology, when microscopic analysis of the tissue samples gives crucial information on the morphology of the cells, tissue structure, and disease progression. The combination of whole-slide imaging and digital pathology into the IoMT-based healthcare systems has facilitated the process of obtaining, storing, and transferring high-resolution histopathological data between the interconnected laboratories and clinical facilities. Digital pathology environments based on the IoMT also increase problems related to stain variability, tissue heterogeneity, and multi-scale morphological complexity, especially in diseases such as breast cancer. IoMT-supported pathology processes have been, at large, subjected to ML and deep learning (DL) applications. Although the initial approaches used handmade feature extraction with subsequent classical classifiers, more recent developments focus on convolutional neural networks (CNNs) to detect and segment nuclei, as well as classify the tissue.

7.3.6 3D Imaging and Advanced Visualization Techniques

Discoveries in 3D imaging have taken the 2D histological methodology further by providing the possibility of quantifying tissue architecture as a volume. Methods like confocal microscopy, light-sheet fluorescence microscopy, and optical projection tomography have given detailed representations of complex tissues in three dimensions with great detail. Through CV-based reconstruction, segmentation, and quantitative analysis, 3D data could be used to properly execute the morphometric evaluation and observation of the spatial cellular relationship. These methods provide better perspectives on tumor morphology, tissue structure, and disease development and progress, helping biomedical research and future clinical practices.

7.3.7 Applications in Clinical Diagnostics

CV and DL allow the processing of histopathological images automatically, detecting cellular and tissue features that are related to malignancy. The techniques assist in the detection, grading, and prediction of biomarkers in breast, prostate, and lung cancers. Computational pathology with AI enhances the accuracy of diagnosis, reduces variation, and efficiently processes pathology by aiding pathologists in the detection of mitosis, classification of subtypes, and detection of lymph node metastasis. Automated cytology image analysis aids in the early detection of cancer by detecting the slightest abnormalities in the cells in their samples (i.e., fine-needle aspiration and exfoliate cytology). Computer-aided systems give objective morphometric measurements that minimize interobserver variability and enhance screening accuracy. AI-based cytology systems result in quicker diagnosis, promote standardized reporting, and improve clinical decisions on applications such as cervical or thyroid cancer, as well as those that are non-gynecological cancers [9].

7.3.8 Video-Based and Sensor-Derived Visual Analysis

Vital signs are usually monitored using contact-based sensors, like chest sensors in pacemakers, pulse analyzers of blood oxygen content, and glue-based sensors for continuous measurements. Nevertheless, being effective, longer-term use of such interface devices could lead to discomfort as well as dermal irritation and a rise in medical waste due to the disposable components. These noninvasive devices could be useful in continuous monitoring and telehealth applications because they are less invasive, convenient, and cost-effective. This section presents an organized taxonomy in which one can study the current developments in DL-based medical video analytics. The taxonomy provides a clear structure for understanding the current findings by categorizing diagnostic medical videos based on imaging mode and clinical use. This chapter contains only diagnostic medical videos; training and surgical videos are not included since they are outside the scope of this investigation.

7.3.9 Medical Video Types

- **Echocardiography:** Echocardiography is a noninvasive, relatively inexpensive imaging modality that is commonly applied to assess the morphology and functionality of the heart. Temporal detail is necessary in an accurate diagnosis since it develops video sequences that show the action of the heart over time.
- **Endoscopy:** This is to test inner organs, such as the colon and stomach, by inserting an endoscope with a camera into the body. Such films, which facilitate the process of recognizing lesions, polyps, and malignancies, are most commonly colorful and high-resolution. Fluids and other obstacles to vision, however, can complicate automatic recognition of the region of interest, which is enabled by motion blur and incomplete frames.
- **Ultrasound:** Ultrasound imaging is commonly applied to vascular, breast, and gastrointestinal diagnosis and makes use of sound waves that are of high frequency in visualizing the organs. Despite it being cheap and safe, DL-based video analysis has challenges with loss of image quality due to noise, low brightness, and limited access to tissues, especially in patients with high body mass.

7.3.10 Vision-Based Approaches for Clinical Gait Analysis

Due to the cheapness of vision-based gait analysis, the lack of intrusion and the ability to provide an automated and research-based evaluation of human mobility, vision-based gait analysis is increasingly being used in medicine. Most of the existing approaches are based on pose-based representations of depth sensors or motion recording devices because there are not many publicly available datasets containing pathological gait films. Transformer-based models, graph convolutional networks (GCNs), and recurrent neural networks (RNNs) are often used to extract stochastic gait dynamics using skeleton data. Even though recent studies have demonstrated that it is possible to combine monocular red green blue (RGB) movies and posture estimation, problems of viewpoint dependence and spatial variability cannot be ignored. There is thus an increasing need to come up with strong, view-invariant, and multimodal vision-based models to facilitate accurate and scalable clinical gait analysis [10]. CV in IoMT Healthcare is shown in Figure 7.1.

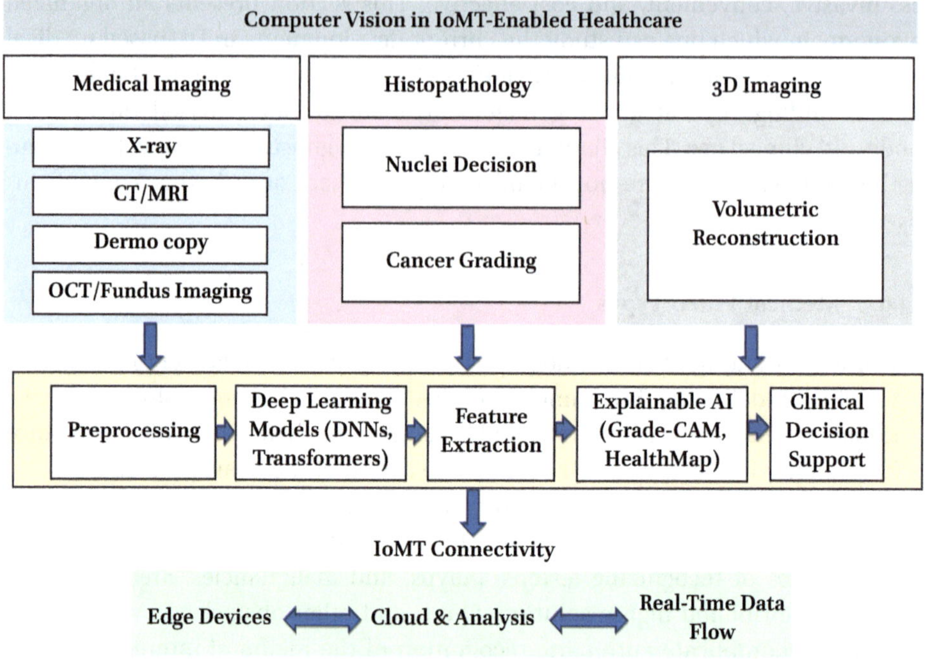

FIGURE 7.1 Computer vision in IoMT healthcare. (Source: Own elaboration)

7.4 NLP in Healthcare

EHR Data Organization and Management: The efficient management and organization of electronic health record (EHR) data is also one of the important applications of NLP in the healthcare sector. Hunting and appraising the huge quantities of unstructured clinical text in EHR systems, including physician's notes, discharge summaries, and progress reports, is difficult. NLP can be used to convert this unstructured language into structured forms of data since it is much easier and more usable by academics and physicians.

Data Extraction of Electronic Health Records: NLP is necessary to extract therapeutically valuable information from free-text data of EHR. Named entity identification and relation extraction are key clinical concepts that can be used to identify diagnoses, drugs, test data, and treatment plans through the use of NLP models. This organized data can assist healthcare professionals in identifying abnormal clinical trends, monitoring symptom progression, and identifying disease-specific trends that can help them make clinical decisions quickly and accurately.

Automated Report Generation and Patient Summarization: The other two applications of NLP in EHRs are automated report generation and automated patient summaries. This saves time by preventing clinicians from spending most of their time going through patient histories and by allowing NLP algorithms to offer summaries of long clinical notes.

NLP in EHR Systems: NLP can be used to automate medical coding and billing processes in EHR systems. Using NLP algorithms, it is possible to successfully apply an International Classification of Diseases (ICD) by identifying diagnostic terms and procedures used by the clinician in a clinical narrative. This technology enhances work within the healthcare organizations, as it reduces the administrative stress and reduces the errors in coding, as well as ensures that the billing and payment are performed correctly [11].

7.4.1 Named Entity Recognition (NER)

Named entity recognition (NER) is a basic NLP task, which entails detecting and categorizing clinically significant entities in medical text. Clinical notes and EHRs are used to identify entities in healthcare that include diseases, symptoms, medications, procedures, lab tests, and anatomical words, among others, through the use of NER. Structured data can be obtained from unstructured text by means of accurate NER, which supports downstream functions like clinical analytics, risk prediction, and decision support.

7.4.2 Clinical Concept Extraction and Normalization

Clinical concept extraction is an extension of NER that maps entities found to standard medical vocabularies and ontologies, including ICD, SNOMED-CT, and Unified Medical Language System (UMLS). This normalization process allows uniformity of clinical documentation as well as interoperability between healthcare systems. Through the association of various expressions in texts with standard concepts, NLP systems allow proper aggregation of patient records, enhance the quality of clinical documentation, and facilitate massive healthcare research and population studies.

7.4.3 Information Retrieval and Summarization

NLP tasks that are important enable clinicians to find information about patients in large amounts of clinical text swiftly and gain a summary of this information, including information retrieval and information summarization. The retrieval systems based on NLP allow quick searching of the treatment and patient records according to symptoms, diagnoses, or treatments. Automated summary systems produce brief patient summaries based on long clinical notes, which help clinicians to reduce workloads and enhance care team communication.

7.4.4 Clinical Coding and Phenotyping

Clinical coding is the process of coding patient records with standardized diagnostic and procedural codes in order to bill, reimburse, and report on patients. This is automated by NLP, which enables the retrieval of the relevant information in clinical narratives and the correct classification of patient encounters. NLP is applied to clinical phenotyping, which involves the recognition of patient groups with certain clinical features based on textual information in EHRs. Such tasks serve in the provision of epidemiological research, precision medicine programs, and population health services [12]. Figure 7.2 depicts the NLP workflow in IoMT healthcare.

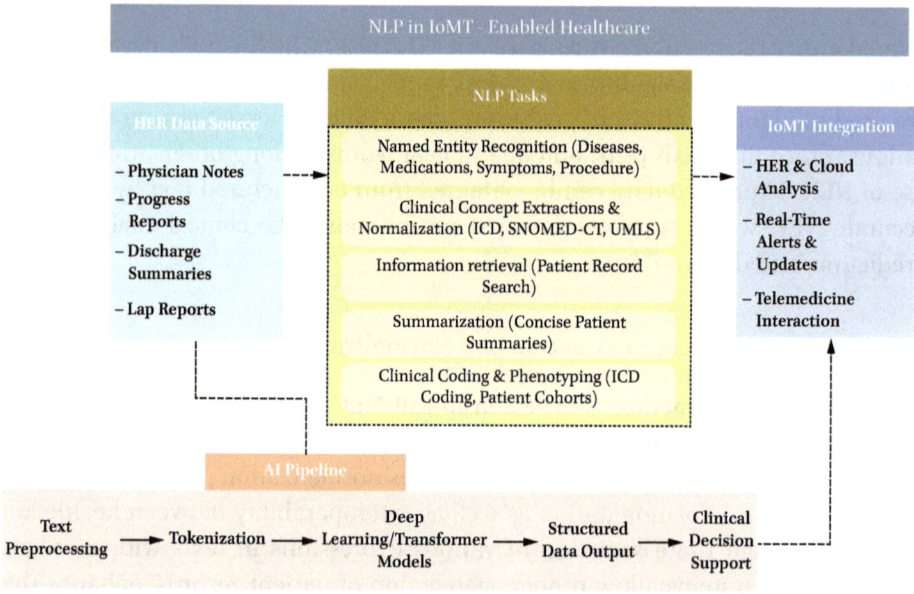

FIGURE 7.2 NLP workflow in IoMT healthcare. (Source: Own elaboration)

7.5 Multimodal Integration of CV and NLP in IOMT-Enabled Healthcare

In an IoMT-based healthcare setting, multimodal information is generated at all times with all the interconnected medical devices, imaging systems, wearable sensors, and clinical information platforms. These data integrate both visual data of medical images and videos and unstructured word data of clinical narratives, diagnosis reports, and patient histories. Although CV and NLP can analyze visual and linguistic health information independently, when combined in a system of IoMT, they would give a more detailed, real-time, and context-specific insight into patient conditions.

Multimodal learning systems can also be well applied in IoMT-based healthcare, when constant streams of data need automated interpretation at a larger scale. One notable task is the radiology report generation in related imaging settings, where vision-language models match medical images with the related reports to assist in automated reporting, abnormality identification, and cross-modal retrieval. This combination enhances the consistency of reporting and minimizes variability in clinical processes powered by IoMT.

Multimodal models in clinical decision support systems are the integration of imaging results with patient history, laboratory results, and physician notes, which are the result of platforms linked together. As an illustration, combining pathology images and histopathology reports can help in cancer grading and the prediction of prognosis, whereas a combination of video-based patient monitoring and text clinical monitoring can ensure continuous disease monitoring of patients in distant and in-hospital environments. More recent developments in vision language transformers add even more power to multimodal analytics, as joint reasoning and explainability can currently be facilitated, allowing dependable real-time decision support through distributed IoMT healthcare systems [13]. Figure 7.3 illustrates the multimodal CV-NLP IoMT framework.

7.6 Deployment of CV and NLP Models in Internet of Medical Things (IOMT) Architectures

The technological basis of the implementation of CV and NLP models in the actual healthcare setting is the IoMT. IoMT systems are architectures of medical sensors, imaging devices, wearable devices, smart cameras, and electronic health record systems that allow continuous data gathering and real-time health monitoring. Implementation of AI-based CV and NLP analytics in these systems enables scalability of clinical services, low-latency, and context-aware services. In IoMT-enabled analytics, edge cloud computing is important. Local processing of physiological and visual data is done on bedside monitors, wearable sensors, and smart imaging systems through lightweight CV models. This minimizes the delay in communication, maintains patient confidentiality, and allows quick decision-making in emergency clinical situations. Cloud infrastructures are usually used to run more compute-intensive activities, including the training of multimodal models and massive NLP semantic analysis of clinical text.

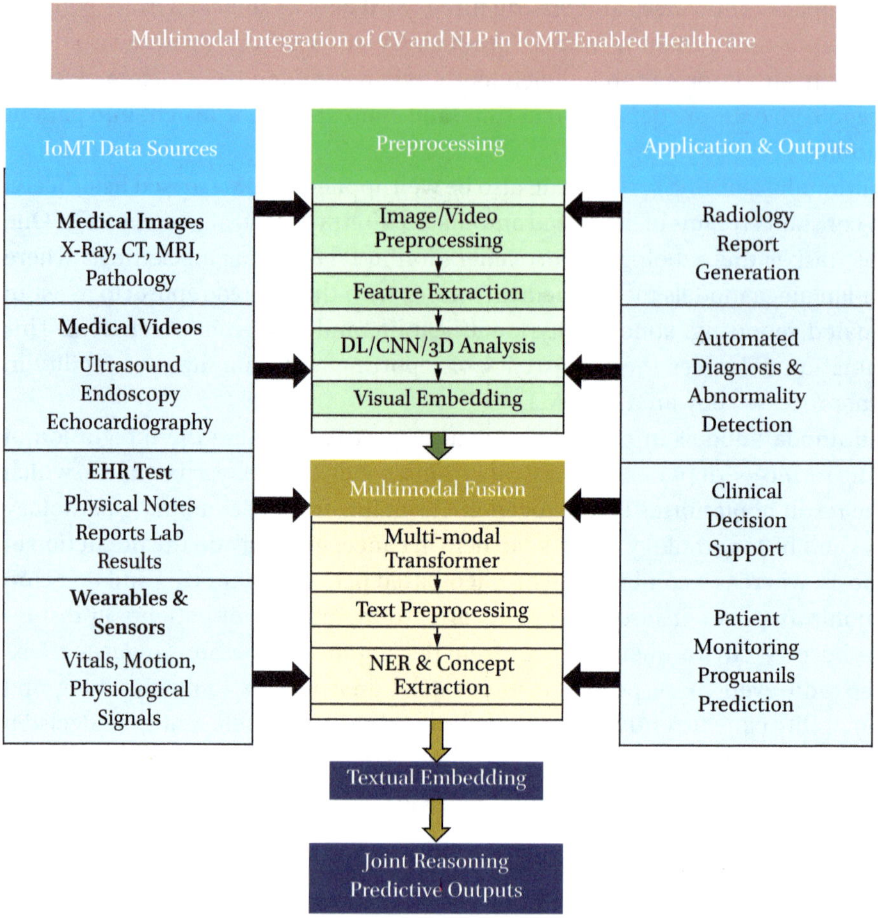

FIGURE 7.3 Multimodal CV-NLP IoMT framework. (Source: Own elaboration)

A combination of CV and NLP in the IoMT architecture facilitates continuous monitoring of the patient, early disease diagnosis, and customized healthcare provision. As an illustration, on-the-edge vision-based vital sign monitoring and remote assessment of patients are backed by NLP analysis of clinical notes and alerts in the cloud (vision-based) and cloud-based NLP configuration. The inclusion of interoperability between IoMT devices and healthcare information systems also promotes the smooth flow of data, longitudinal tracking of patients, and integrated care provision. Irrespective of the deployment issues that could be associated with security, bandwidth, and heterogeneity of devices, sound system design has provided dependable and effective AI-driven healthcare services [14].

7.6.1 Challenges in IoMT-Based CV and NLP Systems

IoMT-empowered CV and NLP systems present various technical and operational challenges because of real-time analytics, heterogeneous devices, and continuous data collection. The quality of data, privacy, interpretability, scalability, and compliance with regulations can influence the quality of clinical safety and system reliability directly. All these issues must be tackled to ensure the credible, scalable, and effective implementation of smart healthcare systems [15]. Table 7.1 provides the challenges in IoMT-based CV and NLP systems.

TABLE 7.1 Extended challenges in IoMT-based CV and NLP systems.

Challenges	Description	Implications
Data quality and bias	Heterogeneous IoMT devices and clinical sources produce uneven and biased data.	Reduced generalizability and unequal performance across patient groups.
Privacy and data security	Continuous sensing and data transmission expose sensitive patient information.	Risk of data breaches, regulatory violations, and loss of trust.
Model interpretability	CV and NLP models often provide limited transparency in real-time decisions.	Hinders clinical validation and safe decision-making.
Real-time decision risk	Automated alerts and predictions could influence urgent clinical actions.	Errors can directly compromise patient safety.
Human–AI collaboration	Unclear roles between AI recommendations and clinician judgment.	Accountability concerns and resistance to adoption.
Interoperability	IoMT devices and hospital systems use diverse data formats and standards.	Limits seamless integration and cross-system analytics.
Scalability	Large-scale IoMT deployments generate massive continuous data streams.	Increased computational load and system performance degradation.
Latency and bandwidth constraints	Real-time CV and NLP processing requires fast and reliable communication.	Delays can reduce responsiveness in critical care scenarios.

TABLE 7.1 Extended challenges in IoMT-based CV (*cont.*)

Challenges	Description	Implications
Edge resource limitations	Edge devices have limited power, memory, and compute capacity.	Restricts deployment of complex DL models.
Model drift	Changing patient conditions and environments affect model behavior over time.	Degraded accuracy without continuous retraining and monitoring.
Regulatory and legal compliance	AI-enabled IoMT systems must meet evolving healthcare regulations.	Slower deployment and higher certification requirements.

Source: Own elaboration.

7.7 Conclusion

This chapter was a review of how CV and NLP can be used to facilitate intelligent analysis of heterogeneous healthcare data. CV methods facilitate automatic interpretation of medical images, videos, and sensor-acquired visual inputs, whereas NLP makes it possible to understand unstructured clinical text found in EHRs in a structured fashion. A combination of these technologies can create the basis of smart healthcare solutions that can facilitate proper diagnosis, constant monitoring, and effective clinical decision-making. The integration of CV and NLP with multimodal learning models helps to improve both contextual interpretation and diagnostic accuracy and is implemented in IoMT models to support scalable and real-time healthcare analytics. Although there are current issues regarding the quality of data, privacy, and explainability, new solutions are emerging in explainable AI, vision-language models, and edge-cloud computing to solve these problems. CV, NLP, and IoMT technologies coming together allow the creation of a new generation of intelligent, interconnected, and patient-centered healthcare systems with the capability to provide real-time, scalable, and reliable clinical intelligence.

References

[1] Elyan E, et al. Computer vision and machine learning for medical image analysis: Recent advances, challenges, and way forward. Artificial Intelligence Surgery. 2022; 2(1):24–45. https://doi.org/10.20517/ais.2021.15.

[2] Olveres J, et al. What is new in computer vision and artificial intelligence in medical image analysis applications. Quantitative Imaging in Medicine and Surgery. 2021;11 (8):3830. https://doi.org/10.21037/qims-20-1151.

[3] Aljabr AA, Kumar K. Design and implementation of Internet of Medical Things (IoMT) using artificial intelligent for mobile-healthcare. Measurement: Sensors. 2022;100499:24. https://doi.org/10.1016/j.measen.2022.100499.

[4] Kanaujia VK, et al. Evaluation of CNN-based computer vision recommended treatments for recognised guava disease. International Journal of Embedded Systems. 2023;16(5–6):354–63. https://doi.org/10.1504/IJES.2023.141930.

[5] Narwaria M, et al. Harnessing the Power of NLP for Mental Health Diagnosis and Treatment. In Demystifying the Role of Natural Language Processing (NLP) in Mental Health 2025 (pp. 71–98). IGI Global Scientific Publishing. https://doi.org/10.4018/979-8-3693-4203-9.ch005.

[6] Ara G, et al. "Deciphering the Digital Healthscape: Unveiling the Intricacies of the Internet of Medical Things (IoMT)." Technoarete Transactions on Advances in Data Science and Analytics 3 (2024). https://doi.org/10.36647/TTADSA/03.01.A001.

[7] Ramalingam M, et al. A comprehensive analysis of blockchain applications for securing computer vision systems. IEEE Access. 2023;11:107309–30. https://doi.org/10.1109/ACCESS.2023.3319089.

[8] Li AL, et al. Automated detection of posterior vitreous detachment on OCT using computer vision and deep learning algorithms. Ophthalmology Science. 2023;3 (2):100254. https://doi.org/10.1016/j.xops.2022.100254.

[9] Dhabliya D, et al. Computer vision: Advances in image and video analysis. In E3S Web of Conferences 2023 vol. 399. EDP Sciences. https://doi.org/10.1051/e3sconf/202339904045.

[10] Ranjan R, et al. Computer vision for clinical gait analysis: A gait abnormality video dataset. IEEE Access. 2025;13:45321–39. https://doi.org/10.1109/ACCESS.2025.3545787.

[11] Sarella PNK, TherissaMangam V. AI-driven natural language processing in healthcare: Transforming patient-provider communication. Indian Journal of Pharmacy Practice. 2024;17(1). https://doi.org/10.5530/ijopp.17.1.4.

[12] Thukral A, et al. Knowledge graph enrichment from clinical narratives using NLP, NER, and biomedical ontologies for healthcare applications. International Journal of Information Technology. 2023;15(1):53–65. https://doi.org/10.1007/s41870-022-01145-y.

[13] Manickam P, et al. Artificial intelligence (AI) and internet of medical things (IoMT) assisted biomedical systems for intelligent healthcare. Biosensors. 2022;12(8):562. https://doi.org/10.3390/bios12080562.

[14] Kazi KSL. Healthcare Monitoring System Driven by Machine Learning and Internet of Medical Things (MLIoMT). In Convergence of Internet of Medical Things (IoMT) and Generative AI 2025 (pp. 385–416). IGI Global Scientific Publishing. https://doi.org/10.4018/979-8-3693-6180-1.ch016.

[15] Khattak WA, Rabbi F. Ethical considerations and challenges in the deployment of natural language processing systems in healthcare. International Journal of Applied Health Care Analytics. 2023;8(5):17–36.

CHAPTER 8

Applications of Machine Learning in Real-Time Monitoring and Diagnosis of Chronic Diseases

Sharon M., Sweta Kumari Barnwal, Avinash Samantra

Abstract Chronic diseases require continuous and adaptive monitoring due to their progressive, variable, and often unpredictable physiological changes. The increasing use of wearable sensors and connected health systems has produced vast amounts of real-time data, often exceeding the capacity of traditional analytical methods. Machine learning (ML) offers advanced computational tools to interpret complex, multimodal, time-dependent data streams, supporting accurate real-time assessment and early detection for chronic disease. The proposed framework combines supervised models, including extreme gradient boosting (XGBoost), random forest (RF), and *K*-nearest neighbors (KNN), with unsupervised techniques such as *K*-means and DBSCAN, alongside deep learning (DL) architectures such as hybrid convolutional neural network (CNN)-long short-term memory. Continuous physiological signals from wearable and IoMT-enabled devices were collected from diverse sources, including public healthcare repositories, UCI ML repository, Kaggle datasets, electronic health records, laboratory results, and sensor streams, ensuring real-world relevance. Comparative evaluation shows high predictive performance, with the hybrid CNN + KNN achieving 96% accuracy, 93% precision, 99% recall, and a 97% F_1-score. Disease-specific models, including DNN for diabetes and XGBoost for hypertension, demonstrated similarly strong results. Key applications include predictive risk scoring, anomaly detection, AI-driven alerts, and integration into clinical decision support systems, enabling proactive interventions and personalized care across diverse populations.

Keywords Machine learning – chronic disease monitoring – real-time diagnosis – wearable sensors – Internet of Medical Things – digital health

8.1 Introduction

Chronic diseases, including diabetes, heart disorders, gastrointestinal disorders, Alzheimer's disease, mental illnesses, and other disorders, are some of the health issues that present a major global health challenge in the world since they are long-term, could present inconsistent symptoms, and require complex management strategies [1]. These conditions tend to occur in a silent form, and therefore early

 | HTTPS://DOI.ORG/10.1515/9783112237083-008

diagnosis, constant checking, and ensuring that the patients are intervened in time are crucial in the reduction of complications, better patient results and reduced healthcare costs. Nevertheless, the traditional methods of clinical follow-up, based mostly on the periodic visits to hospital and self-reporting by patients, do not provide real-time fluctuations in physiological parameters, often leading to the fact that the diagnosis is delayed, and the decisions regarding the method are not optimal [2]. The most recent advancements in wearable biosensors, and the Internet of Medical Things (IoMT) for short, have enabled continuous and noninvasive recording of vital physiological variables, including heart rate, blood glucose, body temperature, physical activity, and sleep behavior, in natural world settings [3]. These technologies produce high amounts of real-time, multimodal health data that view the health of patients comprehensively outside of clinical environments. Simultaneously, machine learning (ML) and deep learning (DL) approaches showed a high level of potential in processing multifaceted and high-dimensional healthcare records to facilitate early disease detection, prediction of its progression, and the development of a treatment plan with personal factors [4]. These developments notwithstanding, there are a number of challenges. The heterogeneity of sensor data, a different sampling rate, noise, and missing data could decrease model robustness and reliability. Also, not all predictive models are easily incorporated into clinical processes and patient care systems, constraining the practical use and clinical adoption [5]. Moreover, the conventional diagnostic methods are not usually scalable and flexible to use with large cohorts or patients with more than one chronic illness. To overcome these drawbacks, the proposed research aims at creating a powerful ML-based system to monitor and early diagnose chronic diseases in an instantaneous fashion using data in produced by IoMT-equipped devices [6]. The suggested system combines wearable technologies with modern data preprocessing, feature engineering, and ML/DL models to provide continuous, automated and personal health insights. The framework seeks to improve disease management, decrease spending on healthcare and ultimately improve the quality of life of patients by contributing to proactive intervention and informed clinical decision-making [7].

The research aims to real-time monitoring, early detection, and diagnosis in ML models of chronic circumstances using wearable and IoMT-based systems. It aims at building a powerful structure that will allow predictive analytics, anomaly detection, and individual clinical decision support for chronic diseases management.

8.2 Machine Learning Models for Disease Prediction and Diagnosis

Continuous data streams created by wearables and IoMT systems are used by ML models to predict the development and evolution of chronic diseases early enough to prevent them. Supervised learning methods like the *K*-nearest neighbors (KNNs), extreme gradient boosting (XGBoost), and random forest (RF) are utilized in classification and regression, in which the disease states, risk level, and clinical outcomes are predicted using labeled physiological data. Unsupervised algorithms, including the K-means and the DBSCAN, are applied to uncover latent structures and hidden patterns in high-dimensional and unlabeled data, allowing the identification of anomalies and clusters effectively. DL models, like the hybrid convolutional neural network-long short-term memory (CNN-LSTM) models, are used to learn both time-dependent and space-dependent features in time-series sensor data. The cross-validation, hyperparameter optimization, and ensemble learning strategies are used to improve model robustness and generalization. The performance of these optimization methods is stable across a wide spectrum of patients as well under a number of data situations, which leads to the accurate, reliable, and scalable predictive models of real-time chronic disease monitoring and clinical decision support.

8.2.1 XGBoost for Monitoring a Disease

The XGBoost classifier is used to evaluate the chronic disease status by learning multifaceted associations to patient health data as in real-time wearable and IoMT surveillance instances [8]. XGBoost is an ensemble boosting algorithm, which utilizes several weak decision trees in order to create a robust predictive system at the minimal stage of overfitting. Its goal function combines the error of prediction and complexity in the model, as in eqs. (8.1) and (8.2):

$$K(\theta) = \Sigma_j k(z^s_{pred}, z_{truth}) + \Sigma_l \Omega(g_l) \tag{8.1}$$

$$\Omega(g_l) = \gamma S + \frac{1}{2}\gamma||\nu||^2 \tag{8.2}$$

where θ denotes the set of trainable parameters of the ML model, and $K(\theta)$ represents the overall objective function to be minimized during training. The $k(\cdot)$ is loss function, quantifies the variation of predicted and factual findings for each patient sample j. The term $z^{(s)}_{\text{pred}}$ indicates the predicted disease status, while z_{truth} corresponds to the absence of early-stage chronic disease. The index l refers to the model layers, and g_l denotes the associated parameter group. The regularization term $\Omega(g_l)$ includes γ as the regularization coefficient, S as the structural complexity constraint, and $||\nu||^2$ as the weight vector. These terms penalize excessive model complexity, reduce overfitting caused by noisy healthcare data, and promote stable and interpretable predictions for chronic disease monitoring.

8.3 Real-Time Monitoring and Predictive Analytics

IoMT devices and wearable sensors constantly send medical data, such as heart rate, ECG signals and glucose levels, in real time. These data analyzed ML models that use sliding window approaches and real-time streaming pipelines to identify the temporal patterns to facilitate continuous and dynamic monitoring. Some of the anomaly detectors used to identify abnormal patterns include isolation forest, One-Class SVM and autoencoders. Predictive risk scoring models approximate the probability escalation of the patient condition, enabling healthcare professionals to introduce clinical interventions in a timely manner. With the help of this combined method, it is possible to contribute to the proactive management of the disease, improve patient safety, and increase the efficiency of real-time systems for chronic disease monitoring.

8.3.1 Autoencoders (AEs) for Monitoring and Predictive Analytics

An unsupervised learning approach that encourages obtaining the normal structure of input data in AE, making it very appropriate to apply to the constant observation of chronic diseases using wearable and IoMT sensor data [9]. The encoder takes the input vector W(e.g., heart rate, glucose level, ECG features) to a compact concealed representation G as shown in eq. (8.3):

$$G = \partial\left(X_{wg}W + \beta_{wg}\right) \tag{8.3}$$

where G represents low-dimensional latent representation capturing normal physiological patterns of patients, W represents input feature vector obtained from wearable and IoMT sensors, X_{wg} is weight matrix, β_{wg} is bias vector associated with the encoder layer, $\partial(\cdot)$ denotes the nonlinear activation function used to model complex physiological relationships. The decoder then reconstructs the original input from the hidden representation, as given in eq. (8.4):

$$\hat{W} = \partial\left(X_{\hat{w}g}g + \beta_{\hat{w}g}\right) \tag{8.4}$$

where the latent representation g is mapped back to the reconstructed input $\hat{W}$, the matrix $X_{\hat{w}g}g$ denotes the decoder weight parameters, and $\beta_{\hat{w}g}$ represents the corresponding bias term. This reconstruction process aims to reproduce the original sensor data as accurately as possible, allowing the model to learn intrinsic structures of normal health behavior. The quality of reconstruction is measured using the reconstruction error in eq. (8.5):

$$\varepsilon = ||W - \hat{W}|| \tag{8.5}$$

where the reconstruction error ε is the absolute variance between the actual input W and the resultant signal $\hat{W}$. This error functions as an anomaly indicator, with low values representing normal physiological conditions and large values indicating abnormal health patterns. This allows for early detection of chronic disease-related anomalies in real-time monitoring systems.

8.4 Early Warning and Alert Systems

The ML-based systems also provide notifications when the risk to the patient is beyond safe levels. Models raise threshold-based alerts based on crossing clinical limits, probabilistic alerts on the basis of model-predicted risk probabilities, and AI-based alerts are based on deep learning to predict events before exceeding threshold levels. These alerts can be embedded in dashboards, mobile applications, or information systems and are used to deliver a visual indicator and take action act on it. Automated responses could be in the form of notifications, reminders, or emergency notifications to caregivers and patients. Such systems improve patient safety and allow timely interventions, and proactive management of chronic conditions, reducing the risk of critical health events because of the combined use of analytical forecasting and continuous monitoring in real time.

8.4.1 AI-Driven Alerts

To process wearable and IoMT stream skeletal physiological data in real-time, DL-based AI-based alert systems use recurrent neural network (RNN) models [10]. These models have the potential to predict possible disease flare-up or clinical deterioration before conventional threshold-sensitive approaches by incorporating time-dependent relationships and trends in vital measurements. Predictive alerts are dispatched to both the patient and healthcare provider, allowing timely interventions to reduce emergency events and enhance chronic disease management. Intelligent alerts can guide personalized real-time monitoring and active clinical support and constantly adapt to the behaviors of individual patients according to AI.

8.5 Clinical Decision Support Systems (CDSS) Integration

Predictions and risk scores based on ML are added to clinical decision support system (CDSS) platforms to assist clinicians in making decisions in a timely manner. Privacy is guaranteed through systems through the General Data Protection

Regulation (GDPR) or Health Insurance Portability and Accountability Act (HIPAA) compliance, anonymization, and consent management. This is done using transfer learning to enhance model reliability and performing validation across sites. edge computing, lightweight models, pruning, and graphics processing unit (GPU) acceleration are used to provide computational efficiency to high-frequency real-time health data streams.

8.5.1 Edge Computing-Based Ensemble Learning for Real-Time CDSS in Chronic Disease Monitoring

A CDSS with edge computing takes ensemble learning models to process real-time wearables and IoMT device physiological and clinical data [11]. The system unites various extreme learning machine predictors to enhance robustness and accuracy at the expense of imbalance in classes and outliers. It can process at the edge and offer low-latency, continuous monitoring and issues alerts when abnormal patterns or high-risk conditions are observed. This framework helps to take clinical decisions promptly, engage in predictive diagnostics, and make proactive interventions, improving the chronic disease management process by making personal and real-time observations and providing early warnings about such conditions as CKD, heart disease, and arrhythmia.

8.6 Real-World Applications

ML-based applications track chronic conditions such as diabetes, cardiovascular, and gastrointestinal conditions with the help of wearable data. Hypoglycemia or arrhythmias are detected by predictive models, whereas GI flare-ups are detected by anomaly detectors. Clinicians are guided on individual interventions by real-time alerts and dashboards. Hospitalization, biomarkers, and patient outcome tracking confirm system effectiveness and facilitate more refined model processes to support personalized and proactive care.

8.6.1 LSTM

The ability to gain sustained chronologies in bodily functions enables the LSTM-based model to be applied to the real-time monitoring of chronic diseases relying on wearer sensor information [12]. The input is the variable(s) of the model, which is the input time-series data divided into a set number of sliding window lengths and could be stated in eq. (8.6):

$$W = \{w_1, w_2, ..., w_S\} \tag{8.6}$$

where $W = \{w_1, w_2, ..., w_S\}$ refers to a segmented time-series sequence of wearable and IoMT sensor data, w_S is the sensor feature vector at time step S, which is a collection of physiological data including heart rate, glucose level, ECG, and activity data. The fixed sliding window size that is used to capture temporal patterns in the progression of chronic diseases is equal to the length S. The LSTM network is created to embrace the continuing dependencies in the form of three gating processes. The forget gate is a control of the past information retention as in eq. (8.7):

$$e_s = \sigma\left(X_e.[g_{s-1}, w_s] + a_j\right) \tag{8.7}$$

where e_s is the entry point for forgetting activation at time s, $\sigma(\cdot)$ is function of sigmoid activation that scales values between 0 and 1, and X_e is the forget gate weight matrix, g_{s-1} is the prior latent state, w_s represents the significant input, while a_j is a biased term. This model rejects superfluous ancient time's physiological data. It is the input gate that identifies the latest data that's recorded and computes the candidate cell state as eqs. (8.8) and (8.9):

$$j_s = \sigma\left(X_j.[g_{s-1}, w_s] + a_j\right) \tag{8.8}$$

where j_s is the ratio of the freshly acquired data that was stored in the state, X_j represents the input gate's weight matrix, and the concatenation of past context and current sensor readings g_{s-1} is used to determine which new health-related information is important. The update of the cell state is given in eq. (8.9):

$$D_s = e_s.D_{s-1} + j_s.\tilde{D}_s \tag{8.9}$$

where D_s denotes the new cell state at time step s,D_{s-1} reflects the previous cell state, and $\tilde{D}_s$ represents the prospective state. The forget gate controls the memory held, while the gate utilized for input j_s regulates the incorporation of the innovative medical data, facilitating one to learn long-term dependency. The hidden state and output gate are given as follows:

$$p_s = \sigma\left(X_p.[g_{s-1}, w_s] + a_p\right)ps = \sigma\left(X_p.[gs{-}1, ws] + cp\right) \tag{8.10}$$

where p_s is the output gate at time step s, which determines the extent to which the information in the current cell state should be revealed to the next hidden layer or output. The role of $\sigma(\cdot)$ is the sigmoid activation role, which limits the values of the gates to be 0 and 1, to guarantee control of the information flow. The term X_p is the weight parameters that are trainable for the output gate. The symbol $[g_{s-1}, w_s]$denotes the composition of the last latent state g_{s-1} and the recent input bit w_s which corresponds to wearable and IoMT sensor data, e.g., heart rate, glucose level, or ECG characteristics. A bias term a_p or c_p is included to enhance the model's flexibility and stability at the training stage. In general, this gate qualifies the learnt temporal health information that is applicable in real-time chronic disease monitoring and diagnosis.

$$g_s = p_s \, . \tanh(D_s) \tag{8.11}$$

The hidden state g_s is computed by modifying the engaged condition of cells tanh (D_s) utilizing the outcome of gate p_s. This hidden representation captures temporal health patterns and is used for coincident prediction and monitoring of chronic disease conditions.

8.7 Data Collection

The datasets utilized in publicly available healthcare repositories, the UCI ML Repository and Kaggle, are mostly in this research. These repositories hold organized records of patients with chronic illnesses such as diabetes, heart diseases, and breathing problems [13]. The information also contains demographics and clinical characteristics, which are age, gender, blood pressure, cholesterol, glucose levels, BMI, smoking history, and family medical history, assisting in the early detection of the disease. Anonymized clinical data, both structured and unstructured, such as symptoms and physician visits, were added to make the data more realistic, without personal identifiers, as this would ensure client privacy [14]. Also, EHRs, laboratory reports, wearable sensors and clinical databases, including MIMIC-III, provided multimodal healthcare data that enhanced the model generalizability and robustness [15].

8.7.1 Evaluation Metrics and Performance Assessment

Accuracy, or overall accuracy, is the percentage of correct predictions, which is susceptible to bias caused by the out-of-true data. Precision is used to determine the level of real positives; low precision minimizes false positives. True positives that were recalled were true positives and high recall reduces false negatives. Preciseness and recall are harmonic means to obtain F_1-score. AUC is used to measure the capability of the model to differentiate between patients that have and do not have chronic diseases, where a greater value of an AUC implies improved discrimination performance and a higher ability to rank the positive cases above the negative ones correctly.

TABLE 8.1 shows a comparative performance evaluation of diverse ML models of chronic disease diagnosis. The conventional learning models feedforward neural network (FNN), logistic regression (LR), RF, and gradient boosting (GB) demonstrate modest accuracy, recall, F_1-score, precision, and AUC scores, with LSTM being better in temporal data processing accuracy 88.7%, F_1-score 85.9%, and

AUC 0.92 [13]. The hybrid CNN + KNN model also performs better, with a detection rate of 96%, specificity of 93%, recalls of 99%, and an F_1 score of 97%, indicating that it is capable of finding local and global feature patterns that could be utilized to monitor chronic diseases in real time [14]. FIGURE 8.1 shows the relative performance of different ML models in early chronic disease identification.

TABLE 8.1 Comparison between achievements of various ML models on chronic disease diagnosis [13, 14]

Model	Accuracy (%)	Precision (%)	Recall (%)	F_1-Score (%)	AUC
FNN [13]	84.2	82.3	81.5	81.9	89
LSTM [13]	88.7	85.5	86.3	85.9	92
RF [13]	83.1	80.2	79.6	79.9	87
GBM [13]	84.0	82.1	81.2	81.6	88
LR [13]	79.3	75.8	73.9	74.8	83
CNN + KNN (hybrid) [14]	96.0	93.0	99.0	97.0	–

TABLE 8.2 shows the results of various ML models in the prediction of various chronic diseases. In the case of diabetes, deep neural network (DNN) scored the best (94.3%) accuracy and F_1-score (92.5%), which is a high predictive ability. XGBoost prediction of hypertension performed with high accuracy (93.1%) and AUC (0.954), and it is capacitated to handle structured clinical data effectively. RF demonstrated a high level of reliability in classification with 91.7% accuracy and 89.8% F_1-score which showed that it is reliable in the case of the recurrence of the cancer. Overall, these findings outline the usefulness of custom ML models in real-time monitoring of diseases and support of diagnosis of various chronic conditions. FIGURE 8.2 illustrates the estimation of multimodal ML frameworks to forecast chronic diseases.

8.8 Discussion

To determine how ML models perform in the continuous surveillance and detection of chronic illnesses, data has been taken from the UCI ML Repository, where all the organized patient data is available in the form of age, gender, cholesterol, blood pressure, blood sugar levels, weight, body mass index (BMI), nicotine history, and family medical history. The older models, like FNN, RF, GBM, and LR spoke of

TABLE 8.2 Performance evaluations of specific ML models for predicting disease, highlighting model effectiveness in chronic disease diagnosis [15]

Disease	Model	Accuracy (%)	Precision (%)	Recall (%)	F_1-Score (%)	AUC
Diabetes	DNN [15]	0.943	0.918	0.932	0.925	0.962
Hypertension	XGBoost [15]	0.931	0.912	0.921	0.916	0.954
Cancer recurrence	RF [15]	0.917	0.901	0.895	0.898	0.945

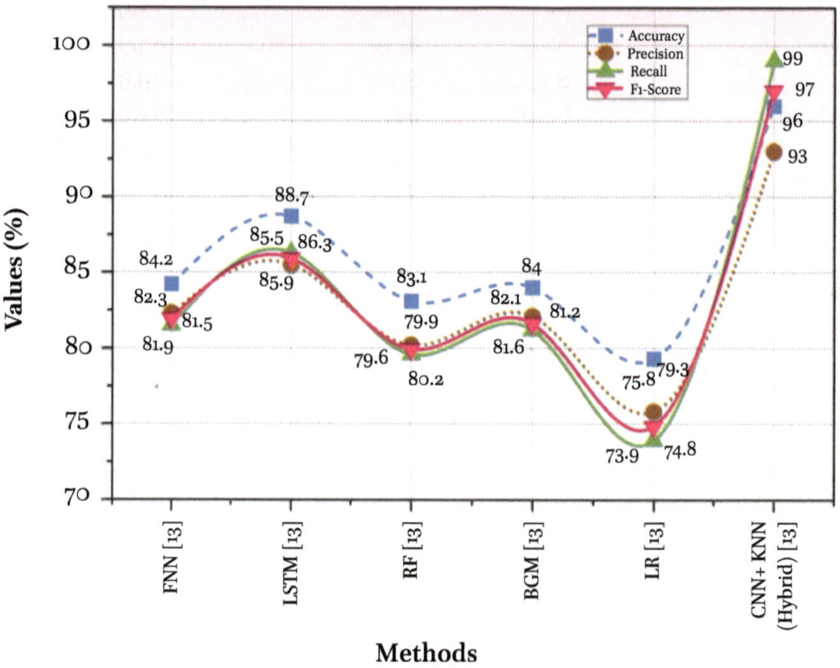

FIGURE 8.1 Comparative performance of ML models for early-stage chronic disease prediction. (Source: Own elaboration)

moderate performance, with LSTM performing better than these in time related data accuracy: 88.7%, F_1-score 85.9%, AUC 0.92. The limitations are the possibility of overfitting on training data and inconsistency of patient data quality [13].

Multimodal information (the combination of anonymized clinical symptoms, physician visits and wearable sensor streams) was added to enhance predictive capability in the chronic disease monitoring. The CNN + KNN hybrid model

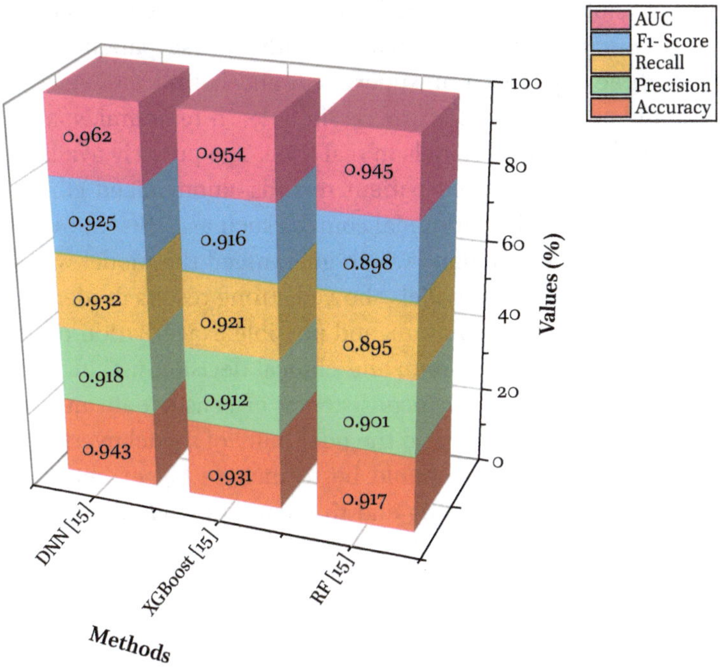

FIGURE 8.2 Evaluation of model metrics across different chronic disease prediction methods. (Source: Own elaboration)

obtained better outcomes with the highest recall of 99%, precision of 93%, greatest reliability of 96%, and F_1 grade of 97%, which effectively identified both local and global characteristics. Weaknesses are missing clinical annotations and wearable data may be inconsistent and thus not applicable in the real world [14].

The datasets taken are EHRs, laboratory reports, and MIMIC-III; disease-specific evaluation was used in the prediction of diabetes, hypertension, and cancer recurrence. DNN diabetes prediction accuracy is 94.3%, F_1-score is 92.5%, XGBoost hypertension prediction accuracy is 93.1%, AUC is 0.954, and RF cancer recurrence prediction accuracy 91.7%, F_1-score 89.8%. These are limitations such as sensor variability, possible overfitting, and incomplete clinical annotations to the continuous monitoring performance [15].

8.9 Conclusion

To facilitate the early detection and diagnosis of chronic diseases, ML systems are constantly tracking and analyzing real-time data from wearable sensors and IoMT-enabled devices. These systems allow early intervention, personalized medical care, and improved monitoring and control of chronic diseases through the analysis of continuous physiological signals, which means more effective medication and a higher success rate of health outcomes with diverse patients. The approach combines all three supervised, unsupervised, and hybrid models, e.g., LSTM, XGBoost, CNN, and autoencoders, to successfully learn temporal, spatial, and abnormal features of physiological signals in real time. The publicly available healthcare repositories, such as structured patient records, anonymized clinical notes, wearable sensor streams, and multimodal sources such as EHRs and laboratory reports, were used to gather the data, which guaranteed the model would be applicable to other populations. The models showed strong results in identifying the appearance of the disease, its evolution, and possible exacerbation to use the predictive analytics, early warning system, and clinical decision-making. The research has various limitations, such as inconsistency of sensor data quality, incomplete or absent clinical annotations, and the possibility of model overfitting during training. The future of this work should be to include bigger heterogeneous datasets and better data preprocessing and to incorporate adaptive learning frameworks to provide increased personalized, scalable, and continuous monitoring of chronic disease management.

References

[1] Hosain MN, et al. IoT-enabled biosensors for real-time monitoring and early detection of chronic diseases. Physical Activity and Nutrition. 2024;28(4):6. https://doi.org/120463/pan.2024.0033.

[2] Wu C-T, et al. A precision health service for chronic diseases: Development and cohort study using wearable device, machine learning, and DL. IEEE Journal of Translational Engineering in Health and Medicine. 2022;10(1-14). http://doi.org/1109/JTEHM.2022.3207825.

[3] Islam MR, et al. Deep learning-based IoT system for remote monitoring and early detection of health issues in real-time. Sensors. 2023;23(11):5204. https://doi.org/13390/s23115204.

[4] El-Rashidy N, et al. Mobile health in remote patient monitoring for chronic diseases: Principles, trends, and challenges. Diagnostics. 2021;11(4):607. https://doi.org/13390/diagnostics11040607.

[5] Narwaria M, et al. Harnessing the Power of NLP for Mental Health Diagnosis and Treatment. In Demystifying the Role of Natural Language Processing (NLP) in Mental Health 2025 (pp. 71–98). *IGI Global Scientific Publishing*. https://doi.org/14018/979-8-3693-4203-9.ch005.

[6] Upadhyay P, Tomar P, Yadav SP. ADNet: DL based model for Alzheimer's disease classification. Iran Journal of Computer Science. 2025;1–13. https://doi.org/11007/s42044-025-00243-x.

[7] Battineni G, et al. Applications of machine learning predictive models in the chronic disease diagnosis. Journal of Personalized Medicine. 2020;1(2):21. https://doi.org/13390/jpm10020021.

[8] Raihan MJ, et al. Detection of the chronic kidney disease using XGBoost classifier and explaining the influence of the attributes on the model using SHAP. Scientific Reports. 2023;13(1):6263. https://doi.org/11038/s41598-023-33525-0.

[9] Rassam MA. Autoencoder-Based neural network model for anomaly detection in wireless body area networks. IoT. 2024;5(4):852–87. https://doi.org/13390/iot5040039.

[10] Scheid MR, et al. Development and validation of a clinical wearable DL based continuous inhospital deterioration prediction model. Nature Communications. 2025;16 (1):9513. https://doi.org/11038/s41467-025-65219-8.

[11] Vincent ACSR, Sengan S. Edge computing-based ensemble learning model for health care decision systems. Scientific Reports. 2024;14(1):26997. https://doi.org/11038/s41598-024-78225-5.

[12] Pavithra D, et al. Application of LSTM networks for continuous patient monitoring and anomaly detection in wearable health devices. Indian Journal of Science and Technology. 2024;17(37):3909–21. https://doi.org/117485/IJST/v17i37.2600.

[13] Akhi SS, et al. Early-Stage Chronic Disease Prediction Using DL: A Comparative Study of LSTM and Traditional Machine Learning Models. Library of Frontline Medical Sciences and Pharmaceutical Journal. 2025;5(07):8–17. https://doi.org/137547/medical.

[14] Alanazi R. Identification and prediction of chronic diseases using machine learning approach. Journal of Healthcare Engineering. 2022;2022(1):2826127. https://doi.org/11155/2022/2826127.

[15] Dias BL. Predictive Analytics for Early Detection of Chronic Diseases Using Multimodal Healthcare Data. International Journal of Humanities and Information Technology. 2022;4(01-03):36–52. https://doi.org/121590/ijhit.04.01-3.05. chronic disease, 1.

CHAPTER 9

Wearable Sensors for Early Detection of Cardiovascular Diseases

Ajeetsingh, Praveena K. N., Amit Prakash Sen

Abstract Cardiovascular diseases continue to dominate worldwide health outcomes, reinforcing the importance of continuous surveillance and prompt detection strategies. It examines how wearable devices enable real-time, noninvasive measurement of key physiological parameters, including heart rate, blood oxygen saturation (SpO_2), heart rate variability, electrocardiogram signals, blood pressure, physical activity, and sleep patterns. Continuous monitoring supports the timely detection of gradual and subtle physiological variations associated with cardiovascular abnormalities such as arrhythmias, hypertension, and heart failure, often before clinical symptoms become evident. The chapter further explores the integration of wearable sensors with cloud-based platforms and mobile health applications to enable fast, remote patient monitoring and clinical decision-making based on data. Additionally, machine learning and sophisticated signal processing approaches for analyzing multimodal sensor data are discussed, highlighting their role in improving diagnostic accuracy and risk prediction. Key challenges, including signal artifacts, measurement reliability, data privacy, and user adherence, are also addressed. Overall, this chapter highlights how recent advancements in wearable sensing, sensor fusion, and artificial intelligence are driving a transition from traditional episodic care toward predictive, preventive, and personalized cardiovascular healthcare, with substantial potential to enhance early diagnosis, patient outcomes, and healthcare efficiency.

Keywords Wearable sensors – cardiovascular diseases – remote patient monitoring – physiological signal analysis – personalized healthcare

9.1 Introduction

Health statistics in the world have always categorized cardiovascular diseases (CVDs) as the leading cause of mortality and morbidity. Arrhythmias, coronary artery disease, as well as heart failure significantly lower life expectancy and general quality of life [1]. It has been shown that every year millions of people die because of CVDs, and it becomes even more widespread among different age groups. In turn, the creation and implementation of efficient and scalable cardiovascular monitoring procedures are of the utmost priority, along with the related

 | HTTPS://DOI.ORG/10.1515/9783112237083-009

healthcare costs and subsequent productivity costs [2]. On-time detection of cardiovascular pathology has a positive effect on the prognosis; long-term monitoring of physiological indicators allows preventing a pathology and reducing the risk of severe clinical complications, as the incipient pathology is identified before the manifestation of the main symptoms [3]. The conventional forms of cardiovascular examination, which include blood pressure (BP) and electrocardiogram (ECG), are the major sources of diagnostic information. However, a comprehensive evaluation of the dynamics of daily cardiovascular is limited by intermittent measures and clinical dependence [4]. Besides clinical applications, wearable sensors technologies can be used to aid noninvasive cardiovascular monitoring continuously. Wearable patches, implantable devices, smartwatches, textile sensors allow obtaining real-time physiological information in the process of daily activities [5]. Wearables offer a holistic measurement of the cardiovascular performance, recording several key physiological parameters during the process, including BP, heart rate (HR), blood oxygen saturation (SpO_2), and activity, heart rate variability (HRV), ECG, and sleep [6].

The system of wearable technologies combined with telemedicine, cloud solutions, and mobile devices, makes the process of data transfer and analysis unproblematic. Longitudinal cardiovascular records and improved clinical insight are beneficial to remote health delivery [7]. Continuous weaknesses are motion artefacts, interoperability, signal noise, and limitations in data reliability. There are limited predictive performances, difficulties in standardization, and privacy issues which limit the successful application of wearable information in detecting cardiovascular early [8]. The chapter refers to wearable sensors as the means of early diagnosis and continuous tracking of CVDs with the emphasis on the key physiological parameters and combined with the mobile health applications, cloud services, and machine learning (ML). The developments in signal processing, artificial intelligence, and sensor technology support the enhancement of predictive, preventive, and personalized cardiovascular care and overcoming the issues of data protection, reliability of the measurement, and the adherence of the user.

9.2 Physiological Parameters for Early CVD Detection

Planning the early diagnosis of CVDs requires constant observation of necessary physiological parameters indicative of cardiac and vascular well-being. Wearable sensors are able to measure these parameters in real time and noninvasively, in the real world, which means that subtle abnormalities can be detected early in life before clinical manifestations occur [9].

Cardiac functions and autonomic regulation are basic parameters of HR and HRV. HR patterns and decreased HRV are considered abnormal and associated with arrhythmias, stress, hypertension, and high cardiovascular risk. Constant surveillance records the dynamic and continuous changes that can be overlooked during the normal clinical evaluation phases. Wearable and cuffless BP monitoring can detect hypertension and abnormal blood pressure fluctuations at the earliest stage, which are risk factors of cardiovascular complications. SpO_2 indicates cardiopulmonary efficiency; pulmonary monitoring of SpO_2 with continuous measurements helps to identify oxygen desaturation of blood that accompanies heart failure and sleep-disordered breathing. Indirect measures of cardiovascular health are provided by physical activity and sleep behaviors measured with motion sensors and physiological devices; a lower level of activity, low sleep quality, and sleep disorders are greatly correlated with high risks of CVD. Together, these physiological parameters provide a multimodal basis of early CVD detection and predictive, personalized, and preventive healthcare using wearable sensing technology.

9.3 Wearable Sensor Technologies and System Architectures

Wearable sensor technologies enable continuous, noninvasive cardiovascular monitoring by capturing physiological signals in everyday environments. Recent advances in sensor miniaturization, wireless communication, and low-power electronics have made these devices suitable for long-term use and clinical integration [10]. Commonly used wearable sensor technologies include:

- **Photoplethysmography (PPG):** HR, HRV, SpO_2, and pulse waveform analysis
- **Electrocardiogram (ECG):** Detection of arrhythmias and electrical conduction abnormalities
- **Inertial Sensors:** Monitoring physical activity, posture, and sleep behavior
- **Bioimpedance Sensors:** Assessment of respiration and fluid-related cardiovascular changes
- **Pressure Sensors:** Support cuffless BP estimation

The wearable devices come in various form factors with each being designed to balance the comfort, signal quality, and the duration of monitoring. The key wearable devices form factors are:

- Wristbands and smartwatches to be used on a daily basis and long-term monitoring.
- Back/plate chest straps and adhesive patches with better ECG signal fidelity.
- Garment sensors made out of textile and embedded to be monitored unobtrusively.
- Medical implantable sensors with high accuracy of clinical use.

An example of monitoring based on a smartwatch is vital ECG, a relatively cheap medical smartwatch that performs a regular check of the health with the help of a simple one-push interface. The gadget facilitates remote monitoring outside of hospital settings and is easy to use, requiring no medical knowledge. Vital ECG measures key vital parameters such as skin temperature, moisture content, HR, SpO_2, ECG signals, and physical activity. Early anomaly identification and real-time reporting to medical professionals via smartphones are made possible by ML algorithms integrated into the system, showcasing the usefulness and clinical significance of contemporary wearable platforms.

Monitoring requires efficient design of hardware to be uninterrupted. Wearable systems are based on low-energy system-on-a-chip and adaptive sampling to reduce energy consumption and maintain data accuracy. The key design factors are miniaturized and flexible electronics, low-power system-on-a-chip based on design, and adaptive sensor sampling to ensure energy-efficient and continuous monitoring. Preprocessing of data on-device lowers the requirements of data transmission and increases the responsiveness of the system.

Internet of Medical Things platforms involve the use of wearable systems so that real-time healthcare can be delivered. Edge computing can be used to perform initial signal processing and quick event detection, and cloud rendezvous can store signals over the long term, analyze them using ML, and combine them with electronic health records. The main aspects of integration include edge-cloud collaboration to perform real-time analysis, cloud storage and ML, and automatic integration with remote patient monitoring and electronic health records. The integration of wearable sensor technologies with smart system architecture provides a strong foundation to early CVD detection and can help to constantly monitor, assess risks on a personalized basis and shift toward predictive and preventative cardiovascular treatment.

9.4 Signal Processing Techniques for Wearable Cardiovascular Data

Wearable cardiovascular signals are continuously obtained in the unstructured setting and are often affected by noise, motion artifacts and physiological variability. Signal processing methods are essential to improve the quality of data and allow its proper interpretation in order to detect CVD at the earliest stage [11]. The signal preprocessing and noise elimination techniques are as follows:
- Bandpass filtering
- Notch filtering
- Adaptive filtering
- Signal quality index assessment

Matters of physiological significance are then obtained after preprocessing and feature extraction, which detects cardiac rhythm and vascular properties. The extraction of features is an important part of converting cleaned signals into infor-

mative cardiovascular functional representations. Commonly extracted features include:
- Time-domain features: HR, inter-beat intervals, HRV metrics
- Frequency-domain features: low- and high-frequency HRV components
- Nonlinear features: entropy measures, Poincaré plot descriptors

For vascular and BP assessment, pulse waveform analysis techniques are widely applied to wearable signals. Three distinctive peaks are usually found in a waveform of the wrist artery's pulse, as depicted in Figure 9.1. These usually contain systolic pressure ($P1$) as a systolic peak (s_p); peripheral wave reflection produces a reflected peak (r_p), which is linked to reflected pressure ($P2$), and a diastolic pressure ($P3$) that corresponds to a diastolic peak (d_p) [12]. Accurately characterizing cardiovascular health requires identifying minute changes in the (s_p), (d_p), and (r_p).

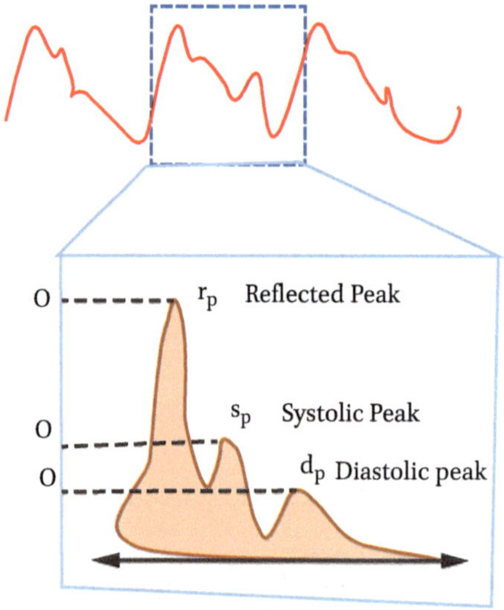

FIGURE 9.1
Wrist-derived arterial pulse waveform illustrating characteristic peaks. (Source: Own elaboration)

Pulse wave morphological features provide information about the arterial stiffness and hemodynamic changes. Pulse-based methods of analysis include:
- Pulse transit time and pulse arrival time estimation
- Waveform morphology features such as amplitude and rise time
- Assessment of arterial compliance and hypertension risk

Multimodal wearable systems require synchronization and fusion of signals from different sensors to improve accuracy and robustness. Simultaneous ECG, photoplethysmography (PPG), and inertial data analysis overcomes sensor-related constraints. Multimodal processing techniques include time alignment and sensor synchronization, cross signal validation to detect artifacts and feature level and decision level data fusion.

To facilitate timely intervention, real-time processing is becoming more and more performed at the edge; thin-sliced algorithms facilitate quick identification of abnormal cardiovascular patterns, whereas cloud platforms are used to offer in-depth analysis and long-term monitoring. Raw wearable sensor data is transformed into quality and clinically useful information by the signal-processing techniques, which are a crucial basis of intelligent cardiovascular monitoring systems.

9.5 Machine Learning and Artificial Intelligence for CVD Risk Prediction

Continuous wearable and clinical data can be transformed to predictive indicators of CVD risk with the help of ML and AI. Through the learning of nonlinear and complicated relationships based on multimodal health data, AI-driven models contribute to early detection, tailored monitoring, as well as proactive cardiovascular care.

9.5.1 Random Forest (RF)

Wearable sensor data can be used to detect CVD using the RF algorithm, a supervised classification method. RF models process an assortment of physiological measures acquired through wearable equipment, such as patterns of BP, activity, HRV, ECG, and sleep indicators. Multiple decision trees (DTs) are used to combine predictions that are used to classify and a majority vote has to decide the final label of CVD risk. The number of trees improves robustness and classification accuracy in particular when dealing with noisy wearable data with high dimensions.

RF allows learning on a large scale using physiological signals constantly obtained due to random feature sampling and bootstrap sampling. The ultimate risk forecast relates to the prevalence of a single most common class in all the DTs which are trained using a randomly chosen sample of both clinical and wearable data. When trees grow without pruning, overfitting is minimized and the diversity of model is increased.

Importance measures of feature relevance such as mean decrease accuracy (MDA) and mean decrease Gini (MDG) can be useful in enhancing clinical interpretation by extracting clinically significant physiological variables. RF is a reliable method of early prediction of CVD in wearable monitoring systems due to the ability to withstand signal artifacts and missing values [13]. Some of the steps involved in the implementation of the algorithm of the RF are as follows:

1. Bootstrap samples of size n are produced from the dataset using random sampling with replacement.

2. Each node in a DT chooses the optimal split from a randomly selected subset of features when the tree is expanded to its maximum depth without pruning ($m \ll p$).
3. A forest of k DT is created by repeating these procedures k times.

9.5.2 Extreme Gradient Boosting (XGBoost)

A popular tree-based ensemble learning algorithm that makes use of wearable sensor measurements to detect CVD in its initial stages. To minimize the error in prediction and retain the efficiency in computation, the method repeatedly improves decision boundaries. XGBoost develops optimized tree structures that enhance generalization and reduce overfitting through a regularized objective value that is imperative in dealing with noisy physiological data including ECG, activity patterns, and HRV.

The common characteristics of continuous wearable monitoring data that have been successfully applied in XGBoost are sparse inputs and missing values. Efficient utilization of cache, parallel calculations, and optimizer memory are used to provide real-time or near real-time identification of cardiovascular risks. Through the analysis of feature relevance with training, the interpretability of the model to a clinical decision assistant can be improved [13]. The learning process combats the complexity of the models by shrinkage as it is formulated additively as in eq. (9.1):

$$G_t(Z) = G_{t-1}(Z) + \alpha h_t(Z), \quad 0 < \alpha < 1 \tag{9.1}$$

where $G_{t-1}(Z)$ and $G_t(Z)$ denote model output from the previous $(t-1)$ and current (t)iteration, h_t indicates a weak learner added at iteration t, and α, Z, and t refer to learning rate, input features from wearable signals, and boosting iteration index. XGBoost uses wearable sensor measures to predict the risk of CVD in its early stages in a scalable and accurate manner.

9.5.3 SmartHeart Framework for Real-Time CVD Risk Prediction

The SmartHeart model combines ML with cloud-based computing to deliver real-time CVD prediction and personalized health guidance. Such a framework breaks the barrier of the older methods of risk assessment relying on the use of infrequent and static clinical measurements but integrates upon constantly acquired data of wearable sensors, electronic health records, survey on lifestyle, and known cardiovascular risk factors [14]. The XGBoost, RF, support vector machine (SVM), neural networks (NNs), and logistic regression (LR) are the types of ML models that were utilized in the SmartHeart framework. This is an all-inclusive, cloud-based approach that encourages early diagnosis, timely intervention, and overall risk evaluation, which encourages the management of cardiovascular health.

9.5.4 Lightweight Machine Learning Models for Cloud-Based CVD Prediction

The models used to predict cardiovascular risks in cloud systems have to be efficient in order to provide a balance between accuracy and speed. Unobtrusive ML models are particularly effective at structured clinical real-time inference as well as wearable-based data [15]. The models chosen include naive Bayes (NB), DTs, versatile adaptive boosting (V-AdaBoost), and LR, which have low latency, less resource consumption, and scalability, thus, suitable in being deployed in cloud-based healthcare systems.

V-AdaBoost: An ensemble learning approach based on wearable and clinical data for reliable and precise CVD prediction. The method integrates several weak pupils into a potent classifier by iteratively focusing on misclassified input. V-AdaBoost is a further enhancement of the standard AdaBoost, which uses adaptive weighting of features and a simplified selection of the sample, thus enhancing the convergence rate, predictive quality and computational efficiency. This algorithm works in the following way:

1. Set all training instances' weights to the same initial values.
2. Use the weighted dataset to train a weak classifier $h_t(x)$ for each round t.
3. Determine the weight α_t and classifier error.
4. Modify instance weights to highlight instances that were incorrectly classified.
5. Construct a final strong hypothesis by combining weak classifiers.

The overall prediction $H(u)$ is expressed in eq. (9.2):

$$H(u) = \text{sign}\left(\sum_{t=1}^{T} \alpha_t h_t(u)\right) \tag{9.2}$$

where $h_t(u)$ denotes the output of the tth weak classifier, α_t indicates the weight assigned to the classifier, and denotes the total number of iterations. V-AdaBoost is effective in dealing with imbalance in medical data, as well as is resistant to noisy measurements. Adaptive feature selection and advanced stopping criteria reduce overfitting and also minimizes the time of training. V-AdaBoost can be used to predict abnormalities, risk stratify accurately, and achieve reliable real-time prediction using wearable devices and cloud services when used in cardiovascular monitoring.

Decision Tree (DT): The use of a nonparametric supervised model is known as a DT, which is used to describe wearable sensor output to include HR, ECG, and HRV, thereby helping to detect CVD at an early stage. The data are divided into decision rules and terminal nodes are class labels and internal nodes feature test.

Impurity metrics such as the Gini index (GI) are used in node splitting $\text{Gini}(S_n)$ as described in eq. (9.3):

$$\text{Gini}(S_n) = 1 - \sum_{i=1}^{M} q_i^2 \tag{9.3}$$

where q_i mentions sample percentage that belongs to the ith class, M denotes the overall class count, and S_n indicates the sample set at a node. These models collectively enable responsive and reliable CVD prediction while maintaining system scalability and real-time performance.

9.5.5 Data-Driven Models for Personalized CVD Risk Assessment

The type of input data utilized for CVD prediction determines the ML model to be employed. Classical and successful ML models, which include SVM, DT, and RF, are useful in structured medical and wearable data [16]. DL systems provide predictive advantages that are significantly stronger because they automatically dig out complex representations based on large physiologic repositories of data. RNNs demonstrate some level of particularity in terms of temporally resolved signals and DNNs help detect more subtle determinants of cardiovascular risks that cannot be identified through standard diagnostic modalities.

Such AI-based models facilitate earlier cardiovascular aberration detecting and support the development of personalized, data-driven intervention regimes, thus, developing the paradigm of personalized and preventive cardiovascular care.

9.6 Results

The results summarize datasets employed in existing CVD studies, together with evaluation metrics and the benchmarking of ML models used to ensure accurate and reliable CVD risk prediction.

9.6.1 Datasets for AI-Based CVD Prediction

A description of the datasets that are regularly utilized to predict CVD with the help of AI is made to emphasize the variety of data sources that can be utilized to estimate the risk accurately. The SmartHeart model [14] combines ongoing data of electrical health records, life survey and wearable to create individualized cardiovascular health profiles. Wearable devices will be used to give real-time measures, whereas the electronic health records would give clinical measurements such as blood pressure, cholesterol levels and medical history. The behavioral factors that are included in the lifestyle surveys are physical activity, diet, and smoking habits.

UCI heart disease dataset [15] is a popular zone to develop and test models; it consists of 303 result cases and has 14 attributes with diagnostic measurements, clinical characteristics, and demographic factors. A binary target label shows the presence or absence of cardiac disease. Healthy prediction systems [16] also include medical imaging data and longitudinal clinical records to aid in the holistic risk assessment, but there are still some issues that surround data integration, interoperability, and security.

9.6.2 Quantitative Metrics for Assessing CVD Prediction Performance

Key indicators (the *F*-score, accuracy, recall, AUC-ROC, and precision) are used to provide the benchmarking of the various ML methodologies to predict CVD. RF support, SVM, LR, NN, XGBoost, DT, naive Bayes, V-AdaBoost and CNN/RNN models are compared to measure the accuracy of classification, sensitivity and discriminating capacity of the various approaches.

Accuracy is the percentage of correct outputs of a classification model. It assesses the ability of the model to differentiate CVD as well as non-disease cases across the whole data.

Precision defines the proportion of confirmed positive results to the total of the cases referred to as high risk. It determines the accuracy of CVD forecasting as well as it assists in reducing the number of false-positive classifications, which might lead to the unjustified clinical action.

Recall is an assessment of how well the model can pick up all true cases of high-risk within the grouping of patients. It highlights the ability of the model to identify the persons at cardiovascular risk, thus aiding in the early diagnosis and clinical decision-making.

F-score is a harmonic mean of recall and precision. It looks at the balance of false negatives and positive predictions giving a more realistic measure of performance in situations where the class distribution is uneven.

AUC-ROC is a measure of the model that is able to accurately classify positive cases as well as negative cases under different threshold levels. It determines the strength and consistency of prediction of CVD in varied classification settings. Table 9.1 and Figure 9.2 show a comparative evaluation of ML algorithms in CVD risks prediction.

Table 9.1 reports that NN [14] achieved 93.2% accuracy, 91.5% precision, 95.0% recall, 93.2% F_1-score, and 96.0% AUC-ROC, indicating excellent overall predictive performance and a strong ability to detect CVD cases.

The V-AdaBoost [15] attained 87.6% F_1-score, 88.2% precision, 87.1% recall, and 89.4% accuracy, indicating reliable predictive performance and effective detection of CVD cases in imbalanced clinical datasets.

TABLE 9.1 Model performance summary for CVD prediction [14–16]

Models	Accuracy (%)	Precision (%)	F_1-Score (%)	Recall (%)	AUC-ROC
RF [14]	92.5	90.1	92.1	94.2	95.3
SVM [14]	89.3	88.2	89.3	90.5	91.8
LR [14]	85.7	84.3	85.7	87.1	88.6
NN [14]	93.2	91.5	93.2	95.0	96.0
XGBoost [14]	91.8	89.9	91.1	92.4	94.1
LR [15]	85.6	84.2	83.8	83.5	-
DT [15]	84.3	82.6	81.8	81.1	
NB [15]	81.2	79.3	78	76.8	
V-AdaBoost [15]	89.4	88.2	87.6	87.1	
LR [16]	0.79	0.77	0.78	0.82	0.85
CNN/RNN [16]	0.91	0.92	0.93	0.94	0.95

The CNN/RNN model [16] recorded 0.91 accuracy, 0.95 AUC-ROC, 0.93 F_1-score, 0.94 recall, and 0.92 precision, reflecting high effectiveness in capturing temporal patterns for CVD detection.

9.7 Challenges and Limitations in Wearable CVD Monitoring

The limitations and weaknesses of wearable CVD monitoring encompass several aspects of paramount imperativeness, such as the privacy of data, model quality, processing of incomplete or abnormal data, interpretation, diversity of different populations, and the incorporation of new sensor technologies. The major issues are as follows:

Data Privacy and Security: The protection of sensitive health information on cloud computing systems would require the application of strong encryption measures and the implementation of strict regulatory measures. Confidentiality of patients and preventing unauthorized access are two of the most important requirements that facilitate extensive use.

Data Quality and Completeness: The ML models will be sensitive to missing and incomplete, or low quality data. It is essential that correct inputs are provided to obtain credible cardiovascular risk estimates hence the need to have stringent pre-processing and imputation approaches.

Model Validation and Generalizability: Predictive models should be consistently tested in a variety of healthcare settings, multiethnic cohorts, as well as additional

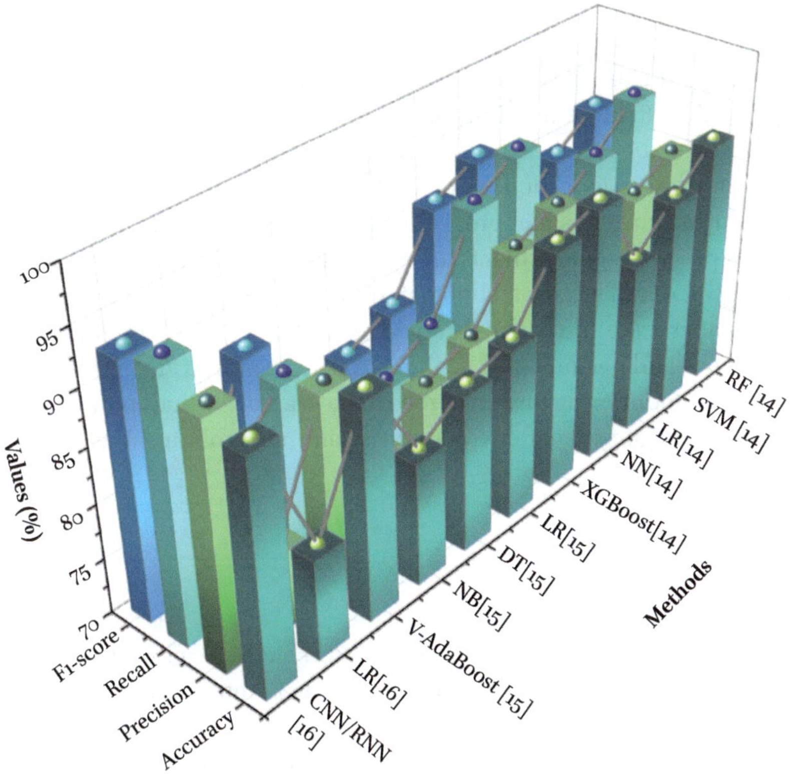

FIGURE 9.2 Visual comparison of ML performance in detecting cardiovascular risks. (Source: Own elaboration)

data. Regional and ethnic, facility-specific differences in cardiovascular risk profile necessitate the constant updating of the models.

Explainability and Interpretability: Modern systems often do not have explainable attributes, and so limit the understanding of model-driven decisions by clinicians. It is vital to augment interpretability, and this will be necessary to achieve trust and effective clinical uptake.

Incorporation of New Data Sources: Many systems do not include the data provided by new wearable or new sensors, which hinders flexibility and reduces the quality of cardiovascular surveillance.

Challenges of Training and Deployment: Structural differences between training and deployment datasets, as well as the combination of heterogeneous data streams, can give rise to reduced model performance, possible data leakage, and reduced generalizability.

9.8 Conclusion

This chapter discusses the use of wearable sensors in the early detection and regular monitoring of CVDs. Remote monitoring and timely, data based, clinical decision making was enabled by the use of cloud based platforms and mobile health applications, and further diagnostic accuracy and cardiovascular risk prediction were achieved using advanced signal processing and ML methods. Although posing the risk of data privacy, signal artifacts, measurement reliability, and user compliance, the improvement of sensor technologies, sensor fusion, and AI has transformed cardiovascular care throughout conventional episodic framework to predictive, preventive, and personalized healthcare. The invention and implementation of the wearable monitoring systems prove to have a tremendous potential in improving the early diagnosis, patient outcomes, and streamlining the overall healthcare functioning.

References

[1] Zang J, et al. A novel wearable device integrating ECG and PCG for cardiac health monitoring. Microsystems & Nanoengineering. 2025;11(1):7. https://doi.org/10.1038/s41378-024-00858-3.

[2] Lopez-Jimenez F, et al. Development and Internal Validation of an AI-Enabled Cuff-less, Non-invasive Continuous Blood Pressure Monitor Across All Classes of Hypertension. Journal of Cardiovascular Translational Research. 2025;18(2):280–90. https://doi.org/10.1007/s12265-024-10589-5.

[3] Awasthi S, et al, eds. AI and IoT-based Intelligent Health Care & Sanitation. Bentham Science Publishers; 2023.

[4] De Sabbata G, Simonini G. Real-Time Forecasting from Wearable-Monitored Heart Rate Data Through Autoregressive Models. Journal of Healthcare Informatics Research. 2025;1–20:9. https://doi.org/10.1007/s41666-025-00191-y.

[5] Guler S, Aslanger E, Yapici MK. Wearable armband with a floating mobile exploratory electrode at fingertip for on-demand touch-and-measure multileadelectrocardiography. Wearable Technologies. 2025;e21:6. https://doi.org/10.1017/wtc.2025.11.

[6] Coste A, Millour G, Hausswirth C. A comparative study between ecg-and PPG-based heart rate sensors for heart rate variability measurements: Influence of body position, duration, sex, and age. Sensors. 2025;25(18):5745. https://doi.org/10.3390/s25185745.

[7] Baigutanova A, et al. A continuous real-world dataset comprising wearable-based heart rate variability alongside sleep diaries. Scientific Data. 2025;12(1):1474. https://doi.org/10.1038/s41597-025-05801-3.

[8] Al-Turjman F, et al, eds. Transforming Management with AI, Big Data, and IoT. Springer; 2022.

[9] Muliawan A, Rizal A, Hadiyoso S. Heart disease prediction based on physiological parameters using ensemble classifier and parameter optimization. Journal of Applied Engineering and Technological Science (JAETS). 2023;5(1):258–67. https://doi.org/10.37385/jaets.v5i1.2169.
[10] Nazir A, et al. Wearable Technology and Its Potential Role in Cardiovascular Health Monitoring and Disease Management. Health Science Reports. 2025;8(11):e71486. https://doi.org/10.1002/hsr2.71486.
[11] Panga NKR. Applying discrete wavelet transform for ECG signal analysis in IOT health monitoring systems. International Journal of Information Technology and Computer Engineering. 2022;10(4):157–75. https://doi.org/10.62647/ijitce.2022.v10.i4.pp62-78.
[12] Chowdhury AH, et al. Monitoring and analysis of cardiovascular pulse waveforms using flexible capacitive and piezoresistive pressure sensors and a machine learning perspective. Biosensors and Bioelectronics. 2023;237:115449. https://doi.org/10.1016/j.bios.2023.115449.
[13] Ansyari MR, et al. Implementation of Random Forest and Extreme Gradient Boosting in the Classification of Heart Disease Using Particle Swarm Optimization Feature Selection. Journal of Electronics, Electromedical Engineering, and Medical Informatics. 2023;5(4):250–60. https://doi.org/10.35882/jeemi.v5i4.322.
[14] Gadhave RT, et al. SmartHeart: A Cloud and Machine Learning Framework for Early Cardiovascular Disease Prediction. International Journal of Environmental Sciences. 11(5):2025. https://doi.org/10.64252/0b2sfy54.
[15] Cenitta D, et al. A Privacy-Preserving Cloud-Based System for Cardiovascular Disease Prediction Using Lightweight Machine Learning Models. Journal of Internet Services and Information Security. 2025;15(3):129–48. https://doi.org/10.58346/JISIS.2025.I3.009.
[16] Sozib HM. Wearable AI for Cardiovascular Health Monitoring: Enabling Early Detection and Prevention. Journal of Computer Science and Technology Studies. 2025;7(2):294–304. https://doi.org/10.32996/jcsts.

CHAPTER 10

Machine Learning-Based Prediction Models for Diabetes Management Using Wearable Sensors

Manivannan Karunakaran, Mahesh Singh Danu, Geetha Arjunan

Abstract Effective diabetes management requires continuous monitoring and timely intervention to reduce health risks. Recent advances in wearable sensor technologies enable noninvasive, continuous collection of physiological and lifestyle data, offering strong potential for intelligent diabetes care. This chapter presents a machine learning (ML)-based framework for predicting blood glucose trends using data from wearable devices, including continuous glucose monitors, heart rate sensors, physical activity trackers, and sleep monitoring systems. Multimodal time-series data are preprocessed through noise reduction, normalization, missing-value imputation, and temporal segmentation to ensure data reliability. Features representing glucose dynamics, physiological responses, and daily activity patterns are extracted and analyzed using multiple ML models, including support vector regression, random forest, extreme gradient boosting, and ensemble approaches. Experimental results demonstrate that tree-based and ensemble models outperform linear methods. Random forest achieves root mean square error values as low as 0.183 with a mean absolute error of 0.102 across age groups, while extreme gradient boosting attains an RMSE of 11.9 mg/dL and an R^2 of 0.73 under free-living conditions. For hypoglycemia prediction, ensemble models reach up to 90% accuracy at a 15-min prediction horizon. Clarke error grid analysis confirms over 99% of clinically acceptable predictions, highlighting the robustness and real-world applicability of the proposed framework.

Keywords Diabetes management – wearable sensors – machine learning – blood glucose prediction – personalized healthcare – predictive modeling

10.1 Introduction

Diabetes is a chronic metabolic condition that requires ongoing monitoring and treatment to prevent serious short-term and long-term effects such as cardiovascular disease (CVD), neuropathy, nephropathy, and retinopathy [1]. Maintaining the patient's blood glucose levels within the safe physiological range throughout the day can be considered one of the most significant issues in diabetes care, as even minor fluctuations may result in acute hypoglycemic or hyperglycemic episodes with severe clinical implications [2]. The physiology of blood glucose is

 | HTTPS://DOI.ORG/10.1515/9783112237083-010

highly complex and depends on a multitude of interconnected factors, including food consumption, physical exercise, sleep, stress, medication use, and individual physiological fluctuations [3]. As a result, continuous glucose monitoring has emerged as a crucial tool in the management of diabetes and the enhancement of the standard of living of patients through the provision of timely feedback as well as interventions [4].

The conventional methods of managing diabetes are mostly founded on regulated plans, frequent blood tests through finger-pricks, and constant clinical principles [5]. Although these techniques offer some basic clinical support, they are limited by nature in their capacity to identify rapid changes in glucose and the intricate interplay between behavioral, environmental, and physiological measures. This means that such conventional systems are not usually adaptable, do not provide much personalization, and do not produce real-time early warning of adverse glycemic events, especially in free-living environments [6, 7]. The constraints emphasize the importance of smart and evidence-based solutions capable of going beyond fixed thresholds and universally applicable suggestions. The latest innovations in wearable sensor technologies have provided the development of new opportunities in smart and personalized diabetes treatment [8].

Continuous glucose monitors, heart rate sensors, activity trackers, and sleep monitors are all examples of devices that can be used to collect high-resolution physiological and lifestyle data on a constant, noninvasive basis in the real world [9]. These multimodal data streams can be analyzed with machine learning (ML) and artificial intelligence (AI) methods to identify complex, nonlinear processes underlying individual glucose behavior and to predict short-term glucose more accurately than conventional methods can [10]. ML models are especially adapted to this, since they can combine heterogeneous data sources, accommodate inter-individual variation, and learn temporal dependencies that are challenging to model in more standard statistical models.

This chapter presents an ML-driven architecture that uses wearable sensor data to perform intelligent diabetes management, motivated by technological and methodological advancements. The goal is to make predictions about the short-term change of the blood sugar level, as well as to make the prediction of the possible hypoglycemic and hyperglycemic episodes more accurate and clinically significant. In addition to the model development, this chapter focuses on the need to apply such predictive capabilities in strong system-level architecture. To operationalize the idea of data-based glucose prediction in a real-world healthcare setting, one needs to have an integrated framework that is able to facilitate continuous sensing, scalable computation, secure data management, and intelligent decision-making. In accordance with this, the following section offers an Internet of Medical Things (IoMT) framework that harnesses the potential of AI to incorporate connected devices, edge-cloud analytics, and predictive models in a systematic manner to promote proactive, customized, and clinically relevant diabetes care.

10.2 AI-IoMT Framework for Intelligent Diabetes Care

Diabetes can be continually and individually monitored by combining AI and an internet of medical things (IoMT). Wearable devices, edge computing, and ML in this framework work together to deliver real-time health monitoring and predictive information:

- Various types of wearable sensors, such as glucose sensors and activity trackers, acquire physiological and lifestyle information at the sensing layer. This data is sent to an edge processing unit that does some basic filtering, noise elimination, and data prioritization in order to maximize the latency and network efficiency [11].
- The cloud layer contains secure data storage, synchronization, and scalable computation, and ensures privacy by encryption requirements and embraces interoperability between disparate devices and electronic health records.
- The intelligence tier uses ML algorithms to predict complicated associations between glucose, physiological indicators, and behavioral trends. Fuzzy logic is incorporated to deal with uncertainty, enhance interpretability, and make early detection of hypoglycemic and hyperglycemic occurrences possible.
- Lastly, the decision-support layer provides actionable information and alerts to patients and clinicians to facilitate timely interventions and customized care and minimize the clinical workload. This multilevel AI-IoMT design proves to be a scalable system to manage intelligent diabetes in the next generation.

In general, the framework is a scalable and flexible design of intelligent diabetes management, which emphasizes the synergistic approach between wearable sensing, edge-cloud analytics, and AI-drawn decision-making in the next-generation healthcare systems. Although the AI-IoMT framework describes a scalable architecture of smart diabetes management, the ultimate success of an implementation of the framework is determined by the work of predictive models trained on the data of wearable sensors. With the aim of evaluating the present condition of such methods, the section below critically reviews the representative studies implementing the use of ML and ensemble methods in glucose estimation and hypoglycemia prediction in controlled and free-living conditions.

10.3 Methodology

The chapter is a critical review of three representative research studies that examine the estimation of levels of blood glucose using AI and ML algorithms and predict hypoglycemia based on noninvasive wearable sensor data. The chosen articles address the controlled and free-living regimes, multimodal sensing, and ensemble learning approaches, which provide a comprehensive perspective on the existing approach and performance patterns of wearable-based diabetes prediction.

10.3.1 Data Collection and Feature Engineering

The research taken into consideration differs in how wearable sensor data is gathered, integrated, and transformed into meaningful features to predict glucose because the quality and representation of the datasets form the foundation of any predictive model. The prediction of glucose based on wearable absorption is based on the incessant accrual of diverse physiological and behavioral indications that are noisy, asynchronous, and incomplete in nature. In the analyzed research papers, the noninvasive wearable sensors, which record both the direct and indirect measures of glucose dynamics, were used to collect data. Physiological indicators used in the research [12] included photoplethysmography, heart rate, physical activity, and skin temperature, which were used to indicate glucose variations relating to metabolic and autonomic reactions. In the investigation [13], data were further acquired under free-living conditions by incorporating continuous glucose monitoring with wearable-obtained heart rate, intensity of activity, sleep patterns, and contextual variables in order to obtain the real-life variability in the glucose behavior. The research [14] concentrated on CGM-oriented datasets that are supplemented with contextual facts, such as meal times and physical activity, to aid in the early identification of hypoglycemia in type 1 diabetic patients.

Due to the variability and flaws of the wearable data streams, systematic preprocessing was utilized to enhance the data quality and model resilience. Raw sensor data were denoised to remove motion artifacts, and missing data due to sensor dropout were addressed through interpolation or imputation methods. Multimodal signals were time-warped and resampled to uniform time windows to make them synchronous with glucose measurements. To reduce interindividual variability, feature normalization and scaling were performed, after which temporal segmentation was performed in order to capture both short-term glucose trends and delayed physiological responses. These preprocessing and feature engineering measures converted raw wearable data into time-series inputs in a structured format, which can then be fed into ML-based prediction. Therefore, the following section describes the ML and AI algorithms that were utilized to forecast the glucose dynamics using such wearable data.

10.3.2 Machine Learning Model and Predictive Strategies for Glucose Monitoring

Various ML and DL (deep learning) approaches have been investigated to monitor and predict glucose based on wearable sensor data, support voting machine, and use decision tree (DT)-based, ensemble learning model, and neural network models. The primary types of ML techniques that have been utilized in glucose monitoring applications in recent years, centered on the rising popularity of tree-based and ensemble models in dealing with noisy and high-dimensional physiological data. Recent research proved the efficiency of ML and AI to forecast the level of blood glucose using the data that was gathered with the help of noninvasive wearable sensors. Such methods allow creating individual and data-driven management of diabetes by modeling nonlinear and complicated relationships between physiological indicators and glucose dynamics [15].

10.3.3 Noninvasive Assessment of Blood Glucose Using Wearable Data

To predict the level of blood glucose parameters based on wearable sensor data, the research [12] employed both DL and ML models. Particular models were developed for each age group. Root mean square error (RMSE) and mean absolute error (MAE) were used to measure performance. RF builds classes and trees in the form of multiple decision classes on a random subset of the features at each part, thereby minimizing overfitting and variance. To estimate blood glucose, the overall result of all three outputs is the average of all three outputs is shown in eq. (10.1):

$$D(s) = \max_{O} F_s \sum_{j=1}^{L} \left(d_j(S) = O\right) \tag{10.1}$$

where $d_j(S)$ is the predicted blood glucose value from the jth tree, and L is the total number of trees. The final estimate $D(s)$ is obtained by aggregating the outputs of all trees for sample s, representing the glucose output space F_s; max indicates the selection of the output value O with the highest aggregated contribution across all trees. In eqs. (10.2) and (10.3), the averaging reduces variance and improves the stability of predictions across high-dimensional wearable data:

$$j(s) = 1 - e_1^2 - e_0^2,\ e_i = \frac{m_i}{m} \tag{10.2}$$

$$\delta j(s) = j(s) - e_o j(s_o) - e_r j(s_r) \tag{10.3}$$

where $j(s)$ denotes the parent node and s_o and s_r represent the left and right child nodes, respectively. $e_i = \frac{m_i}{m}$ is the proportion of samples belonging to class i, where m_i is the number of samples in class i and m is the total number of samples at node s. Minimizing $\delta j(s)$ ensures homogeneous glucose predictions:

$$J_H(\Theta) = \sum_q \sum_r \delta j_\Theta (s, S) \tag{10.4}$$

In eq. (10.4), δj_Θ is the set of all nodes where feature Θ is used. q indexes the DTs in the ensemble, and r indexes the split nodes within each tree. Higher $J_H(\Theta)$ indicates a stronger influence of the feature on glucose predictions.

10.3.4 Continuous Glucose Prediction in Free-Living Conditions

Under the research conditions, a multimodal, tree-based regression model of continuous noninvasive glucose was developed [13]. Their method adopted a combination of physiological and contextual data from wearable sensors and compared various regression models, such as ridge regression (RR), linear regression (LR), eXtreme gradient boosting regression (XR), and random forest regression (RFR). Training of the models was done with repeated train-test splits so as to make the models robust. Xr, an ensemble of boosted trees, minimized prediction loss by regularized objective functions, and was able to learn high-dimensional time-series wearable data.

10.3.5 Hypoglycemia Prediction Using Ensemble Learning

The XGBoost and SVR were selected as the predominant base models in research [14] devoted to the prediction of hypoglycemia in type 1 diabetes. This was combined into an ensemble model, which significantly reduced prediction error in the hypoglycemic region at the cost of comparable accuracy throughout the glucose range. XGBoost repeatedly uses this to create an ensemble of K DTs, with each tree fixing the errors made by the previous ones to enhance overall prediction. The value of glucose Y_i, which the prediction algorithm predicts at an observation x_i, is determined as the sum of all predictions of all K trees as expressed in eq. (10.5):

$$\hat{Y}_i = \sum_{k=1}^{K} f_k(x_i) \tag{10.5}$$

An impartial function $L(\varphi)$, which comprises a loss factor l assessing error in forecasting and a regularization term Ω managing model complexity, is minimized repeatedly for each tree f_k:

$$L(\varphi) = \frac{1}{n}\sum_{i=1}^{n} l\left(Y_i, \hat{Y}_i\right) + \sum_{i=1}^{n} \Omega(f_k) \tag{10.6}$$

In eq. (10.6), Y_i is the observed glucose assessment, $\hat{Y}_i$ is the projected significance, and the variance of the squared error loss is represented by $l\left(Y_i, \hat{Y}_i\right) = \left(l\left(Y_i, \hat{Y}_i\right)\right)^2$. This iterative refinement and regularization allow XGBoost to achieve robust and accurate glucose predictions even with high-dimensional, multimodal input features. To ensure that these models provide not only statistical accuracy but also clinical safety, standardized evaluation metrics are required to quantify prediction errors and assess medical relevance. The next section, therefore, summarizes the performance metrics used across studies to evaluate glucose prediction and hypoglycemia detection.

10.4 Model Evaluation and Performance Metrics

Experiments have been conducted all in Python on regular ML packages, such as Scikit-learn to run regression and ensemble models, and XGBoost to run boosted tree implementations. Training of the model, validation, and performance evaluation were performed in a workstation environment to facilitate the reproducibility and consistency between experiments. The quantitative findings introduced into the three researches, in general, prove that wearable-based prediction of glucose is best accomplished by both ensemble and tree-based methods of learning, especially in a complex, real-world scene. Measurement of evaluation across the studies comprised:

- **Regression Measures:** Accuracy metrics consist of MAE) and root mean square error (RMSE) in glucose prediction by numerically quantifying deviations of the model in predicting short-term glucose dynamics of the model.
- **Measures of Classification:** Assess the ability of the algorithm to detect hypoglycemia, including prediction accuracy, sensitivity, specificity, and F1-score.
- **Measures of Clinical Relevance:** Clarke error grid analysis to assess the accuracy of glucose predictions in a clinically relevant scenario.

10.4.1 Non-invasive Glucose Estimation Using Wearable Sensors

To forecast the amount of glucose in a noninvasive wearable device, the research [12] evaluated different ML techniques. As shown in Table 10.1 and Figure 10.1, the models' performance was assessed using RMSE and MAE with two age groups of participants: young people and adults.

Random forest (RF) had the lowest RMSE across the young group as well as the adult one, which implies the most stable predictive accuracy. ANFIS ran with the worst MAE in both groups, which indicated the worst absolute predictions. The results of all the models were better when they were subjected to feature selection and pre-processing, with little difference between classical and advanced models. The range of RMSE was from 0.183 to 0.197, and the range of MAE was from 0.097 to 0.112, which indicated high accuracy compared to the previous research.

TABLE 10.1 Machine learning model reliability for glucose prediction across age groups [12]

Model	Young group		Adult group	
	RMSE	MAE	RMSE	MAE
SVR [12]	0.195	0.106	0.188	0.104
MLP [12]	0.193	0.112	0.186	0.108
ANFIS [12]	0.197	0.097	0.193	0.099
RF [12]	0.189	0.102	0.183	0.102

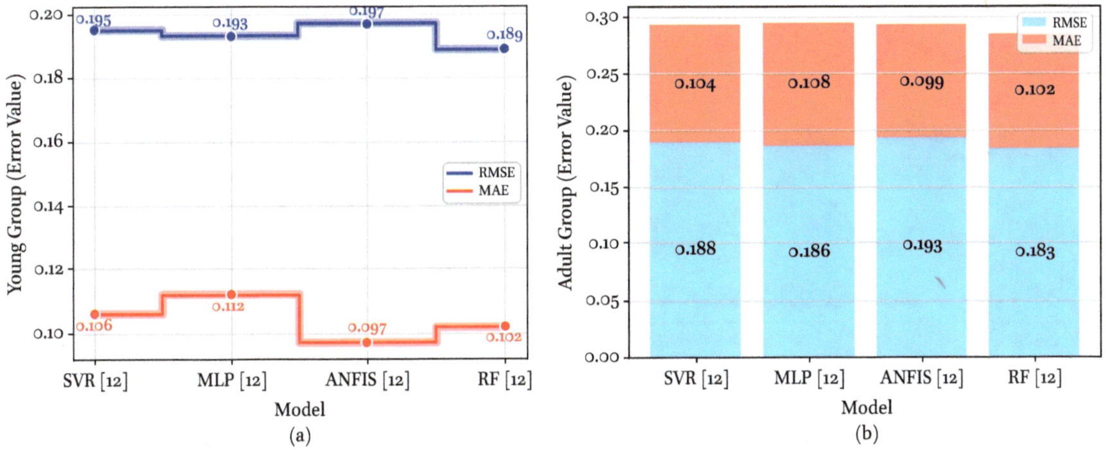

FIGURE 10.1 Comparison of machine learning models for glucose prediction across age groups (a) young and (b) adult group. (Source: Own elaboration)

10.4.2 Continuous Glucose Prediction in Free-Living Conditions

A multimodal tree-based regression model was presented in the investigation [13] and used for continuous noninvasive glucose prediction under free-living circumstances. Their procedure combined the usage of physiological and contextual information of the wearable sensors and tested multiple regression models, such as the gradient boosting (GB), logistic regression (LR), RF, and ensemble voting (EV), which were measured against standard prediction metrics and evaluated across repeated train-test splits to assure stability and reliability of the results, as presented in Table 10.2 and Figure 10.2.

The proposed XR model achieved the best result, the highest R^2 (0.73), the lowest RMSE (11.9 mg/dL), NRMSE (0.52 mg/dL), and MARD (7.1%). RFR was also a

TABLE 10.2 Regression model performance for wearable sensors for prospective glucose estimation in healthy lifestyle environments [13]

Model	RMSE (mg/dL)	R^2 (mean ± SD)	MARD (%)	NRMSE (mg/dL)
RFR [13]	15.6 ± 0.3	0.53 ± 0.01	9.7 ± 0.1	0.68 ± 0.01
RR [13]	21.3 ± 0.3	0.13 ± 0.01	13.7 ± 0.1	0.93 ± 0.00
LR [13]	21.3 ± 0.3	0.13 ± 0.01	13.7 ± 0.1	0.93 ± 0.00
XR [13]	11.9 ± 0.3	0.73 ± 0.01	7.1 ± 0.1	0.52 ± 0.01

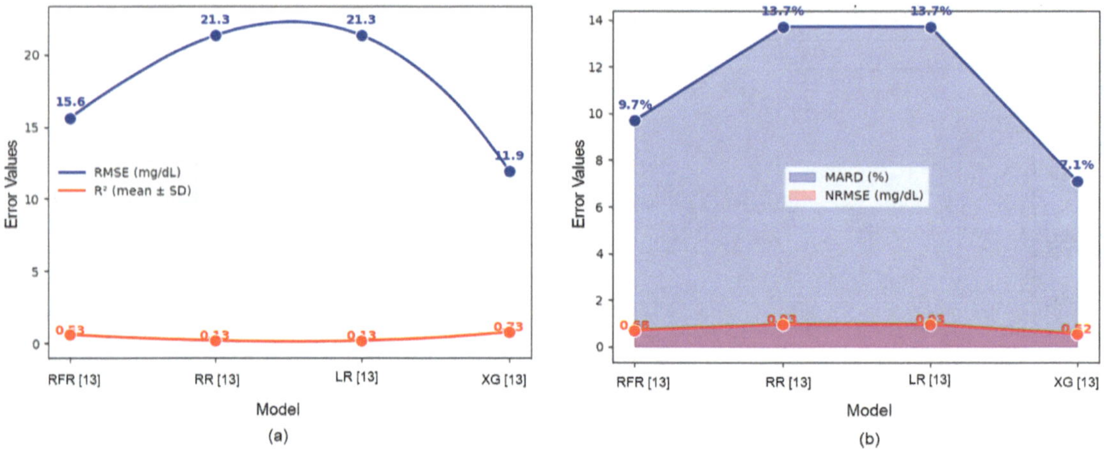

FIGURE 10.2 Regression model performance for wearable sensor-based constant glucose prediction: (a) RMSE and R^2, and (b) MARD and NRMSE (Source: Own elaboration)

good model that demonstrated high accuracy estimates on all measures. The data are not well fit by linear models (LR and RR) with low R^2 and high error rates. All models had high clinical accuracy (Clarke error grid) with more than 99%, which showed better clinical reliability.

10.4.3 Hypoglycemia Prediction Using Ensemble Models

With a focus on hypoglycemia detection, the research [14] developed an ensemble ML model that combines XGBoost and SVR to predict the short-term blood sugar levels of individuals with type 1 diabetes.

Table 10.3 illustrates how the ensemble model's performance was assessed over many prediction horizons (15, 30, and 60 min) in comparison to its component basis models (XGBoost and SVR).

TABLE 10.3 Short-term glucose prediction performance of ensemble and constituent models for hypoglycemia detection [14]

Model	Prediction horizon	RMSE (mg/dL)	MAE (mg/dL)	MAPE (%)	Time lag (min)	Hypoglycemia accuracy (%)
SVR [14]	15 min	21.92 ± 9.13	15.57 ± 4.72	10.82 ± 3.59	21.97 ± 6.49	70
	30 min	29.72 ± 9.80	19.68 ± 5.66	15.57 ± 4.62	32.45 ± 8.75	30
	60 min	42.95 ± 12.11	22.91 ± 6.23	20.43 ± 6.34	51.28 ± 10.99	53
XGBoost [14]	15 min	21.91 ± 5.93	15.63 ± 5.34	10.33 ± 3.41	21.80 ± 5.98	80
	30 min	23.65 ± 5.99	21.97 ± 6.49	15.48 ± 4.64	32.45 ± 8.75	22
	60 min	30.97 ± 8.37	32.45 ± 8.75	22.95 ± 6.36	51.28 ± 10.99	22
Ensemble [14]	15 min	16.55 ± 4.55	11.33 ± 3.03	7.85 ± 2.14	9.21 ± 3.49	90
	30 min	27.32 ± 8.68	15.57 ± 4.72	13.67 ± 3.80	22.21 ± 6.93	81
	60 min	41.48 ± 11.86	19.68 ± 5.66	21.24 ± 6.93	49.90 ± 10.10	58

The XGBoost and the SVR were chosen to be the foundation models of the ensemble because of their high performance in the hypoglycemia region. The ensemble model greatly mitigates the erratic forecasts caused by hypoglycemia with an EP of 19 at a 30-min horizon, as opposed to single basis models such as SVR (30%) and XGBoost (22%). The overall performance of the error metrics (RMSE, MAE, MAPE, and time lag) across the entire glucose range is also robust and similar to the basis models. This shows that the ensemble can make predictions that are balanced under hypoglycemic, euglycemic, and hyperglycemic conditions.

10.5 Discussion

This chapter analyzed the recent developments in ML and AI algorithms in blood glucose prediction based on noninvasive wearable sensor measurements, particularly continuous glucose prediction and hypoglycemia prediction. The three chosen studies all show that data-driven models can be used to effectively model the nonlinear, complex interactions of physiological signals and glucose dynamics in controlled and free-living settings. Investigation [12] demonstrated that classical ML and neuro-fuzzy algorithms are very competitive in the noninvasive estimation of glucose. Their findings emphasized the soundness of the tree-based models, specifically the RF, in their capability to deal with noisy and high-dimensional wearable data, in addition to ANFIS, which possesses the capacity to adopt fuzzy logic, resulting in lower absolute prediction errors. The similarity in the performance of the age groups is another pointer that suggests the possibility of personalized models to be adopted in predicting more accurately, without the use of invasive sensing technologies. Research [13] further projected glucose in real-world free-living conditions and combined multimodal wearable and contextual information. They established that boosted tree ensembles, particularly the eXtreme gradient boosting regression, perform much better than linear and regularized regression models. This affirms that the dynamics of glucose in free day-to-day life are very nonlinear and can be optimally learned by complex temporal interaction models. Notably, the paper emphasized sound methodology with repeated train-test splits instead of algorithm novelty, which contributes to the feasibility of the practical use of ensemble learning in real-life diabetes management. Research [14] dealt with a clinically critical issue by focusing on early hypoglycemia prediction in type 1 diabetes. Ensuring that XGBoost and SVR are incorporated into an ensemble system, the research employed significant error reduction in the hypoglycemic range, especially under a short-term perspective of prediction. The capability of the ensemble model to reduce the erroneous predictions without compromising the model's consistent performance between the euglycemic and hyperglycemic ranges emphasizes the importance of the hybrid models of modeling in safety-critical applications. These findings highlight the fact that the importance of the optimization of performance in clinically sensitive glucose regions is compared to the minimization of global error metrics.

These results collectively confirm that noninvasive and clinically meaningful glucose predictions may be obtained by combining multimodal wearable data with cutting-edge ML methods. These solutions aid in the early identification of risks, self-reliant decision-making, and preemptive diabetes care. Although their deployment may have issues that affect the quality of data, variation between individuals, and overall implementation in the real world, with the increasing numbers of wearable sensing technologies, explainable AI, and mass validation studies, the resilience and clinical integration of intelligent diabetes monitoring systems will likely continue to be improved. These reflections are a clear basis for generalizing the implications of wearable-based ML models and finding essential follow-ups to the development of intelligent systems for managing diabetes.

10.6 Conclusion

To control diabetes with noninvasive wearable sensors, this chapter reviewed and synthesized the literature on ML-based prediction models. The reviewed articles reveal that when wearables are combined with physiological and contextual data and trained with the help of advanced methods of ML and AI, accurate, continuous, and personalized blood glucose prediction becomes possible. Data-driven models are more appropriate in modeling the complicated, nonlinear dynamics of glucose linking activity, sleep, and variability in physiological regulation as opposed to traditional rule-based or linear methods. In all of the reviewed publications, ensemble and tree-based models, including RF, XGBoost, and hybrid ensembles, performed better than linear and single-model methods in statistical and clinical analysis measures. The experiments on the free-living features validated the strength of increased regression models in estimating glucose in the real world, whereas hypoglycemia-biased ensemble models drastically minimized the critical errors in prediction without affecting the general performance. The safety and applicability of these models to diabetes care were further proven by clinical relevance tests, such as Clarke error grid assessments. In general, the results indicate the high potential of ML in conjunction with wearable sensing devices to assist in non-invasive, continuous glucose monitoring and early risk detection. The chapter provides a strong starting point for further studies that will enhance the model's generalizability, interpretability, and real-time implementation, which will eventually lead to intelligent and patient-centric diabetes management systems.

References

[1] Gregg EW, et al. Multiple long-term conditions as the next transition in the global diabetes epidemic. Communications Medicine. 2025;5(1):42. https://doi.org/10.1038/s43856-025-00742-9.

[2] Awasthi S, et al, eds. AI and IoT-based Intelligent Health Care & Sanitation. Bentham Science Publishers; 2023.

[3] Veeranki V, Prasad N. Utilising continuous glucose monitoring for glycemic control in diabetic kidney disease. World Journal of Diabetes. 2024;15(10):2006. https://doi.org/10.4239/wjd.v15.i10.2006.

[4] Ardelean A, et al. Personalized and predictive strategies for diabetic foot ulcer prevention and therapeutic management: Potential improvements through introducing Artificial Intelligence and wearable technology. Medicine and Pharmacy Reports. 2024;97(4):419. https://doi.org/10.15386/mpr-2818.

[5] Salama R, et al. (Benefits of Internet of Things (IoT) Applications in Health care-An Overview). In 2023 International Conference on Computational Intelligence, Communication Technology and Networking (CICTN) 2023 IEEE. https://doi.org/10.1109/CICTN57981.2023.10141452.

[6] Fraser RA, et al. Integration of artificial intelligence and wearable technology in the management of diabetes and prediabetes. npj Digital Medicine. 2025;8(1):687. https://doi.org/10.1038/s41746-025-02036-9.

[7] Huang X, et al. Digital biomarkers for interstitial glucose prediction in healthy individuals using wearables and machine learning. Scientific Reports. 2025;15(1):30164. https://doi.org/10.1109/JSEN.2023.3319360.

[8] Ahmed S, et al. Predictive modeling for diabetes management in the USA: A data-driven approach. Journal of Medical and Health Studies. 2024;5(4):214–28. https://doi.org/10.32996/jmhs.2024.5.4.24.

[9] Shi B, et al. Assessing Elevated Blood Glucose Levels Through Blood Glucose Evaluation and Monitoring Using Machine Learning and Wearable Photoplethysmography Sensors: Algorithm Development and Validation. JMIR AI. 2023;2(1):e48340. https://doi.org/10.2196/48340.

[10] Alhaddad AY, Aly H, Gad H, Al-Ali A, Sadasivuni KK, Cabibihan JJ, Malik RA. Sense and learn: Recent advances in wearable sensing and machine learning for blood glucose monitoring and trend-detection. Frontiers in Bioengineering and Biotechnology. 2022;10:876672. https://doi.org/10.3389/fbioe.2022.876672.

[11] Maqbool S, et al. A smart sensing technologies-based intelligent healthcare system for diabetes patients. Sensors. 2023;23(23):9558. https://doi.org/10.3390/s23239558.

[12] Ahmed A, et al. Performance of artificial intelligence models in estimating blood glucose level among diabetic patients using non-invasive wearable device data. Computer Methods and Programs in Biomedicine Update. 2023;3:100094. https://doi.org/10.1016/j.cmpbup.2023.100094.

[13] Karunarathna TS, Liang Z. Development of Non-Invasive Continuous Glucose Prediction Models Using Multi-Modal Wearable Sensors in Free-Living Conditions. Sensors. 2025;25(10):3207. https://doi.org/10.3390/s25103207.

[14] Katsarou DN, et al. Optimizing hypoglycaemia prediction in type 1 diabetes with Ensemble Machine Learning modeling. BMC Medical Informatics and Decision Making. 2025;25(1):46. https://doi.org/10.1186/s12911-025-02867-2.

[15] Alhaddad AY, et al. Sense and learn: Recent advances in wearable sensing and machine learning for blood glucose monitoring and trend-detection. Frontiers in Bioengineering and Biotechnology. 2022;10:876672. https://doi.org/10.3389/fbioe.2022.876672.

CHAPTER 11

Real-Time Monitoring and Diagnosis of Chronic Respiratory Diseases with Wearable Sensors

Ayesha Taranum, Shalinee Pareek, Soumya Sharma

Abstract Chronic respiratory diseases (CRDs) like asthma, chronic obstructive pulmonary disease, and bronchitis are major causes of global morbidity and mortality. Timely detection and continuous monitoring are vital for preventing exacerbations and improving patient outcomes. Traditional methods, such as periodic clinical visits and spirometry tests, fail to capture real-time physiological variations. This chapter explores wearable sensor technologies for CRD monitoring, enabling continuous, noninvasive tracking of parameters like respiratory rate, SpO_2, heart rate, lung sounds, and airflow. Integrated with IoT platforms, these devices collect multi-modal data for analysis using deep learning and machine learning techniques, enabling predictive modeling, detection of anomalies, and real-time diagnosis. AI models, including long short-term memory, deep neural networks, and convolutional neural networks, show impressive accuracy of 89.10%, 89.10%, and 97.02%, respectively. Advanced signal processing algorithms and predictive analytics provide actionable insights for personalized treatment plans, reducing hospitalizations. Wearable sensor-based monitoring offers a transformative approach to CRD management by combining continuous data acquisition with intelligent analytics, significantly enhancing patient quality of life and supporting data-driven healthcare strategies.

Keywords Wearable sensors – chronic pain management – physiological signal monitoring – machine learning analytics – personalized healthcare systems

11.1 Introduction

Chronic respiratory diseases (CRDs), asthma, chronic obstructive pulmonary disease (COPD), and bronchitis are also the most common factors that result in morbidity and mortality in all countries. Such conditions are managed by continuous monitoring and detecting any changes in the respiratory abilities in time [1]. Some of the more traditional methods of diagnosis, such as clinical inspection, spirometry, and self-reporting by the patient, only provide intermittent data and do not analyze the changes in health of a patient in real-time, delaying the given action to be taken and increasing the risk of severe complications [2]. The wearable sensor technologies are going to become a potential solution to the real-time monitoring

 | HTTPS://DOI.ORG/10.1515/9783112237083-011

and timely warning of breathing issues. Respiratory parameters that the sensors of wearable devices could continuously measure include breathing rate, oxygen saturation, airflow, and lung sounds [3]. Real-time data from these devices turns out to be able to detect deviant respiratory rates in real- time, eliminate potential aggravations in time, and provide an opportunity to render medical assistance in time. The personalized care was also aided by constant monitoring, as the approaches to treatment and management could be adapted to the needs of the particular patients, encouraging active self-management and adherence [4].

Integrating the information of a wearable sensor with computational algorithms, the deviations of respiratory patterns would be accurately monitored, and a potential threat of a health problem would be predicted. This integration makes the analysis automated and therefore offers actionable insights to the patients and the healthcare provider. Wearable-based monitoring lessens the need to be hospitalized, improves patient interaction rates, and enhances the overall quality of life, resulting in the provision of full-time monitoring and a proactive approach to the management of data rather than a reactive one [5]. On-demand oversight and computer-controlled analysis allow individuals with respiratory health issues and clinicians to respond immediately to any respiratory health change, which ultimately results in the suppression of the illness, early diagnosis, and improved results in the long term [6].

11.2 Smart Wearable Systems for Real-Time Respiratory Disease Detection

An entire framework of real-time tracking of respiratory wellness was grounded in wearable gadgets and IoT-connected intelligent devices, which are employed to oversee respiratory health in real time. In addition to the chemical and biological factors that possibly might impact the respiratory system, including hydrogen peroxide, gas molecules, and infections, the system records the environmental factors, including temperature, humidity, and airflow. The data collected by the connected technology are transmitted to a central processing unit that examined the trend of breathing and physiological cues [7]. The information is input into sophisticated AI analysis, predicting, and real-time monitoring and early detection of disease. The framework allows the introduction of timely interventions and improved quality of life of the potentially vulnerable people to respiratory diseases by incorporating insights of personal data, which facilitates the proactive healthcare management, improved preventative strategies, and overall physical health.

The wearable sensors are used to diagnose and monitor chronic respiratory conditions in real time and are capable of measuring the respiratory rate, oxygen saturation, heart rate, lung sounds, and airflow. Machine learning (ML) and deep learning (DL) are used to evaluate continuous and noninvasive data collection and personalized care, predictive models, and early diagnosis. The solution is amenable to proactive data-based respiratory care, fewer hospitalizations, and distance accessibility to work with the IoT systems.

11.3 Wearable Sensor and Ml Technologies for Respiratory Health

To design wearable sensor technology for diagnosing and real-time monitoring of CRD, such as bronchitis, asthma, and COPD. Ideally, it becomes feasible to monitor critical respiratory functions that would otherwise be necessary in identifying the early exacerbation and timely remedies, such as respiratory rate, airflow, oxygen saturation, lung sounds, and chest motion, it is hoped. The conventional approaches of clinical surveillance, such as periodic spirometry tests and hospitalization, are not appropriate for obtaining long-term and continuous observation, since they are both costly and cumbersome, and sometimes the data are not gathered continuously [8]. The investigation used simple, light, and comfortable wearables to overcome these limitations, which use the most recent sensing materials that capture the strain, pressure, and humidity variation throughout respiratory cycles. These resources include graphene, reduced graphene oxide, conducting polymers, CNTs, and cellulose-based substrates. Such sensors, together with IoT applications and smart data analytics such as ML and prediction models, might give the system real-time fault detection, actionable insights, and individualized monitoring. The main intention could be to use continuous, precise, and remote respiratory tracking to promote active disease management, patient survival, reduction of hospitalization, and quality of life.

11.4 Ml Models for Real-Time Monitoring and Diagnosis of CRDs Using Wearable Sensors

The wearable technology has the capability to detect physiological parameters of heart rate and oxygen saturation by relying on ML and DL algorithms and predicting the existence of breathing issues in real-time. These models are feasible and may be implemented to identify early exacerbation recognition by navigating through the complicated sensor data to point out the significant patterns and trends. Besides, both performances of the two models have been appropriately evaluated, and Section 11.8 provides the exact measurements about how each of the models is applied in active health management.

LSTM: To investigate the technologies in wearable sensors to diagnose and monitor the CRDs in real time. The network applied was LSTM networks that were trained on sequential data of wearable devices that continuously measured physiological data of the breathing rate, SpO_2, pulse rate, and lung sounds, as in Figure 11.1. Such networks could be useful in assisting CRD patients to achieve individualized treatment and early detection of the issues by detecting the long-term correlation of the time of variable of blood pressure and heart rate [9]. The element of AI-modeled data, coupled with the advanced signal processing and predictive analytics, was supposed to provide patients with better outcomes, fewer hospitalizations, and closer management of the diseases in a more efficient manner. In eqs. (11.1)–(11.3):

$$e_s = \sigma\left(Z_e.\left[g_{s-1},\ e_s\right] + C_d\right) \tag{11.1}$$

$$n_t = \sigma\left(X_e.\left[g_{s-1},\ w_s\right] + G_d\right) \tag{11.2}$$

$$D_s = e_s * D_{s-1} + j_s * \tanh\left(X_d.\left[D_{s-1}, w_s\right] + a_d\right) \tag{11.3}$$

e_s and j_s represent gating functions at step s, regulating the extent to which the current input and earlier state affect the current state, respectively. The gate e_s is calculated using a sigmoid activation σ over the weighted combination of the preceding hidden state g_{s-1} and itself, with the bias term an C_d, whereas n_t is calculated similarly from g_{s-1}. The input that is now w_s with bias G_d. The concealed state D_s is updated by keeping some of the prior state D_{s-1}. The scaled by e_s and adding the new input's contribution, scaled by j_s, is the following transformation using weights X_d and a hyperbolic tangent function $\tanh(\cdot)$, coupled with bias a_d. In this case, Z_e and were weight matrices that are applied to the concatenated vectors, $\left[g_{s-1},\ e_s\right]$,$\left[g_{s-1},\ w_s\right]$, $\left[D_{s-1}, w_s\right]$and $\left[D_{s-1}\right]$. The non-linear dynamics of memory update are controlled by the activations σ and tanh, which are used to simulate interactions between past and current data:

$$R_s = \sigma\left(G_p.\left[g_{m-1},\ w_s\right] + a_q\right) \tag{11.4}$$

$$H_s = Q_s * \tanh(E_O) \tag{11.5}$$

In eqs. (11.4) and (11.5), R_s indicates the activation of the update/output gate at time step s; the sigmoid activation function is denoted by $\sigma(\cdot)$, and G_p is the gate's weight matrix; $\left[g_{m-1},\ w_s\right]$ denotes the preceding hidden state's concatenation. Q_s is the particular system amount or source rate of flow that generates H_s. g_{m-1} and the vector of current input, w_s; a_q is the bias vector; H_s is the output at time step s; E_o shows the internal memory vector or cell state at time step O.

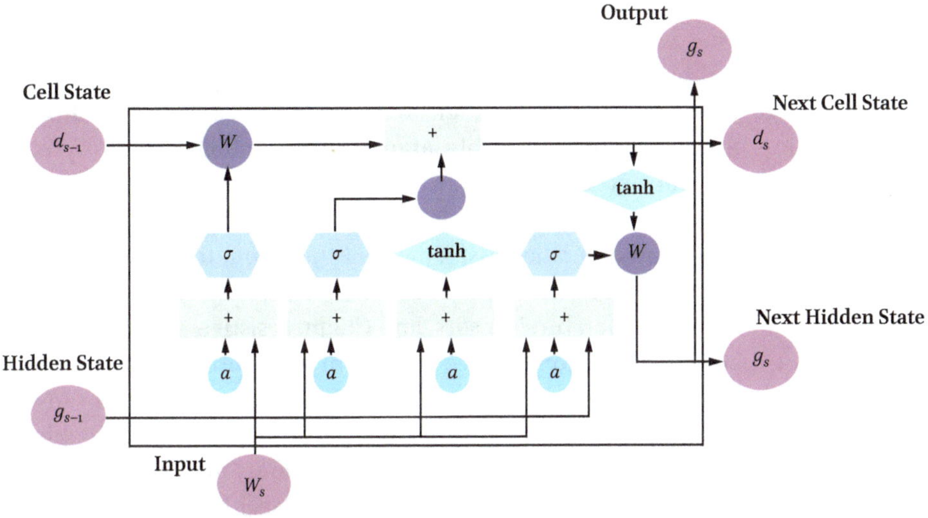

FIGURE 11.1 Comparative analysis of detection method performance metrics for accurate and reliable wearable sensor-based CRD monitoring. (Source: Own elaboration)

11.5 Integration of IoT and AI for Continuous Respiratory Health Tracking

Wearable sensors, along with AI, should be suggested to be used as a CRD monitoring and diagnostic system in real time, such as for bronchitis, asthma, and COPD. Complex and multidimensional data obtained through wearables, smart inhalers, pulse oximeters, environmental sensors, and electronic health records could be appropriately analyzed by advanced AI tools, including ML and DL. The small physiological infractions that could lead to a sign of early exacerbation or disease progression could be detected by AI algorithms, often before clinical manifestation, through constant monitoring of respiratory parameters such as airflow, lung sounds, oxygen saturation, and respiratory rate. Predictive models assist in simplifying the categorization of risk, customized modifications to personal therapy, and early interventions. Additionally, natural language processing provides a holistic view of the factors that determine lifestyle, medication adherence, and environmental stimuli due to the extraction of information in clinical notes and patient-reported outcomes [10]. The use of wearable-based monitoring and AI-driven analytics could complement self-management, lower the rate of hospitalization, allow patients and healthcare providers to treat patients based on the data, and provide patients and physicians with valuable feedback. It is intended toward improving CRD individuals' general level and existence, optimization of the treatment plans, and disease management.

11.6 Physiological Parameters Monitored

Asthma, COPD, and bronchitis are examples of CRDs that might be tracked and diagnosed in real time by the wearable sensor devices. The constant monitoring and evaluation of the vital physiological parameters, including respiratory rate, respiratory volume and oxygen saturation rates (SpO_2), airflow dynamics, lung sounds, and exhaled and blood gases (CO_2, O_2, NH_3, nitric oxide (NO), and volatile organic compounds) allow a detailed assessment of the respiratory condition and metabolism of a patient. The airflow-based sensors, wearable spirometry, thermal imaging, humidity and temperature sensors, and Chemi resistive and colorimetric gas detection methods make noninvasive long-term monitoring of noninvasive procedures adaptable to real-world conditions [11]. In addition, blood gas sensors that are noninvasive, such as transcutaneous and pulse oximetry, are capable of continually measuring oxygen and carbon dioxide without invasive procedures. High-level signal processing, predictive analytics, and algorithms powered by AI could be useful in proactive management of illness, development of an individual intervention plan, and identification of an exacerbation at an early stage. The constraints of traditional clinical examination are overcome by the multiparametric, continuous examination through the provision of total real-time respiratory health. The use of such a system leads to enhancement of the patients' standard life quality with CRDs through facilitation of patient-centered, data-driven respiratory health care, reducing the hospitalization rates, and enhancing the quality of clinical decision-making.

11.7 The Platforms of Internet of Things (IoT) Integration

Bronchitis and breathing disorders such as COPD and asthma are some of the CRDs that could be tracked and diagnosed immediately on the wearable sensor technologies and IoT platforms. The respiratory parameters could be continuously and noninvasively monitored, such as lung sounds, oxygen saturation, airflow, and respiratory rate, which could assist in individual treatment and make it possible to detect the exacerbation early. Simple, cheap, and easy to-operate systems make use of audio, strain pressure, and humidity-sensitive sensors to track physiological information [12]. Through telemedicine systems, IoT-based devices, cloud computing services, edge computing, secure data communications, and predictive analytics, actionable insights and transmission to patients or healthcare professionals are made in low-latency data processing. It further aids in the sustained surveillance in regard to the treatment of the patients, both at home and in the assisted living facility. The wearable-IoT systems could positively impact the disease's proactive management, decrease hospitalizations, increase patient survival, and use data-focused strategies to enhance the usual lifestyle of individuals suffering from CRD due to the shortage of proximate data as a result of the capability to register the data, intelligent analysis, and make it easily accessible offsite.

11.8 Results

An approach to an AI-monitored CRD was the use of wearable sensors and ML methods, 2D convolutional neural networks (CNN), long short-term memory (LSTM), deep neural networks (DNN), and time-frequency analysis to identify the disease and act on its development in time. Various training options and datasets were also detected to evaluate the performance of the pre-trained CRD detection models, such as hybrid CNN-recurrent neural network (RNN). The most performing has been the Hybrid CNN-RNN (ICBHI dataset), and the NA model presented high sensitivity and low specificity. They rely on multi-modal data (e.g., respiratory rate, SpO_2, and heart rate) to enhance the disease's treatment, which could lower rates of hospitalization, and enhance the outcomes for patients. Such measures as F_1-score, sensitivity, precision, specificity, accuracy, TN, FP, TP, and FN were used in measuring the model performance, and 2D CNN and LSTM models were demonstrated to produce higher results in CRD diagnosis. The techniques offer active and continuous surveillance in contrast to the conventional intermittent ones.

11.8.1 Datasets Used in AI-Driven Monitoring of CRD Using Wearable Sensors for Real-Time Diagnosis

Multi-modal data are applied to treat CRD through the assistance of AI-based surveillance, which includes respiratory rate, SpO_2, heart rate, lung noises, and airflow to treat CRD as asthma, COPD, and bronchitis. They are realized through real-time detection, prediction, and early intervention with the assistance of AIs, which are trained to detect, predict, and intervene early with CRD data (time-frequency analysis [13], 2D CNN [13], LSTM [15], CNN [15], and DNN [15]). Various training strategies and datasets were used to test the trained models that have been presented to detect CRD, Hybrid CNN-RNN [14]. The integration of AI with wearable sensors has addressed the inability of traditional episodic monitoring methods to offer proactive continuous monitoring to enhance disease management, lower hospitalization rates, and improve patient outcomes:

- **Accuracy**: The general ability of the model to identify CRD-related anomalies such as abnormal respiration, oxygen saturation, and abnormal heart rate. The DL models, like 2D CNN, could be applied in the enhancement of the model accuracy because the model could identify critical respiratory features with high accuracy in the context of real-time detection.

- **Precision:** The capability of the model to identify CRD-related abnormalities, including abnormal respiratory patterns, oxygen saturation problems, and abnormal heart-beat rates. Another potential of the model is the improvement of the level of the model accuracy since the deep learning models, such as 2D CNN, could detect the essential respiratory features with high reliability as far as real-time detection aspects are concerned.
- **Recall:** The fraction of the occurrence of CRD that actually took place that was successfully recognized using the model. Recall was the manner in which the model would respond to real occurrences of CRD, it was incapable of ignoring major events.
- **Sensitivity:** The ability of the model to recognize the real incidences of CRD, which are true positives. The high sensitivity means that the real cases of CRD are detected in time to take action aimed at eradicating the possibilities of the unidentified cases.
- **Precision:** It is the percentage of right positive model predictions. This accuracy could reduce the cases of false and render the medical process effective because only the real cases of CRD could be investigated.
- **F_1-Score:** A balanced score of accuracy and recall that is expected to present a more centered appearance of the model action when CRD occasions are not well-balanced across the information. It makes sure that the occurrences of CRD and the occurrences of non-CRD are identified and addressed accordingly.
- **Specificity:** The model must find a way to correctly detect cases of non-anomalous events (true negatives) without false alarms. The accuracy of the 2D CNN model on the accelerator patch reduced the number of wrong interventions.
- **True Positives (TPs):** The cases where the model is capable of detecting a CRD event. It is actually a good phenomenon, and the techniques anticipate it to be.
- **True Negatives (TNs):** These are where the model detects the presence of non-CRD. This is not a very good trend that is being projected by the model, as usual.
- **False Positives (FPs):** These are the instances where the model has mistaken a non-CRD event for a CRD event, therefore leading to a false alarm.
- **False Negatives (FNs):** The false negatives are the cases of instances in which the model does not identify the true CRD event and falsely identifies it as normal.

The accuracy, sensitivity, and specificity are the major parameters that were used in determining the performance of the detection systems in Figure 11.2 and Table 11.1. The time-frequency analysis method was used to establish the accuracy of the accelerator patch, with sensitivity and specificity to be 87.45%, 81.39%, and 92.53%, respectively. The use of the 2D CNN model on the accelerator patch improved accuracy to 94.52%, sensitivity to 93.45%, and specificity to 95.72%, which proves the effectiveness of DL with reference to determining the contents of interest to make improved predictions. Similarly, the accuracy, sensitivity, and specificity of

TABLE 11.1 Performance metrics comparison for real-time COPD diagnosis with wearable sensors: accuracy, sensitivity, specificity, and score [13, 14]

Method	Sensitivity (%)	Specificity (%)	Score (%)	Accuracy (%)
Accelerometer patch 2D CNN on test set [13]	93.45	95.72	–	94.52
Digital stethoscope time–frequency analysis [13]	82.98	87.52	–	86.83
Digital stethoscope 2D CNN on test set [13]	91.36	86.65	–	88.95
Accelerometer patch time–frequency analysis [13]	81.39	92.53	–	87.45
Hybrid CNN-RNN (TF speech dataset) [14]	43.01	74.28	58.4	–
Hybrid CNN-RNN (UrbanSound8K dataset) [14]	34.73	88.45	61.59	–
Hybrid CNN-RNN (ICBHI dataset) [14]	56.91	86.70	71.81	–

the time-frequency analysis digital stethoscope were 86.83%, 82.98%, and 87.52%, respectively. The 2D CNN model applied to the digital stethoscope had a high accuracy of 88.95%, sensitivity of 91.36%, and specificity of 86.65%, and it indicates that CNNs could enhance the performance of the detectors. That kind of assessment indicator proves the stability and effectiveness of the combination of the latest DL solutions with the traditional signal processing in detecting the CRDs in real time. The Hybrid CNN-RNN models, and especially those that were trained using the ICBHI dataset, showed indicators of good results with reduced sensitivity (56.91%) and increased specificity (86.70%).

In Table 11.2, the CNN, LSTM, and DNN models were tested based on the measures of F_1-score, recall, precision, and accuracy to ascertain the results of ML models in identifying CRD. Using a dataset, the correct classification ability of these models into TP, TN, FP, and FN results was determined. The CNN model, having the best accuracy of 97.02%, precision of 97.02%, and recall of 97.02%, and F_1-score of 97.02%, was the most accurate and proved that it could predict respiratory diseases with high reliability. The LSTM model with the accuracy of 89.10% and a recall of 89.10% was slightly lower, meaning that the results were a rather good compromise between precision and recall. The DNN model was also very precise, with a score of 94.05%, and its recall was even a bit lower, so it could be seen that

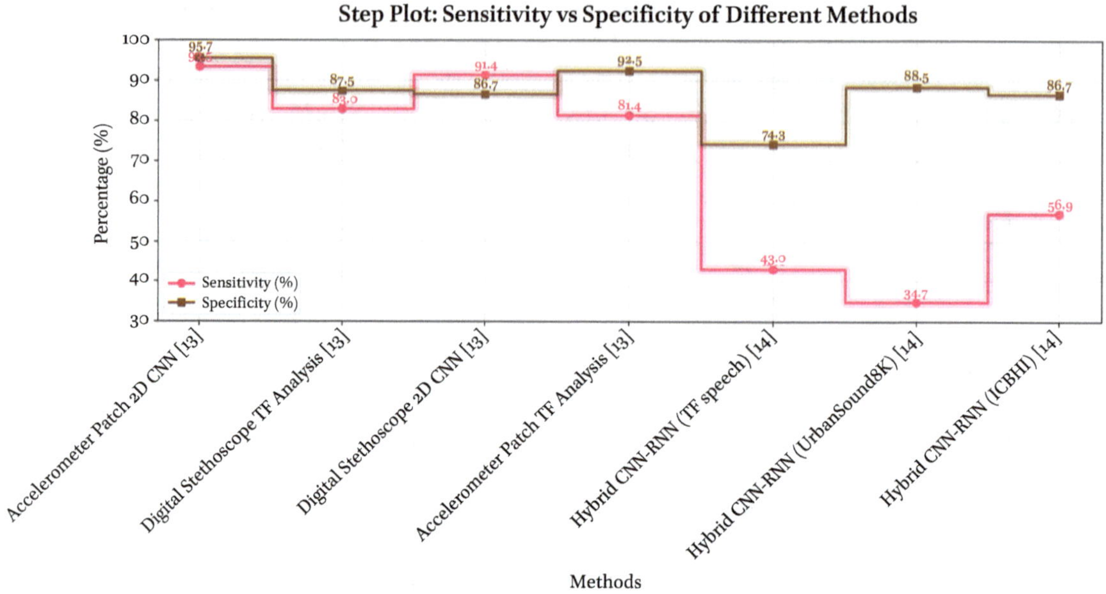

FIGURE 11.2 Bidirectional LSTM model for real-time CRD monitoring using wearable sensor data. (Source: Own elaboration)

TABLE 11.2 Evaluation of CNN, LSTM, and DNN models for real-time diagnosis of CRDs based on performance metrics [15]

Method	Precision (%)	F_1-Score (%)	Accuracy (%)	Recall (%)	FP	TN	FN	TP	FN
CNN [15]	97.02	97.01	97.02	97.02	4	64	2	132	2
LSTM [15]	89.14	88.87	89.10	89.10	16	52	6	128	6
DNN [15]	94.05	94.01	94.05	94.05	8	60	4	130	4

the model could have been more efficient in recognizing all possible cases. Such results indicate that the CNN model was more effective than the LSTM and DNN models in CRD detection.

11.9 Future Trends and Challenges

The future of wearable sensors, AI, and IoT integration development will determine the future of CRD monitoring. Continuous, real-time monitoring systems will be improved to provide an even more accurate reading, which will allow for regulating and providing individual treatment. AI models will also be stronger, and the predictive accuracy will be enhanced, as well as the false alarm rate. Nonetheless, the issues related to information protection and confidentiality, handling the immense volumes of information produced, and standardization across devices and platforms are also challenges. Moreover, its extensive implementation will involve overcoming such obstacles as cost, adoption by users, and incorporation of such systems into healthcare systems. The future investigations will be oriented towards the optimization of these technologies for wider use and their success in identifying the exacerbations at the initial stages.

11.10 Conclusion

Wearable sensors and AI-based models together to monitor CRD recommendations in real-time represent significant contributions to the quality of care received by patients. The systems identify and interrupt exacerbations in time by continuously, noninvasively, and continually monitoring physiological factors, including the saturation level of oxygen, the rate of breathing, and sounds from the lungs, that are capable of causing less hospitalization. Such ML and DL algorithms are especially the CNN, LSTM, and DNN algorithms, which have shown promising results in CRD prediction and diagnosis with better accuracy. Despite the mentioned developments, certain problems, such as the necessity to ensure the safety of the data, to standardize, and to address the constraints tied to the cost, still exist. The further development of the technologies is probably going to be oriented on refining these technologies to be applied on a wider basis, the optimization of predictive efficiency, and the absence of privacy. Wearable technology, IoT, and AI are preventive strategies for controlling breathing problems in people with asthma, which could have an important effect on their quality of existence and performance.

References

[1] Mamdiwar S, et al. Recent advances on IoT-assisted wearable sensor systems for healthcare monitoring. Biosensors. 2021;11(10):372. https://doi.org/10.3390/bios11100372.

[2] Yadav SP, Yadav S. Fusion of medical images in wavelet domain: A hybrid implementation. Computer Modeling in Engineering & Sciences. 2020;122(1):303–21. https://doi.org/10.32604/cmes.2020.08459.

[3] Sanchez-Perez JA, et al. A wearable multimodal sensing system for tracking changes in pulmonary fluid status, lung sounds, and respiratory markers. Sensors. 2022;22 (3):1130. https://doi.org/10.3390/s22031130.

[4] Santoni A, et al. Improving Care for People with Chronic Respiratory Diseases: Taking a Policy Lens. Advances in Therapy. 2025;42(6):2569–86. https://doi.org/10.1007/s12325-025-03191-x.

[5] Jung T, Vij N. Early diagnosis and real-time monitoring of regional lung function changes to prevent chronic obstructive pulmonary disease progression to severe emphysema. Journal of Clinical Medicine. 2021;10(24):5811. https://doi.org/10.3390/jcm10245811.

[6] Salama R, et al. Benefits of Internet of Things (IoT) Applications in Health care-An Overview. 2023 International Conference on Computational Intelligence, Communication Technology and Networking (CICTN). 2023;1(1):778–84. https://doi.org/10.1109/CICTN57981.2023.10141452.

[7] Kim D, et al. Recent developments in wearable breath sensors for healthcare monitoring. Communications Materials. 2024;5(1):41. https://doi.org/10.1038/s43246-024-00480-w.

[8] Yin Z, et al. Wearable respiratory sensors for health monitoring. NPG Asia Materials. 2024;16(1):8. https://doi.org/10.1038/s41427-023-00513-9.

[9] Hamayel MJ, Owda AY. A novel cryptocurrency price prediction model using GRU, LSTM and bi-LSTM machine learning algorithms. Ai. 2021;2(4):477–96. https://doi.org/10.3390/ai2040030.

[10] Gowda M. AI-Enhanced Monitoring Systems for Managing Chronic Respiratory Diseases. Journal Publication of International Research for Engineering and Management (JOIREM). 2025;5(06):1–8.

[11] Vitazkova D, et al. Advances in respiratory monitoring: A comprehensive review of wearable and remote technologies. Biosensors. 2024;14(2):90. https://doi.org/10.3390/bios14020090.

[12] Chauhan R, Mishra A, Singh D. RAPT: AI–Powered IoT Framework for Real-Time Respiratory Disorders Monitoring and Prediction. In International Summit on the Global Internet of Things and Edge Computing 2025 (vol. 2719, pp. 159–74). Cham: Springer Nature Switzerland. https://doi.org/10.1007/978-3-032-09555-8_10.

[13] Sang B, et al. An accelerometer-based wearable patch for robust respiratory rate and wheeze detection using deep learning. Biosensors. 2024;14(3):118. https://doi.org/10.3390/bios14030118.

[14] Acharya J, Basu A. Deep neural network for respiratory sound classification in wearable devices enabled by patient specific model tuning. IEEE Transactions on Biomedical Circuits and Systems. 2020;14(3):535–44. https://doi.org/10.1109/TBCAS.2020.2981172.

[15] Dahane A, et al. An Innovative Low-cost IoT-Based Asthma Exacerbation Prediction System Using Federated Learning. IAM. 2023;3694:12–20.

CHAPTER 12

Wearable Sensor Technologies for Chronic Pain Management

Swadhin Kumar Barisal, Sudharsan M, Manivannan Karunakaran

Abstract Chronic pain presents a persistent health challenge, affecting physical function, psychological well-being, and quality of life. Conventional assessments rely on subjective, episodic evaluations, which fail to capture dynamic fluctuations in pain intensity. Wearable sensor technologies provide a solution by continuously monitoring physiological and behavioral signals, including heart rate variability, electromyography (EMG),accelerometry, and skin conductance in real-world settings. Multimodal datasets collected from cloud-based chronic pain management systems, longitudinal electronic health records (EHRs), and hospital databases enable comprehensive analysis of physiological signals such as, heart rate, EMG, and electrocardiogram. Machine learning models were trained using these physiological signals alongside patient-reported numeric pain rating scale scores. Deep learning approaches, including random forest, gradient boosting, XGBoost (XGB), transformer models, bidirectional encoder representations from transformers (BERT), and hybrid term frequency-inverse document frequency-enhanced frameworks, effectively captured temporal dependencies and nonlinear relationships. These models achieved peak performance with accuracy and F_1-scores (0.92) in predicting mild, moderate, and severe pain episodes. Limitations include missing sensor data, heterogeneous EHR formats, and preprocessing complexity. Overall, integrating wearable sensors with AI-driven analysis offers a scalable approach for real-time, personalized chronic pain assessment, enabling proactive management, and supporting improved, patient-centered clinical decision-making.

Keywords Wearable sensors – chronic pain management – physiological signal monitoring – machine learning analytics – personalized healthcare systems

12.1 Introduction

Chronic pain is a common health condition and encompasses continuous or repeated pain that goes beyond the healing duration of the normal tissue and has a severe influence on physical functioning, mental health, and life quality of persons across the globe. The self-report questionnaires and rare clinical evaluations, which are subjective, dependent on patient perception, and cannot keep up with

 | HTTPS://DOI.ORG/10.1515/9783112237083-012

dynamic changes in the intensity of symptoms over time, have been the leading methods of measuring and treating chronic pain chronologically [1]. These limitations contribute to the unfinished clinical knowledge, timely interventions, and the best personalization of treatment methods. The recent progress in wearable sensor technologies offers a possible solution to such gaps offering the possibility to track physiological and behavioral signs of pain and functionality in real-time and continuously [2]. Objective data of heart rates, movements, muscle activities, and autonomic responses in the real world could be collected using smartwatches, accelerometers, electrocardiographic (ECG), electromyography (EMG), and other biosensors. Such sustained biosensing can be used to substitute the previous approach to chronic pain management that was intermittent, subjective, and based on a personalised, proactive approach, which is now data-driven and can be personalised, tailored, and implemented [3]. Irrespective of such a promise, the modern research and clinical practice is full of challenges. The bulk of the existing literature has been focused on aligning wearable-based signals with traditional pain scales, and not as fully as possible taken advantage of predictive modeling techniques to anticipate change in pain or customize treatment. Moreover, most of the studies are conducted under controlled conditions and not under natural conditions and thus, are not generalizable and cannot be used in the long run [4]. Inconsistency of sensors and the quality of the data is also an issue that restricts mass clinical implementation and there are also concerns of privacy and security of personal health data [5]. Speaking about providing a detailed description of wearable sensor technologies in the management of chronic pain, it is crucial to mention this potential of overcoming the constraints of the current technologies by the means of the acquisition of multimodal data and the application of sophisticated analytical approaches such as, machine learning (ML) and deep learning (DL) [6]. The plan would make the process more reliable in terms of diagnosing pain, enable the mitigation of the impact of pain outbreaks over time, and enable a personal care path through the integration of objective data streams and predictive analytics. The research indicates the importance of interdisciplinary innovation to improve the utilization of technology as far as clinical practice is concerned in an attempt to get improved patient outcomes in chronic pain management [7].To develop an integrated model of continuous and personalized chronic pain treatment through a combination of wearable sensor devices with the latest ML and DL models. This is to acquire multimodal physiological and behavioral signals, align them with patient-self-report pain scales, and predictive analytics to differentiate and forecast pain levels with ease. The ultimate task is to have the ability to track chronic pain real-time and help make a decision based on the data and individualized intervention to guarantee improved patient outcomes.

12.2 Chronic Pain: Types and Significance

Chronic pain was categorized into four frequent types, which were neuropathic pain, musculoskeletal pain, inflammatory pain, and postsurgical pain, which were determined by the analysis of medical records, pain questionnaires, and such standardized scales as the visual analog scale (VAS) or numeric pain rating scale (NPRS). The duration, intensity, and functional limitation of patients in terms of pain were recorded to determine the baseline features. This procedure allowed tracing patterns related to particular types of pain and guided the further locations and data gathering of wearable sensors. The clinical relevancy was determined to inform individual pain management interventions.

12.2.1 Use of NPRS for Wearable-Based Pain Prediction

This utilized a standardized measure of persistent pain with the respondents rating pain from 0 (having no pain) to 10 (the greatest amount of suffering they can have) [8]. The self-reported NPRS scores were taken daily using a custom smartwatch app and served as ground-truth labels for ML models to estimate the level of pain in response to passive wearable data. Physiological and behavioral measures that the wearable sensors on the Apple Watch captured included heart rate variability, number of steps, and stand time, correlated with the NPRS pain levels. It was shown that wearable-derived objective digital biomarkers were able to forecast categorical pain severity (mild, moderate, and severe) at high accuracy in training on NPRS scores. This method directly corresponds to chronic pain management through validation of wearable sensor data to standardized clinical pain measures, which allows real-time monitoring of pain and individualized pain management.

12.3 Wearable Sensor Technologies

Wearable sensor technologies were utilized to continuously track physiological and biomechanical indicators that are related to chronic pain. Electromyogram (EMG) sensors, accelerometers, gyroscopes, electrocardiogram (ECG) monitors, and skin conductance sensors were put on the parts of the affected body. The data were recorded during day-to-day operations in order to record immediate changes in movement, muscle activity, heart rate, and autonomic changes. The signals were sent to a mobile or cloud platform to store them. The deployment was directed to noninvasive, comfortable gadgets permitting a prolonged observation. The choice and placement of sensors were informed by previous investigations and the nature of pain such that high-quality signals were obtained and made available to be analyzed through an AI-based tool.

12.3.1 Wearable ECG and EMG Monitoring for Chronic Pain Assessment

The use of ECG and EMG sensors to monitor physiological signals was applied to a wearable device to record these signals during everyday activities [9]. ECG sensors record electrical signals that are produced by the heart, such as the R-R interval and the heart rate variability, whereas EMG sensors record the electrical activity of the muscles, which gives information about autonomic and musculoskeletal reactions. The biosignals are filtered, through amplification and digitization and sent wirelessly on bluetooth to a smartphone or a cloud platform to store and analyze in realtime. The results of the study showed that it is possible to monitor ECG and EMG continuously during the day in a comfortable manner and maintain the quality of signals and long-term stability. This real-time data acquisition framework is physiological data, which can be applied to chronic pain treatment directly, and wearables are able to track the cardiac and muscular activity of the organism. The system will be able to track perceptual differences in heart and muscle reactions to determine pain, and therefore provide personalized pain management by examining physiological alterations related to pain with the help of AI. Such a strategy helps to facilitate ongoing monitoring and objective information gathering and incorporates decision support systems to enhance therapeutic interventions and lifestyle advice.

12.4 ML Models for Pain Assessment

The ML models were required to identify the different levels of pain based on the data they received using the wearable sensors. The *k*-nearest neighbors (k-NN), random forest (RF), and decision trees (DTs) were trained using the assistance of the labeled datasets of the physiological features and patient-reported pain scores. Heart rate variability feature vectors, EMG amplitudes and movement measures, and skin conductance were present. The models were evaluated based on cross-validation and accuracy, F_1-score, and mean absolute error. All these ML techniques allowed the classification of the pain intensity and the identification of patterns that were predictive of the different types of pain, which allowed predictive models and clinical decision support.

12.4.1 RF for Pain Assessment

In wearable sensor-based biomedical signal classification, RF offers a strong ensemble learning framework by combining predictions from multiple DTs, which enhances classification accuracy, reduces overfitting, improves robustness to noisy sensor signals, and ensures reliable pain level prediction. [10]. A bootstrapped subset of the training data is used to obtain each tree $S_i(W)$ which then produces a predicted class given an input feature vector w. The final predicted class z is determined using majority voting across all l trees as follows:

$$\hat{z} = \text{mode}\left(S_1(w),\ S_2(w) \cdots S_l(w)\right) \tag{12.1}$$

where $S_i(W)$ is the forecast of the type of the i-th DT with input w, and $\hat{z}$ is the empirical probability of a model with parameters w to predict the outcome d on all l samples:

$$O(Z = d|w) = \frac{1}{l}\sum_{i=1}^{l} J(S_i(w) = d) \tag{12.2}$$

In eq. (12.2), $O(Z = d|w)$ represents the observed probability of the outcome Z being d given the model parameters w, $Z = d$ represents the event/outcome of interest (e.g., a specific pain level), w denotes parameters of the model (e.g., weights in a neural network or classifier), l is the total number of samples or observations, $S_i(w)$ is the predicted output for the ith sample based on the model w, $J(S_i(w) = d)$ represents indicator function that equals to 1 if the prediction matches d, 0 otherwise.

12.5 Deep Learning Approaches for Pain Prediction

To assess chronic pain, transformer-based DL models, including bidirectional encoder representations from transformers (BERT) and hybrid DL architectures, are used to predict the temporal and spatial trends of wearable sensor information. These models are automatic and derive complicated features of the physiological signals, such as, EMG, ECG, and motion, to accurately classify the intensity of the pain. Continuous real-time monitoring is supported by their capability to capture understated temporal relationships and nonlinear dynamics that are usually overlooked by the conventional ML techniques. These models make it possible to manage pain at a personal level with a large and high-dimensional dataset, provide insights into the personal responses of patients with chronic pain, and enhance personalized responses to interventions.

12.5.1 Transformer/BERT Method

The transformer architecture, which is the foundation of BERT, is a nonrecurring, nonsequential DL architecture capable of efficiently processing sequential data with minimal memory requirements and is used to predict chronic pain intensity using wearable sensor time-series data, including heart rate, EMG, and motion metrics [11]. The scaled dot-product attention is presented as follows:

$$\text{Attention}(P, J, U) = \text{softmax}\left(\frac{PJ^S}{\sqrt{c_l}}\right)W \tag{12.3}$$

where P, J, and U are the query, key, and value matrices, respectively, derived from processed wearable sensor signals such as, heart rate, EMG, and motion data, S denotes the sequence length or time steps of the sensor data, c_l is the scaling factor (dimension of the key vectors) to stabilize gradients, W represents trainable weight matrices in the attention layer. This computes a weighted representation of sensor signal features, focusing on the most relevant temporal patterns for predicting pain intensity. Multihead attention utilized several such concentration techniques so that learned various features of the pattern of sensor signals in parallel as follows:

$$\text{head}_j = \text{Attention}\left(PV_j^P, JV_j^J, UV_j^U\right) \tag{12.4}$$

$$\text{MultiHead}(P, J, U) = \text{Concat}\left(\text{head}_1, \ldots, \text{head}_g\right)V^N \tag{12.5}$$

Each head_j corresponds to one attention mechanism learning distinct aspects of physiological and behavioral patterns from the wearable data, enabling the model to capture multiple dependencies in the signals that correlate with pain severity. Outputs from all attention heads are concatenated to form a comprehensive feature representation, with V^N as the trainable projection matrix. This combined representation integrates multiple signal contexts for robust pain prediction. The feed-forward network (FFN) applies nonlinear transformations to the output as follows:

$$\text{FFN}(w) = \max(0, wV_1 + a_1)V_2 + a_2 \tag{12.6}$$

where w is the multihead attention output, V_1 and V_2 are weight matrices, a_1 and a_2 are biases. The Rectified Linear Unit (ReLU) activation $\max(0, \cdot)$ introduces nonlinearity, allowing the model to learn complex relationships between sensor features and pain levels, enabling continuous, real-time prediction of mild, moderate, or severe chronic pain episodes.

12.6 AI-Based Pain Forecasting and Decision Support

AI-based pain forecasting utilizes past wearable sensor data on future pain attacks and assists clinical decision-making. Time-series models of gradient boosting (GB) and XGBoost (XGB) were used to determine the time-dependent features of physiological indicators of heart rate, EMG activity, and the measures of movement, which enables one to predict an episode of pain exacerbation in advance. Ensemble learning techniques improve the accuracy in prediction by fusing several model predictions. Reinforcement learning systems also make interventions more individualized by tailoring the recommendations to the response of an individual patient. The clinical decision support systems integrate the predictive results, enabling clinicians to optimize treatment plans, alter drugs, and propose personalized lifestyle changes. This strategy will minimize the use of subjective evaluation it will help prevent active pain management and it will improve the results due to constant data-driven monitoring.

12.6.1 XGBoost

The boosting algorithm enhances model accuracy through the co-combination of numerous simple weak learners. The successive learner then concentrates on the errors left behind in the last model, and gradually a reduction in the difference between the predicted and the actual values is made, and methods like regularization ensure that over fitting does not occur [12]. Figure 12.1 demonstrates an intricate transformer encoder architecture developed to analyze multimodal medical inputs from wearable sensors in chronic pain management. A new tree $e(y)$ is trained on the residual errors of the prior ensemble, successively reducing bias. In eq. (12.7) the prediction at the sth iteration is shown:

$$w_j^s = \sum_{l=1}^{s} e_l(y_j) = w_j^{s-1} + e_s(y_j) \tag{12.7}$$

Where w_j^s represents the cumulative predicted pain score for the patient j at iteration s in the XGB model, derived from wearable sensor data such as heart rate, EMG signals, and activity metrics. The term $e_l(y_j)$ denotes the contribution of the l-th decision tree (weak learner) to the prediction of the patient's pain level, capturing patterns in physiological or behavioral signals corresponding to different pain intensities. w_j^{s-1} is the cumulative prediction of the last iteration which means that each new tree $e_s(y_j)$ is aimed at reducing the residual error made by the previous predictions. This recurring updating also allows the model to continuously improve the forecast of pain and offer relevant and precise predictions of the chronic pain episodes that can be used to facilitate real-time monitoring and machine-assisted decision-making in wearable sensor-based pain management frameworks. The XGB model is represented by the training loss $M(\theta)$ and a regularization term $\Omega(\theta)$ in order to eliminate overfitting as follows:

$$\text{Obj}(\theta) = M(\theta) + \Omega(\theta) \tag{12.8}$$

Using wearable sensor-based chronic pain management, XGB is capable of modeling the physiological indicators (heart rate, EMG, and activity) to precisely predict the future intensity of pain and enable proactive measures and care plans.

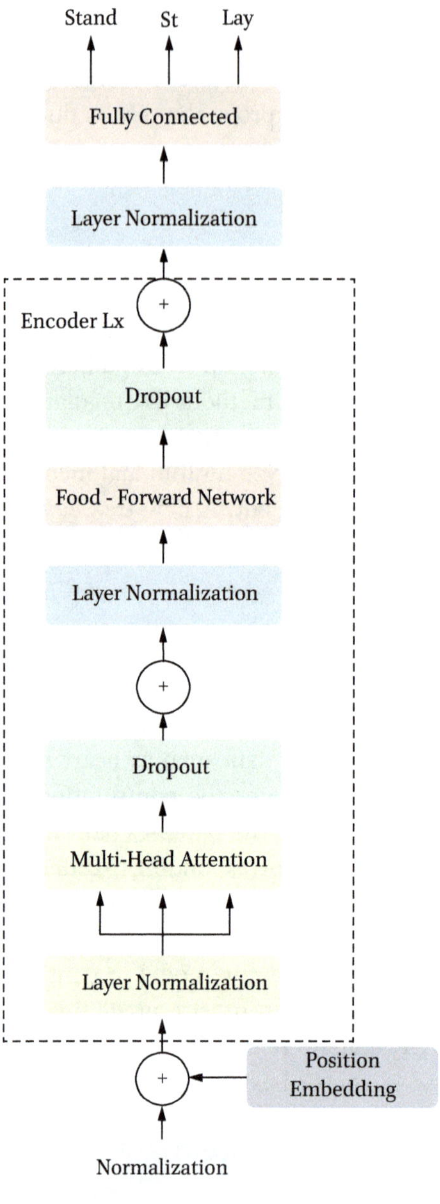

FIGURE 12.1
Transformer-based DL architecture for wearable sensor-driven chronic pain intensity prediction. (Source: Own elaboration)

12.7 Data Gathering

The data was gathered through cloud-based chronic pain management systems and hospital data, such as, structured patient data, demographics, clinical history, pain score, medication history, and wearable sensor data. This multimodal dataset allowed ML models to learn the pattern of longitudinal pain in patients remotely to optimize remote healthcare [13]. The data that was used included vitals, activity levels, opioid usage, surgical data, and patient-reported pain scores, which were longitudinal electronic health records (EHRs). The data used were sequential, time-stamped attributes that can be used to predict the level of postoperative pain and opioid needs, and the data remained anonymous of patients [14].They used real-world medical data in the form of hospital EHRs, lab reports, and wearableness, with demographics, physiological measurements, pain scales, and treatment history. This intermodal data enabled comparative assessment of AI algorithms based on chronic pain detection, which improves the robustness, and generalizability of the model to patient populations [15].

12.8 Performance Evaluation Metrics

The percentage of correct predictions was accuracy, or overall accuracy, which can be biased by out of true data. The level of real positives was determined by precision; low precision reduced false positives. True positives, which were remembered, were true positives and false negatives are minimized by true positives. Preciseness and recall were harmonically combined to give F_1-score. The ability of the model to differentiate between patients with and without chronic conditions is measured as AUC, with high AUC indicating high discrimination ability and greater ability to rank positive cases higher than negative cases.

The widespread comparison of various ML and AI models used to predict chronic pain based on the wearable sensor data, EHRs, and text embeddings based on clinical datasets. The ML models such as, logistic regression (LR), DT, RF, GB, and KNN have good predictive performances with GB having the highest accuracy (0.88) and F_1-score (0.87) of the classical approaches, meaning that they are a robust model to work with structured physiological and clinical features [13]. XGB, RF, LR, and stacking ensembles were compared in terms of prediction of the postoperative pain. The stacking ensemble was superior to single models in precision (0.91), accuracy (0.89), and F_1-score (0.89), indicating the advantage of having multiple learners to achieve predictive reliability and to manage heterogeneous sources of data [14]. Master AI models based on text embeddings were also experimented on for chronic pain detection. XGB using term frequency-inverse document frequency (TF-IDF) embeddings made the best results (accuracy and F_1-score 0.92 each), followed by GB using TF-IDF (0.91 and 0.91, respectively). Moderate performance was achieved with transformer-based models, such as BERT (Bio-Bit), with an accuracy of 0.85 and an F_1-score of 0.84, which is mainly driven by the appropriateness of embedding representations and the effectiveness of domain-specific adaptation [15]. Table 12.1 reflects the steady increase in the performance

TABLE 12.1 Comparison of the performance of ML and DL models in predicting pain in multiple studies [14, 15]

Model	Accuracy	Precision	Recall	F_1-Score	AUC
DT [13]	0.78	0.82	0.75	0.78	0.82
RF [13]	0.85	0.87	0.82	0.84	0.88
GB [13]	0.88	0.90	0.85	0.87	0.90
KNN [13]	0.80	0.83	0.77	0.80	0.84
LR [13]	0.83	0.86	0.80	0.82	0.86
XGB [14]	0.89	0.90	0.89	0.88	0.87
RF [14]	0.74	0.76	0.74	0.73	0.81
LR [14]	0.68	0.68	0.68	0.67	0.74
Stacking [14]	0.89	0.91	0.89	0.89	0.90
XGB + TF-IDF [15]	0.92	0.94	0.91	0.92	–
GB + TF-IDF [15]	0.91	0.93	0.89	0.91	–
BERT (BioBit) [15]	0.85	0.84	0.85	0.84	–

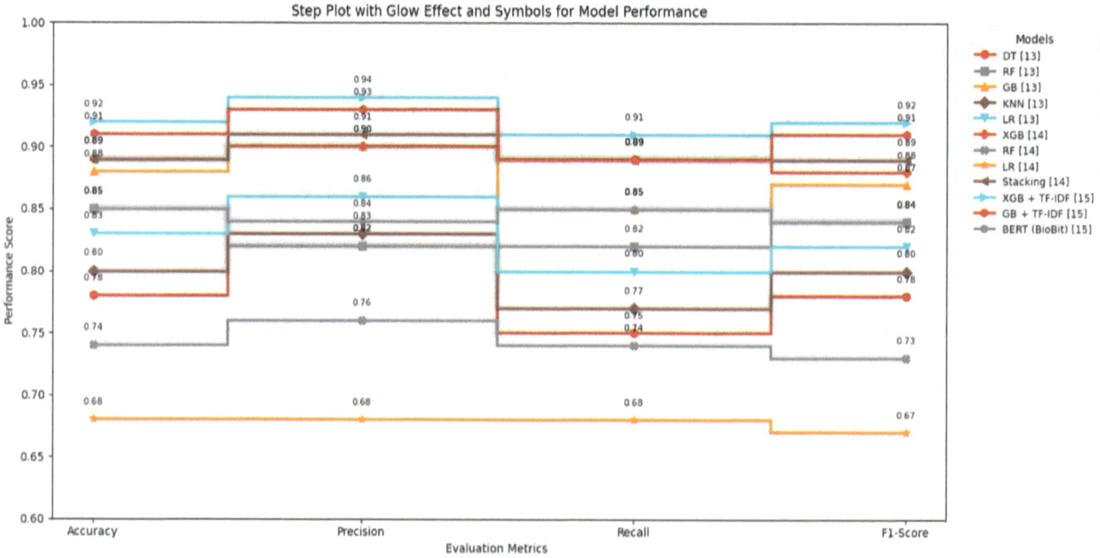

FIGURE 12.2 Comparison of ML and DL models for chronic pain prediction using wearable sensor and clinical data (Source: Own elaboration.)

of classical ML and DL models, with a focus put on the possible potential of accurate and personalized chronic pain management. Figure 12.2 represents the comparative performance analysis of the ML and DL models for chronic pain prediction with wearable sensor and clinical information.

12.9 Discussion

To predict chronic pain intensity using cloud-based records and wearable sensor data, combining heart rate, EMG, and activity logs. The dataset included multimodal patient information, enabling model training for AI-assisted monitoring. GB achieved F_1-score (0.87), precision (0.90), accuracy (0.88), and recalls (0.85). Limitations included small sample size and missing sensor data [13]. On predicting postoperative pain and opioid use using longitudinal EHRs and wearable sensor data. The dataset included vitals, lab results, medications, and activity data. XGB and stacking: the accuracy of both the models was 0.89, an F_1-score between 0.88 and 0.89, a recall of 0.89 and with a precision of 0.90 to 0.91. Limitations involved incomplete EHR documentation and sensor gaps [14]. AI models for chronic pain detection using structured EHRs, lab reports, and wearable sensor data. TF-IDF textual features were included for hybrid models. XGB + TF-IDF outperformed the BERT (BioBit) and GB models, scoring 0.92 accuracy, 0.94 precision, 0.91 recall, and 0.92 F_1-score. Limitations included preprocessing complexity and data imbalance [15].

12.10 Conclusion

The effectiveness of incorporating the wearable sensor technologies into AI-based analysis frameworks to manage chronic pain continuously and individually. Based on the multimodal datasets consisting of cloud-based chronic pain management systems, longitudinal EHRs, and hospital records, which include physiological (heart rate, EMG, and ECG) as well as activity metrics, and patient-reported NPRS of pain, ML and DL models are trained to predict the intensity of pain with a high degree of accuracy. Techniques were RF, GB, XGB, BERT, and hybrid models based on TF-IDF with some additions and alternatives, which were capable of capturing nonlinear trends and temporal aspects in the data. Ensemble and hybrid models especially, XGB + TF-IDF, were the most accurate as a result of detail, recollection, accuracy, and F_1 scoring and were found to be predict mild, moderate, and severe pain events. There were numerous limitations to the research, such as, missing wearable sensor data, variation, and inconsistency in EHR documentation, imbalance in classes in pain categories, and complexity of preprocessing data, which can restrict the strength and externalizability of the results. Future studies are advised to include bigger heterogeneous datasets, effective data preprocessing pipelines, and adaptive learning models to enhance model scalability, sensor variability, and real-time, patient-specific pain monitoring to address diverse patient groups to improve clinical decision-making and intervention planning.

References

[1] Ahmed Kamal M, et al. Telemedicine, E-health, and multi-agent systems for chronic pain management. Clinics and Practice. 2023;13(2):470–82. https://doi.org/10.3390/clinpract13020042.

[2] Lewczak Z, Mitchell M, Mitchell MG. Wearable technology and chronic illness: balancing justice and care ethics. Cureus. 2024;16(11). http://doi.org/10.7759/cureus.73686.

[3] Orzabal M, et al. A forehead wearable sensor for the objective measurement of chronic pain. International Journal of Environmental Research and Public Health. 2022;19(24):17041. https://doi.org/10.3390/ijerph192417041.

[4] Leroux A, et al. Wearable devices: current status and opportunities in pain assessment and management. Digital Biomarkers. 2021;5(1):89–102. http://doi.org/10.1159/000515576.

[5] Rani P, et al. Simulation of the lightweight blockchain technique based on privacy and security for healthcare data for the cloud system. International Journal of E-Health and Medical Communications (IJEHMC). 2022;13(4):1–15. http://doi.org/10.4018/IJEHMC.309436.

[6] Upadhyay P, Tomar P, Yadav SP. ADNet: Deep learning based model for Alzheimer's disease classification. Iran Journal of Computer Science. 2025;1–13. https://doi.org/10.1007/s42044-025-00243-x.

[7] Cricco R, et al. Advancements in sensor technology for monitoring and management of chronic coronary syndrome. Sensors. 2025;25(15):4585. https://doi.org/10.3390/s25154585.

[8] Patterson DG, et al. Objective wearable measures correlate with self-reported chronic pain levels in people with spinal cord stimulation systems. NPJ Digital Medicine. 2023;6(1):146. https://doi.org/10.1038/s41746-023-00892-x.

[9] Jeong J-W, Lee W, Kim Y-J. A real-time wearable physiological monitoring system for home-based healthcare applications. Sensors. 2021;22(1):104. https://doi.org/10.3390/s22010104.

[10] Karim A, Ryu S, Jeong I. Ensemble learning for biomedical signal classification: a high-accuracy framework using spectrograms from percussion and palpation. Scientific Reports. 2025;15(1):21592. https://doi.org/10.1038/s41598-025-05027-8.

[11] Dirgováluptáková I, Kubovčík M, Pospíchal J. Wearable sensor-based human activity recognition with transformer model. Sensors. 2022;22(5):1911. https://doi.org/10.3390/s22051911.

[12] Ma R, et al. A XGBoost-based prediction method for meat sheep transport stress using wearable photoelectric sensors and infrared thermometry. Sensors. 2024;24(23):7826. https://doi.org/10.3390/s24237826.

[13] Mohankumar N, et al. Advancing chronic pain relief cloud-based remote management with machine learning in healthcare. Indonesian Journal of Electrical Engineering and Computer Science. 2025;37(2):1042–52. https://doi.org/10.11591/ijeecs.v37.i2.pp1042-1052.

[14] Soley N, et al. Predicting postoperative pain and opioid use with machine learning applied to longitudinal electronic health record and wearable data. Applied Clinical Informatics. 2024;15(03):569–82. https://doi.org/10.1055/a-2321-0397.

[15] Comito C, et al. Comparative analysis of AI algorithms on real medical data for chronic pain detection. International Journal of Medical Informatics. 2025;203:106002. https://doi.org/10.1016/j.ijmedinf.2025.106002.

CHAPTER 13

Machine Learning Techniques for Real-Time Detection of Neurological Disorders Using Wearable Sensors

Jaimeel Shah, Santosh Kumar Behera, Asif Mohamed H. B.

Abstract The timely identification of neurological conditions is a crucial factor in treating patients and preventing a delay in the treatment process. The research examines the wearable sensor technology, and machine learning (ML) can be used to track and detect neurological disorders in real-time. Constant signals of physiology and movement are taken by wearable sensors. The high-level ML algorithms, such as, support vector machines, random forests, and decision tree models are used to identify the neurological disorders and forecast the abnormal patterns with high accuracy. The system is installed to do noninvasive continuous monitoring to act as early warning signals that could help clinicians in decision-making and intervention. Experimental analysis proves that the ML strategy applied certain metrics such as, accuracy and reliability to conduct a real-time and patient-centered assessment of neurology. The wearable technology with smart algorithms makes the system an active healthcare solution that will help to decrease clinical workload, improve patient monitoring, and overall outcomes. The research displays the potential of Internet of Medical Things-powered, ML-based models at reshaping neurological services to enable early diagnosis, continuous evaluation, and timely replacement of various neurological conditions.

Keywords Neurological disorders – wearable sensors – machine learning – patient-centric healthcare – Internet of Medical Things (IoMT) – patient-centric healthcare

13.1 Introduction

Initial investigations in this field focused on tracking vital signs and mobility with physical sensors. Instead of employing wearable technology to track physical activity, investigators are currently focusing on resolving important issues in medical technologies. The cornerstone of innovative healthcare is Internet of Medical Things (IoMT), a group of technological equipment and applications that are networked to medical information technology (IT) platforms. The sensors gather data, which is then processed by machine learning (ML) algorithms to enable predictive

 | HTTPS://DOI.ORG/10.1515/9783112237083-013

and adaptive healthcare applications. Smart materials improve the responsiveness and usefulness of sensors [1].

13.1.1 Wearable Sensor Technologies for Neurological Monitoring

One kind of neurosensory wearable technology for intelligent biomedical systems is offered for mass consumer electronic (CE) products. A specific kind of personal electronic device is a medical-grade wristwatch for neurological illnesses that alerts caregivers, when a patient is having an epileptic seizure. Wearable sensors are a part of the broader goal of combining and connecting detectors and gauges with automation and intelligent components. These sensors are increasingly being used in the medical field to measure different biomechanical factors. Wearable sensors are easy to establish together, affordable, and applied in immediate terms because the implementation stage is much quicker than the amount of computation required by some common platforms [2]. However, wearable sensors cannot be routinely used in everyday practice because the extremely massive amounts of data they collect require complex signal processing to convert them into information that is useful for science and medicine. Early detection and treatment of postural instability would be possible if doctors had access to quantitative and automated assessments of motor and postural function via portable sensors.

13.1.2 ML Techniques for Real-Time Detection

ML algorithms have demonstrated amazing effectiveness in producing accurate forecasts for complicated situations, including healthcare difficulties, by automatically processing vast volumes of data. ML solutions, which are frequently trained using data collected from multiple wearable sensors, have played a significant role in recent advances in healthcare research by enabling the prediction of various patient health conditions [3].

13.1.3 IoMT-Driven Healthcare Systems and Patient-Centric Applications

Wearable sensors are now one of the important technologies of human-machine interaction (HMI) and personalized healthcare, which ensure continuous monitoring of health indicators and exercise. These devices support applications such as, remote patient monitoring, sports performance analysis, and health management, and the potential to capture such metrics as blood pressure, cardiac rate, and exercise quantities [4]. Wearable sensors provide real-time data that would help identify health problems early, provide personalized feedback, and make informed decisions. In spite of these benefits, the implementation of AI on wearable systems still faces difficulties because of large amounts of data, noise, and unavailability of computing capabilities. ML-based methods of sensor calibration promote the use of IoMT-based, patient-centred health care systems that ensure timely intervention, proactive monitoring, and better clinical outcomes [5].

13.1.4 Purpose and Advancement of the Research

The main purpose of the research is to design and evaluate the ML-based wearable sensor-based real-time monitoring and detection of neurological illness. The aim of the research is to provide the precise, constant, and noninvasive evaluation to combine the information about physiological and motion-based measures of multiple wearable sensors with complex algorithms, such as, support vector machines (SVM), random forests (RF), and decision tree (DT). The research is also aimed at solving problems such as, noise, choice of features, and heterogeneity of data besides provide a patient-centered solution to support proactive care, early intervention, and improved clinical decision-making:
- To create an ML-based framework that makes it possible to use wearable sensors continuously and in real time and diagnose neurological diseases
- To combine and evaluate motion-related and physiological data from several wearable sensors for precise and noninvasive evaluation
- To enhance patient-centered care and early intervention by offering accurate, trustworthy, and useful information for clinical decision-making

13.1.5 Structure of the Chapter

The research comprises six parts: the purpose, history, and objectives of using wearable sensors and ML to detect neurological illness are discussed in Section 13.1. Neurological conditions, monitoring concerns, and the role of wearable sensors in continuous physiological assessment are explained in Section 13.2. Section 13.3 presents the ML techniques employed in the data collection, integration of IoMT, and implementation of RF, SVM, and DT models in neurological illnesses detection. The main technological obstacles and new prospects in wearable healthcare systems driven by ML are described in Section 13.4. In Section 13.5, the effectiveness of the proposed approaches is assessed using common measurements, and the findings are compared and discussed. Section 13.6 concludes the research with the results summarized and the limitations highlighted as well as suggestions on future research.

13.2 Neurological Disorders and Wearable Sensor-Based Monitoring

Neurological diseases that affect the brain, spinal cord, and nerves cause cognitive, physical, and emotional deficiencies that determine the quality of life. The lack of resources and the complex nature of etiologies make detection at an early stage a challenge due to overlapping symptoms. Sophisticated sensors and signal processing increase accuracy, comfort, and real time data acquisition. The combination of these devices with AI and ML solutions allows making personalized assessment, early detection, and enhanced patient-specific therapies possible.

13.2.1 Neurological Conditions and Monitoring Difficulties

Neurological disorders involve conditions of the brain, spinal cord and nerves. They have the potential to extensively reduce life quality through physical, cognitive, and emotional impairment. These conditions offer challenging healthcare issues since they lack resources, a multiplicity of illness etiologies, and also have overlapping symptoms, which often can result in ineffective or delayed interventions. In such disorders as amyotrophic lateral sclerosis, neuronal intranuclear inclusion disease, multiple sclerosis, Alzheimer's disease, and mild cognitive impairment, recent advancements, such as, genetic screening, radiomics, electroencephalography (EEG), and ML solutions, have contributed to the early detection, classification, and risk assessment. The importance of immunological predictors, functional assessment, and postintervention observation of the prognosis of glioblastoma, stroke, and rare diseases, such as Erdheim-Chester illness is also highlighted by clinical research. The examples of new technologies that can potentially provide tools to guarantee proper monitoring, patient-specific treatment, and the enhancement of diagnostic proficiency are DL models and virtual reality applications that can bridge the existing gaps in neurological care in real life [6].

13.2.2 Wearable Sensors for Continuous Physiological Monitoring

Real-time, noninvasive monitoring of vital physiological signals is made possible by wearable sensors, which offer insights into the activities of molecular biomarkers, the heart, respiratory system, and sweat [7].

Brain activity, heart rate, respiration, and sweating are all impacted by the parasympathetic and sympathetic nervous systems' reactions to both internal and external stimuli. Wearable sensors, such as, electrocardiograph (ECG), electroencephalography (EEG), strain/impedance, and electrodermal activity (EDA) sensors, are recorded to monitor this during real-time and to assess the level of stress:

- **Cardiac Activity:** This is monitored with ultra-sensitive ECG sensors, paint-on, textile-based threads, wireless patches, and soft e-tattoos and materials, including graphene, gold, and poly(3, 4-ethylenedioxythiophene)-poly (styrenesulfonate) (PEDOT-PSS).

- **Respiratory Activity:** It is estimated on the basis of piezoresistive and triboelectric textile sensors to the chest and abdomen to produce resistance or voltages signals associated with breathing soundness and lung volume; advanced signal processing isolates cardiac and respiratory signals to be correctly evaluated.
- **Perspiratory Activity:** It is measured using either EDA or galvanic skin response electrodes which measure a change in voltage in response to sweat, indicating the activity of the sympathetic nervous system and stress.
- **Molecular Biomarkers:** Molecular bimarkers such as cortisol are detected by electrochemical, aptamer-based, and microfluidic sensors integrated into patches, bandages, or textiles allowing real-time and individualized tracking. The combination of these systems assists in the constant measurement of physiological, cognitive, and emotional conditions.

13.3 ML Techniques for Wearable Sensor-Based Neurological Disorder Identification

Wearable sensors are used to determine physiological and motion related information to track neurological disorders such as the Parkinson disease (PD) that gathers real-time measurement (blood pressure, heart rate, oxygen levels, and walking patterns) and integrates with hospital medical records. This multimodal data is processed by ML models to classify it accurately and identify it at an early stage. Other techniques include feature relevance and sliding window segmentation that improve interpretability and real-time detection, which allow customized and predictive neurological monitoring and evaluation using AI-based practices.

13.3.1 Neurological Disorder Wearable Sample Collection

Physiological data were collected using wearable sensors, including respiration flow, internal humidity, pulse count, blood glucose, oxygen saturation, and physical activity and was sent through short-range IoMT networks. Wi-Fi facilitated swift exchange of information in the hospitals and all information gathered were stored in a central repository in the internet of things to be analyzed. Additionally, the hospital database provided the actual medical records of 50 patients aged 50–70 [8]. Over four days, the movement patterns and neurological variations of 31 PD participants were tracked using wearable accelerometers on the wrists and a smartphone attached to the hips, allowing both motor symptom assessment and the development of data-driven detection models [9]. Real-time patient monitoring and ML-based neurological evaluation are supported by this combined dataset.

13.3.2 RF-Based Classification for Neurological Disorder Detection

The RF algorithm, consisting of numerous DTs, was employed to classify patient physiological and neurological data. Each DT recursively splits the dataset when a predetermined depth is achieved, measuring node contamination at the node using the Gini coefficient as indicated in eq. (13.1), each DT iteratively separates the dataset until a predetermined depth is reached [10]:

$$\text{Gini}(w_i) = \sum_{j=1}^{l} o_j(1 - o_j) = 1 - \sum_{j=1}^{l} o_j^2 \tag{13.1}$$

where $\text{Gini}(w_i)$ represents the node w_i Gini impurity value,w_i is DT's ith node, l is the total amount of labels for classes, j is index of classes, and o_j is likelihood of samples in a class j. Equation (13.2) computes the variable importance measure (VIM) as the decrease in node impurity as follows:

$$\text{VIM}_{\text{im}}^{(\text{Gini})} = \text{GI}_m - \text{GI}_k - \text{GI}_q \tag{13.2}$$

$\text{VIM}_{\text{im}}^{(\text{Gini})}$ features m's variable importance measure at the node i based on Gini impurity, i denotes the RF's DT index, m is the DT node index the Gini impurity of the parent node prior to the split on feature m, the left child node, and the right child node following the split are represented by GI_m, GI_k, *and* GI_q, respectively. Equation (13.3) states that the total VIM of feature w_i for a single DT is the sum of its VIM over all pertinent nodes

$$\text{VIM}_{\text{ji}}^{(\text{Gini})} = \sum_{m \in M} \text{VIM}_{\text{im}}^{(\text{Gini})} \tag{13.3}$$

$\text{VIM}_{\text{ji}}^{(\text{Gini})}$ feature i's variable importance measure at node j based on Gini impurity, M is a collection of every node in the DT, $\sum_{m \in M}$ representing total VIM values over all pertinent nodes. The total of the feature's VIM scores for each tree in the RF with s trees yields the feature's VIM in eq. (13.4)

$$\text{VIM}_i^{(\text{Gini})} = \sum_{j=1}^{s} \text{VIM}_{\text{ji}}^{(\text{Gini})} \tag{13.4}$$

Here s is the total quantity of features and $\text{VIM}_i^{(\text{Gini})}$ denotes the overall significance of every feature in the ith DT. Finally, eq. (13.5) is used to normalize the VIM scores so that they can be compared across features:

$$\text{VIM}_i = \frac{\text{VIM}_i^{(\text{Gini})}}{\sum_{j=1}^{n} \text{VIM}_j^{(\text{Gini})}} \tag{13.5}$$

where VIM_i represents normalized VIM, $\sum_{j=1}^{n} \text{VIM}_j^{(\text{Gini})}$ denotes the total of all features' raw VIM scores, and n is the dataset's total feature count. The RF finds the most important characteristics for classification by computing the Gini impurity at each node, adding up the impurity decreases over all nodes and trees, and then normalizing. Neurological disorders can be accurately and comprehensibly detected thanks to this methodical technique, which guarantees that the features that contribute most to lowering uncertainty in the DT are identified.

13.3.3 SVM-Based Classification for Neurological Disorder Detection

SVM is used in this research to categorize patterns of neurological disorders using physiological and motion-related signals gathered from wearable sensors. To provide the best possible separation of disorder-related classes, the SVM seeks to identify support vectors, or data points that are closest to the decision hyperplane, and maximize the margin between these vectors and the hyperplane. By maximizing the boundary distance between each class and every other class, the SVM expands the binary classifier for multiclass situations. The objective function outlined in eqs. (13.6)–(13.8) is used to formalize this optimization process, allowing for accurate real-time detection of aberrant neurological signals [11]:

$$\min \frac{1}{2} x^S x + D \sum_{j=1}^{M} \max\left\{0, 1 - z_j\left[x^S \partial\phi\left(w_j + a\right)\right]\right\} \tag{13.6}$$

$$\min \frac{1}{2} \|x^2\| + d \sum \|\xi_j + \xi_j^*\| \tag{13.7}$$

$$\text{Subject to} \begin{cases} w_j - \langle x, \phi(w_j)\rangle - a \le \varepsilon + \xi_j, \\ \langle x, \phi(w_j)\rangle, + a - z_j \le \varepsilon + \xi_j, \\ \xi_j, \xi_j^* \ge 0, \end{cases} \tag{13.8}$$

where x is the classifier's weight vector, S denotes the matrix of regularization, D is the penalty amount for incorrect classification, M is the total quantity of training samples, z_j is the label of the jth sample (+1 or – 1), w_j is the vector of features of the jth sample, a is the bias term, and $\partial\phi(w_j + a)$ is the transformed feature in kernel space. $\|x\|^2$ represents regularization phrase to prevent overfitting, d is the penalty parameter that regulates the slack variables' weight, ξ_j and ξ_j^* are the slack variables for faults in the epsilon-insensitive tube above and below. $\langle x, \phi(w_j)\rangle$ represents the scalar component of the mapped matrix of features and the weight matrix, ε is the tolerance margin. In this research, both nonlinear and quadratic kernel functions are applied to improve the classification of gait patterns from wearable sensor data, enhancing the SVM's ability to distinguish between neurological conditions.The SVM algorithm employs classifier models using both the linear and polynomial kernel functions, which are defined in eqs. (13.9), (13.10), and (13.11), respectively:

$$e(w) = \sum_{j=1}^{n} \alpha_j z^{(j)}(w_1, w_2) + a \tag{13.9}$$

$$l(w_1, w_2) = \langle w_1, w_2 \rangle \tag{13.10}$$

$$l(w_1, w_2) = (\langle w_1, w_2 \rangle + Q)^c \tag{13.11}$$

where $e(w)$ is the SVM decision function's output for input w, each training sample's index, n is the total quantity of training samples, α_j is the Lagrange multiplier connected to the jth sample, $z^{(j)}$ is the class label (+1 or − 1) of the jth sample, w_1 *and* w_2 represent the sample's input feature vectors, Q denotes an unchanging phrase, and c is the degree of polynomials. A sliding window technique was used to separate wearable sensor data into training and testing sets.

13.3.4 DT-Based Classification for Neurological Disorder Detection

DT, a supervised ML technique, uses a hierarchical, tree-like structure to split the input space into smaller groups. The algorithm starts at the root node, traverses the inner nodes and then leaves depending on decision rules by attributes. These rules guide new samples through the tree in order to accurately predict the class labels in the process of testing [12].

The path a sample follows in the tree is determined by this equation. The sample of data is transferred to the tiny cluster on the left if the feature value is below or equal to the threshold instead, the correct subsidiary branch is reached. To classify the sample, this procedure is repeated recursively at every node. DTs are suitable for feature significance analysis and are highly interpretable. The data obtained through wearable devices such as a smartphone placed in the hips and a wrist-mounted accelerometer were analyzed using the DT algorithm to gather physiological and gait data. A sliding window of 50 samples with a step of 1 s was used to divide the gait cues. Characteristics that were used to classify the characteristics included a maximal angular acceleration and actual momentum in both the dual period along the *Z*-axis, which was determined using fast Fourier transform (FFT) in the 4D frequency domain. These features were then subjected to the DT model to determine the patterns that are indicative of neurological conditions and classify them. The method can be used to interpret the main characteristics clearly, track patients in real-time, and corresponds to the goal of the research, that is, to detect the onset of neurological diseases and evaluate them thoroughly.

13.4 Technical Obstacles and New Prospects for ML-Powered Wearable Healthcare

Several factors, including limited power, storage, and battery, make the implementation of ML on the device in a wearable sensor system a complex issue. The sensor noise and insufficiently diversified training data reduce prediction accuracy and additional problems are raised by the gaps in theory and the rapid development of ML. Security and privacy of data are crucial issues especially when handling data locally. The complexity of the technical and practical challenges in the application of ML to healthcare applications can be demonstrated by the fact that such techniques as model compression and transfer learning offer solutions, but implementing them without accuracy loss is complex. In comparison to these challenges, there are several development opportunities in ML applications in wearable sensors and healthcare [13]:

- **Real-Time Data Analysis:** ML models can examine and deduce wearable sensor information in immediate effect, enabling evaluations and forecasts of reserve or decompensation status, activity, and context. The consolidated data can then be sent to the user.
- **Identification of Symptoms:** ML can assist in diagnosing or treating symptoms by identifying the telltale indicators of neurological conditions or movement issues.
- **Better Healthcare Services:** The functioning of healthcare technologies depends on ML. Productivity can be raised by allocating time, fulfilling medical requirements, and automating hospital operations. It is possible to develop algorithms with the help of ML that would correspond to the availability of medical providers in the area with the required clinical skill sets.
- **Diagnosis and Treatment Planning:** ML can be used in detecting cases of high-risk patients, preventive, diagnosis, and treatment planning. The diagnostic procedure is facilitated by offering an ingenious model that helps medical professionals to make the most accurate diagnosis.

13.5 Performance Assessment of ML Models for the Identification of Neurological Disorders

This section compares the SVM, DT, and RF models of identifying neurological illnesses with the data of wearable sensors. The results demonstrate that in the case of noisy and heterogeneous sensor signals, RF models based on ensembles are more robust and predictive. Although SVM is effective in capturing complex multidimensional patterns, decision trees can have an explanatory result, which can be used in clinical decision-making. In general, the findings confirm the reliability of wearable devices that have ML functions in patient real-time neurological health monitoring.

13.5.1 Comparative Analysis of ML Classifiers for Neurological Disorder Detection

The experimental assessment was done with the help of wearable sensor data to test how the SVM [14], DT [14], and RF [14] classifiers detect the neurological diseases. The research considers the evaluation of some ML models to be used in clinical settings in wearable sensor-based systems, which emphasize the strength of RF, SVM, and DT. RF achieved higher results than others because the ensemble learning methodology can deal with noisy and heterogeneous physiological data by reducing variance and improving generalization to different patient factors. This renders RF especially useful in clinical use when there are unpredictable sensor data. SVM was a good decoder of more intricate patterns in neural data, it could identify minor changes in heart rate and gait that are indicative of emerging neurological conditions, which is essential to diagnosing diseases such as, Parkinson disease or mental impairment. The flexibility it has in handling the nonlinear relationships places it as an effective instrument of such complex analyses. DT, though a little less accurate, and offered useful interpretability, enabled clinicians to learn about the feature splits which resulted in predictions. This openness facilitates clinical decision-making, which fosters the introduction of AI with patient care. The discussion has concluded that SVM, DT, and RF are to be used to select a model based on the clinical setting and type of data (complex), interpretability (treatment planning), and high-performance (noisy condition) monitoring. The research lays emphasis on the significance of interpretability and robustness coupled with accuracy in order to have successful real-time monitoring of neurological conditions.

13.5.2 Evaluation of ML Model Effectiveness

Table 13.1 comprises the common metrics in this research employed in the evaluation of the performance of the classifiers in terms of accuracy, precision, recall, and the F1-score. The criteria were used to comprehensively determine the level at which the models identified the patterns of the neurological disorders with wearable sensor data, giving an indication of how they performed and were reliable [13].

In eqs. (13.12)–(13.15), true positives (TPs), true negatives (TNs), false positives (FPs), and false negatives (FNs) are employed as a base for analysis. The comparative analysis highlights the need of matching model selection to the particular clinical context DT for interpretability and clinical understanding, SVM for managing complex multidimensional information, and RF for reliable, high-performance

TABLE 13.1 Performance measures among several classifiers.

Evaluation metrics	Definition	Equation	Equation number
Accuracy	Determines the proportion of every collection that is accurately categorized	$\text{Accuracy} = \left(\frac{\text{TP} + \text{TN}}{\text{TP} + \text{FP} + \text{TN} + \text{FN}}\right)$	(13.12)
Precision	Calculates the ratio of accurately predicted desired samples out of all actual samples that were expected	$\text{Precision} = \left(\frac{\text{TP}}{\text{TP} + \text{FP}}\right)$	(13.13)
Recall (sensitivity)	Calculates the proportion of accurately anticipated desired outcomes out of all actual ones	$\text{Recall} = \left(\frac{\text{TP}}{\text{TP} + \text{FN}}\right)$	(13.14)
F1-score	Positive predictive value and sensitivity metrics are optimized by the harmonic mean	$F_1 = \left(\frac{2 \times \text{recall} \times \text{precision}}{\text{recall} + \text{precision}}\right)$	(13.15)

Source: Own elaboration.

monitoring. Furthermore, since the quality and preprocessing of wearable sensor data greatly affect classifier efficacy, these findings highlight the significance of information representation. Ensemble and multimodal approaches offer a more balanced and dependable framework for real-time neurological disorder detection, supporting proactive patient monitoring and informed clinical decision-making, according to variations in accuracy, precision, recall, and F_1-score across various models and datasets shown in Table 13.2 and Figure 13.1.

TABLE 13.2 Evaluation of ML models for the classification of neurological disorders [14]

Models	Accuracy (%)	Precision (%)	Recall (%)	F_1-Score (%)
SVM [14]	94.97	94.86	94.81	94.84
DT [14]	95.66	94.34	96.72	95.52
RF [14]	96.03	95.09	97.23	96.21

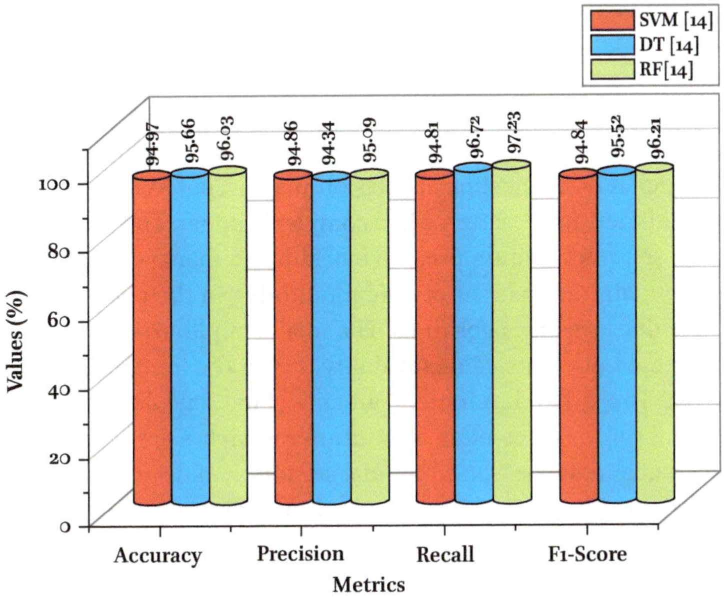

FIGURE 13.1 Comparison of SVM, DT, and RF for different information configurations. (Source: Own elaboration)

13.6 Discussion

ML approaches can effectively identify neurological disorders using wearable sensor data, as demonstrated by the evaluation of SVM, DT, and RF classifiers. Ensemble methods, such as RF, showed strong performance in managing noisy and heterogeneous physiological and motion data. The given DT models gave understandable outputs, especially in the detection of positive results, which can be applied in clinical decision-making. The SVM proved to be useful in the separation of multidimensional sensor signal and complex patterns. The difference in performance between datasets indicates the effect of the characteristics of participants, the placement of the sensor, and the selection of features. On the whole, these findings suggest that wearable sensors and ML can help to provide reliable, real-time, and patient-centered monitoring of neurological conditions, which facilitates timely intervention and proactive healthcare. Ensemble and multimodal methods seem most promising to have robust and scalable detection.

13.7 Conclusion

The research established that wearable sensor technology and ML algorithms could be utilized to monitor and detect neurological issues in real-time. An experimental assessment was done using SVM, DT, and RF. Ensemble learning proved resilient as the RF model was more successful than the others in dealing with noisy and varied physiological and mobility data. The interpretable results of DT models were particularly good in the aspect of recall and overall balance between precision and sensitivity, which is useful in clinical decision-making, but SVM was effective in identifying multidimensional patterns in a complex manner. These findings indicate that wearable sensors facilitate proactive healthcare management, early intervention, and continuous and patient-centric monitoring in the combination with sophisticated machine learning algorithms. The small sample size and limited scope of the research could affect its generalizability due to the limited scope of neurological disorders. The other hindrances are real time calculation limits, patient compliance, and sensor placement. The future research needs to explore advanced ML models, expand datasets, multimodal sensors, and develop scalable and privacy guaranteed models that can be used in a broad clinical practice.

References

[1] AlArnaout Z, et al. Exploiting heart rate variability for driver drowsiness detection using wearable sensors and machine learning. Scientific Reports. 2025;15(1):24898. https://doi.org/10.1038/s41598-025-08582-2.

[2] Bhatt MW, Sharma S. An IoMT-based approach for real-time monitoring using wearable neuro-sensors. Journal of Healthcare Engineering. 2023;2023(1.3):1066547. https://doi.org/10.1155/2023/1066547.

[3] Sagar PK, et al. Using support vector machines for enhancing cancer prediction in recommender systems. In International Conference On Artificial Intelligence Of Things For Smart Societies Vol. 10.2 2024. Cham: Springer Nature Switzerland. https://doi.org/10.1007/978-3-031-63103-0_14.

[4] Xiao X, et al. Advances in machine learning for wearable sensors. ACS Nano. 2024;18(34):22734–51. https://doi.org/10.1021/acsnano.4c05851.

[5] Bajaj M, et al. Study on Degenerative Parkinson's Disease Using Various Machine Learning Algorithms. In 2023 3rd Asian Conference on Innovation in Technology (ASIANCON) 15.2 2023. IEEE. 10.1109/ASIANCON58793.2023.10270216
[6] Ningrum DNA, Kung W-M. Challenges and perspectives of neurological disorders. Brain Sciences. 2023;13(4):676. https://doi.org/10.3390/brainsci13040676.
[7] Kang M, Chai K. Wearable sensing systems for monitoring mental health. Sensors. 2022;22(3):994. https://doi.org/10.3390/s22030994.
[8] Castelli Gattinara Di Zubiena F, et al. Machine learning and wearable sensors for the early detection of balance disorders in Parkinson's disease. Sensors. 2022;22(24):9903. https://doi.org/10.3390/s22249903.
[9] Dahan F, et al. A smart IoMT based architecture for e-healthcare patient monitoring system using artificial intelligence algorithms. Frontiers in Physiology. 2023;14 (1):1125952. https://doi.org/10.3389/fphys.2023.1125952.
[10] Ding J, et al. A novel two-stage feature selection method based on random forest and improved genetic algorithm for enhancing classification in machine learning. Scientific Reports. 2025;15(1):16828. https://doi.org/10.1038/s41598-025-01761-1.
[11] Wang F, Yan L, Xiao J. Human gait recognition system based on support vector machine algorithm and using wearable sensors. Sensors & Materials. 2019;31(4). doi: 10.18494/sam.2019.2288.
[12] Adi GS. Emerging machine learning in wearable healthcare sensors. 2023;32(6):378–85. https://doi.org/10.46670/JSST.2023.32.6.378.
[13] Rahman A, et al. Machine learning and deep learning-based approach in smart healthcare: Recent advances, applications, challenges and opportunities. AIMS Public Health. 2024;11(1):58. doi: 10.3934/publichealth.2024004.
[14] Gedam S, Dutta S, Jha R. Analyzing mental stress in Indian students through advanced machine learning and wearable technologies. Scientific Reports. 2025;15 (1):20610. https://doi.org/10.1038/s41598-025-06918-6.

CHAPTER 14

Challenges and Opportunities in Implementing Wearable Sensor-Based Systems for Chronic Disease Management

A. R. Chayapathi, Dimpal Khambhati, Sanjay Kumar Jena

Abstract Cardiovascular diseases, diabetes, respiratory diseases, and nerve diseases, are chronic diseases that demand constant check-ups and proactive care to avoid complications and enhance patient outcomes. This chapter also looks at the issues and opportunities of adopting wearable sensor-based systems to manage chronic diseases. Wearable sensor-based should be seen as a potential answer as it enables the provision of real, noninvasive nursing of physical, biochemical, and activity-related limitations. This type of systems encourages personalized healthcare, detection of disease exacerbations early, and monitoring of patients remotely to minimize hospital visits and healthcare expenses. Nonetheless, although they have a huge potential, various obstacles make them difficult to be adopted in the treatment of chronic diseases. The main problems are sensor reliability and accuracy as well as short battery life, interoperability of data, privacy and security issues, patient compliance, and absence of unified clinical validation models. Also, user acceptance and accessibility are affected by such socioeconomic factors and digital literacy. The development of sensor technology, artificial intelligence, cloud computing, and telemedicine provides significant opportunities to remove these obstacles. This chapter summarizes the issues concerning the implementation of wearable sensor-based systems to plan long-lasting illness organization.

Keywords Wearable sensors – Internet of Medical Things (IoMT) – chronic disease management – remote patient monitoring (RPM) – personalized healthcare

14.1 Introduction

Chronic disease is a term that is used to describe a wide group of diseases that have an insidious onset, extended incubation, a protracted healing process, the absence of specific biological evidence, and a vague indicator of cure. There are chronic diseases such as, diabetes, cancer, chronic lung disease, and cardiovascular disorders that pose serious challenges to the enhancement of health and life expectancy of people. Chronic disease management is a term which is used to denote incurable, long-time illnesses, and predicting variables to routine and constant testing

 | HTTPS://DOI.ORG/10.1515/9783112237083-014

and evaluating, inclusive measures in the control of the therapeutic process [1]. Wearable technology has been used in the healthcare industry in the recent past, with more devices being developed and sold to support individuals in dealing with their chronic health conditions. These devices such as, fitness tracking and smart watches have the capability to monitor and show a variety of strength limits, such as, breathing, rest, exercise, and compliance with treatments that can be useful to an individual with a chronic condition [2]. The benefits of wearable technology are numerous, yet, the challenges include, in particular, data security and privacy. Manufacturers of devices should ensure high security in order to safeguard the collected data [3].

Mechanization is not a fast-developing subject in computer science and computer vision. For many years, efficiency improvement has been achieved through task automation. Managers in corporations would plan to carry out a specific assignment back in the prehistoric era [4]. Innovations in technology, particularly IoT, are revolutionizing healthcare by enhancing efficiency, services, and trust [5].

Research Objective: This chapter examines the obstacles and possibilities to deploying wearable sensor-based systems for managing chronic illnesses, emphasizing intellectual, clinical, and user-specific barriers while also exploring new developments and research plans to enhance precision, usability, adaptability, and real-world medical integration.

14.2 Various Types of Wearable Heart Monitoring Gadgets

Wearable cardiac detection devices use a variety of technologies, each created specially to fulfill distinct health tracking needs:

Smartwatches: The majority of the contemporary digital wrist watches that Samsung, Google, Apple, Fitbit, and other companies have introduced to the market are equipped with optical sensors that simultaneously measure heart rate (HR) and HR inconsistency. Some of those plans can detect variations in the heart's normal beats that can indicate conditions such as, AFib, along with providing standard electrocardiogram (ECG) results. Numerous individuals who like ongoing, inconspicuous monitoring can use digital watches because they are convenient.

ECG Patches: iRhythm and Zio are the manufacturers of adhesive coverings that are worn on the chest and offer far more reliable long-term ECG data gathering. These devices become more prevalent in medical care for patients who need a thorough assessment of their circulatory system since they can constantly provide the best ECG information intended for times or workweeks, giving a general image of the electric signals of the heart.

Chest Straps: Although chest straps, such as those made by Polar or Garmin, are frequently used during exercise and training, they contain electrodes that let the gadget detect HR accurately. Because of their great accuracy, typical photoplethysmography (PPG) sensors are helpful to everyone who enjoys physical exercise and needs accurate HR tracking from a medical perspective.

Blood Pressure Monitors: The Omron Heart Guide smart watch is one of the wrist-wearable gadgets that use oscillometric technology. Unlike conventional devices that simply measure blood pressure once, these devices monitor changes in blood pressure over time, alerting users to their overall cardiovascular health, including the two types of pressure.

Biosensor-Embedded Clothes: The recent developments allow wearable external cardiac monitors to be carefully incorporated into the fabric of clothes including bras or shirts. Such devices normally lump together several biometrics, including ECG and PPG, and so on, into one reading. The integration of sensors into clothing has provided new possibilities of unobtrusive monitoring of health in a consistent way.

14.2.1 Wearable Heart Devices measure Various Rudimentary Measurements as Signs of Circulatory Health

Cardiac Activity – Heart Rate and Heart Rate Variability **(HRV):** HR is a simple indicator of cardiac activity, and HRV provides information about stress pattern and the percentage of the autonomic of sensory nervous system of the air national guard (ANS). The degree of flexibility is a significant contributor to irregular beating of the heart and other heart diseases. An ECG is the examination that could detect any abnormality in HR like atrial or HRV, which is also a symptom of some forms of coronary disease.

Temperature: A rise in plasma pressure has been associated with high temperature, which is a potential cause of heart attack and stroke [6]. Cholesterol level is one of the key factors of cardiac health. Such monitoring can enable the customers and practically every member of the medical community to monitor or treat pressure.

14.2.2 The Wearable Technology and How It Affects Chronic Diseases

Wearable technology has several benefits compared to traditional solutions in relation to the treatment of chronic medical diseases. Since wearable technologies are always worn and provide real-time information on one's health condition, increased ease and accessibility can be considered one of their most significant benefits. Moreover, wearable technology may offer personalized and targeted treatment in accordance with feedback and health information.

Diabetes: Diabetes management is one of the aspects that wearable technology has demonstrated. Millions of individuals around the world have diabetes which is a chronic condition where proper treatment is very important to avoid complications such as, blindness, kidney failure, and broken leg. Continuous glucose monitoring (CGM) devices and insulin drives have significantly enhanced the control of glycemic and reduced complications in diabetics. Moreover, research demonstrates that glucose pumps alongside CGM make type 1 diabetes patients live better and manage their glucose levels, which means that wearable technology can be used to provide more personalized and effective care to patients with diabetes.

Cardiac Illness: Heart disease is a severe chronic condition which involves coronary artery disease, heart failure, and cardiac arrhythmias. Wearable technology is very promising in the management and treatment of heart disease. HR monitoring and early identification of abnormal rhythms, especially atrial fibrillation, becomes achievable with the help of such devices as fitness trackers and smartwatches, particularly in individuals who are at risk. Additionally, it has been demonstrated that wearable technology-based remote monitoring helps treat heart failure by lowering rates of return to hospitals and enhancing patient outcomes. These tools provide preemptive treatment, individualized treatment, and ongoing cardiac surveillance in a scalable and economical manner.

Chronic Obstructive Pulmonary Disease (COPD): Lots of people globally suffer with COPD, a chronic breathing condition with a significant morbidity and mortality rate. Therefore, minimizing exacerbations and enhancing quality of life (QoL) require appropriate COPD therapy. For instance, the possibility of activity trackers and spirometers to improve COPD therapy has been investigated. Individuals with COPD had better exercise capacity and QoL when they used a wearable device for home-based pulmonary rehabilitation. Furthermore, utilizing an activity monitoring device successfully identified changes in physical activity levels and predicted exacerbations in people with COPD. Wearable technology can provide a noninvasive and economical way to monitor and treat COPD.

14.2.3 Enhancing Compliance in Chronic Care through Wearable Technologies

- By continuously helping people manage complicated chronic care routines, wearable technology promotes treatment adherence.
- Medication reminders are one feature that makes it easier for patients to continuously adhere to recommended dosage schedules.
- Activity monitors enhance adherence to wellness and lifestyle adjustment strategies by promoting frequent physical activity.
- Wearable-based feedback fosters long-term adherence to treatment objectives by raising patient involvement and confidence [7].

- All things considered, wearable technology is a useful and efficient way to enhance consistency and health results in the organization of long-lasting illnesses.

The diseases, sensors, advantages, difficulties, enabling technologies, and potential future paths of wearable sensor-based chronic healthcare systems are shown in Table 14.1:.

TABLE 14.1 Key dimensions of wearable sensor-enabled chronic care systems

Aspect	Description
Target diseases	Cardiovascular disorders, diabetes, respiratory, and neurological conditions
Key sensors	ECG, PPG, glucose, motion, temperature, and biochemical sensors
Clinical benefits	Continuous monitoring, early risk detection, and personalized treatment
Major challenges	Data accuracy, power limitations, privacy, and user compliance
Enabling technologies	Smart materials, machine learning, edge/cloud computing
Future outlook	Personalized medicine, preventive care, and scalable remote monitoring

Source: Own elaboration.

14.3 Technologies for Remote Patient Monitoring (RPM) and Feedback

Machine learning (ML)-powered remote patient monitoring (RPM) devices have revolutionized the management of chronic diseases by providing ongoing, individualized care outside of typical clinical settings. These systems track vital signs, adherence to drugs, exercise routine, and symptom development in real-time by utilizing data from wearable sensors, mobile health apps, and home-based devices.

Supervised and unsupervised learning methods have common uses of modeling patient behavior and health trends. Just to name a few, anomaly detectors can identify changes in HR, oxygen consumption, or glucose level, which may be early warning signs of an exacerbation of a disease such as, diabetes, heart failure, or COPD. Subsequently, they provide their patients with personalized feedback with the help of online resources that assist them in implementing lifestyle and physical activity or pharmaceutical changes [8]. Such strategies allow people to have an active role in their very own care and at the same time enhance clinical outcomes. Moreover, reinforcement learning is becoming more popular in ML-based feedback systems to adjust the suggestions based on the preferences, responses, and practical effectiveness of the user.

14.3.1 Smart Healthcare Systems Using Edge Computing and Electronic Fencing Technologies

Processing data close to the network's edge, closer to the point of data generation, rather than in a centralized data center, is known as edge computing. Edge computing is becoming increasingly important as the internet of things (IoT), which is made up of many connected devices generating massive amounts of data, grows. Devices can function more effectively and with less delay by processing data at the edge, which is crucial in applications such as, industrial automation and driverless cars.

Cloud computing issues such as, reaction time and location perception are addressed by the combination of fog and mobile edge computing, especially in the healthcare industry. Using a hybrid model of type 2 schizophrenia and VIKOR (VIšeKriterijumska Optimizacija I Kompromisno Rešenje) techniques, an IoT-based system was presented for the discovery and maintenance of type 2 diabetes, improving patient survival rates and reducing execution time by 9.8%. An emotion-aware framework used for user interaction combines an edge cloud, smart clothes, and a personal robot.

Wandering Prevention: Automated fences protect older people living with dementia or cognitive disorders from becoming lost by limiting movement outside of designated safe zones.

Chronic Disease Management: Electronic fences facilitate tracking activity and early detection of medical problems for individuals with long-term diseases (such as, diabetes or cardiovascular disease).

Behavioral Surveillance: By identifying odd or potentially dangerous behaviors in homes or care facilities, the system improves patient security.

Elderly Care: Higher-quality care for the elderly is made possible by technological fencing, which also improves assurance, lessens the workload for caregivers, and reassures families [9].

The combination of medical devices and the IoT is referred to as the Internet of Medical Things (IoMT). IoMTs are "things" that have software or sensors for data sharing with other devices over the internet and are monitored by medical experts. The dynamics of the game have been drastically altered by the extensive use of these gadgets in wellness delivery. A few of the RPM, respiratory surveillance, anxiety assessments, personal hygiene evaluations, glucose measurements, regulated drugs, Parkinson's illness observation, and uniform robotic operations are some of the frequent applications of IoMTs. Almost every facet of fitness, from appointment reminders to treatment compliance, health customer service, tracking patients, monitoring, and wellness, is covered by mobile health (mHealth), the extensively used type of IoMT by portable earpieces [10].

14.3.2 AI-Driven Early Health Risk Detection through RPM

AI models use algorithms for early health problem identification to examine patient data, such as, vital signs and individual health characteristics. They set baselines, keep an eye out for variations, and spot indicators of decline. AI can anticipate possible problems by using past data trends, allowing medical practitioners to intervene promptly to enhance patient outcomes and allow early illness identification through RPM integration:

- Quick reaction and intervention when a patient's health status shows even slight changes, which can stop health issues from getting worse and lower the chance of complications.
- Decreasing the need for individuals to be admitted to hospitals because they can obtain medical care at home, saving money for both patients and healthcare systems.
- Enhancing patients' QoL by preventing potential medical emergencies.
- Generating dependability of the patient's state, assisting medical practitioners in more accurate arranging therapy, as well as helping clients make recommendations for their habits and medications [11].

Portable and mobile gadgets that transmit patient data to cloud servers for predictive evaluation, remote clinician access, rescue efforts, and ongoing healthcare tracking.

14.3.3 RPM: Enhancing Chronic Care beyond Clinical Settings

- **Description and Scope:** RPM collects, processes and stores patient generated health data that is not in a traditional clinical environment, through the use of electronic equipment, sensors and wearables, and smartphones and cameras.
- **Creative Assistants:** The ongoing gathering of patient information is enabled by advancements in digital health technologies, which can effectively respond to it and monitor it remotely.
- **Clinical Advantages** RPM helps to provide continuous monitoring of health, prevent unnecessary visits to the hospital, and helps to notice the initial signs of the worsening state of chronic diseases.
- **Availability and Ease:** RPMs can be particularly useful to individuals with low mobility, or to take healthcare services to underdeveloped or remote locations [12].

14.4 Ethical Concerns of Wearable Technology

The necessity of the integrated multi-condition RPM systems has been justified by the fact that sensors have been designed in certain instances to be specific to a particular health condition, and hence integration has been difficult, and the processing complexity is increased by the multisensory use. Although wearable technology has possibilities in the field of medicine to enhance treatment of chronic illnesses and preventive services, one of the challenges that the technology will definitely face is technological issues, privacy issues, and legal issues.

14.4.1 Technical Difficulties

Sensor Data Accuracy and Reliability: This is the primary technical challenge in wearable health technology, as it is necessary to ensure the accuracy and reliability of sensor data. Although these devices measure several physiological indicators, such as, blood sugar and HR, most of the sensors do not have clinical accuracy. Abnormal readings can disadvantage medical staff in making a decision because of pigmentation of the skin, temperature, movement, and fit of a device. Manufacturers are striving toward making sensor responsiveness and measurement better, but they must tackle such problems as battery life and shrinking which can be costly. The gadgets that are used daily should meet the medical-grade requirements of accuracy to be useful in the medical sector.

Power Saving and Battery Life: Battery life is a major technical issue facing wearable technology. Continuous health data monitoring requires constant power consumption which can quickly exhaust its power especially when sensors are collecting vast amounts of data. The wearable technology should be able to offer a longer battery life as most wearable technology is designed to be used continuously. The solution to this issue requires lightweight detectors to be made, and the level of power usage to increase. Enhancing battery performance and maintaining device functions and sensing efficiency is a continuous problem.

14.4.2 Security and Privacy of Data

Risks of Unauthorized Access and Data Leakages: Wearable technology may generate sensitive health information that is vulnerable to security issues. The more associated these devices are with the internet, the more important the issue of hacking and unauthorized access to private medical data (PHI) is becoming. Health data contain specific details regarding the physiological parameters of a person, exercises, and medical history which is of concern to not only medical professionals but also to attackers and other unauthorized organizations.

Usability and User Adoption: Comfort and Design of Wearable Devices: The effectiveness of wearable medical devices is determined by their accessibility, convenience, and effectiveness. To allow effective round-the-clock monitoring, the wearables should be miniature, small, and durable. Yet, much of the wearable technology is uncomfortable or irritating, especially when worn over an extended period of time. In the long-run, they may lead to allergic reactions or discomfort, which would decrease consumer satisfaction and compliance. To make their goods aesthetic and enjoyable to wear, wearable manufacturers must focus on ergonomics. This can make their wide acceptance possible.

User Involvement and Compliance: User participation necessitates constant use of wearable technology, but it remains hard to keep the interest. A lot of individuals consider that these devices are not mandatory but optional, and this is one of the discouraging factors. As a way of circumventing this challenge, manufacturers have to develop wearables capable of delivering the most pertinent information, customized health tips, and real-time responses, thus enhancing usability and promoting more engagement.

Data Use and Management Problems Are Ethical: The wearable technology raises concerns about confidentiality and usage, particularly to the makers of the technology, consumers, or healthcare providers, whose information is being gathered. It is upon the users to ensure their medical files are used to their benefit but not to profit. Providing wearable data to unknown individuals, including insurance companies, may lead to discrimination in the insurance and job market. To eliminate these concerns, there should be clear privacy laws that enable the consumer to control their data [13]. To protect the privacy and dignity of the customers in the wearable technology industry, there is a need to have ethical standards and rules.

14.5 Obstacles and Prospects for the Future

There are many barriers and limitations that are yet to be out of the way to make the most of the potential of intelligent clothing detection schemes which are made possible by intelligent resources and ML, notwithstanding tremendous progress. A user-friendly interface, usability design, data accuracy, clinical capacity, organization in terms of medical needs, cost effectiveness, and compatibility with existing therapeutic schemes should remain a priority because it makes it easier to be accepted by individuals within the health and wellness segments. Discussing data privacy and safety concerns, overcoming the limitations of wearable sensors technologies, such as, accuracy and reliability, meeting the needs of users, motivating and complying, and dealing with ethical and legal challenges are all the factors which should be considered. These variables should be put under strict control in order to implement successfully.

Wrapping Problem: The wearable technology has a problem with packaging as it introduces various features in small sizes that are vital to the comfort of a user. To clear the miniaturization of the sensor would not impair the performance of the sensors or the battery life, it is necessary to optimize the parts that use less power and the energy management methods used. Also, machines should be in a position to endure different conditions, including humidity and temperature changes. The need to ensure constant innovation in the wearable sensor packaging processes is also emphasized by the fact that the packaging structure materials must be biocompatible, strong, portable, beautiful, and simple to afford.

User Compliance and Behavior: To have accurate and useful wearable device data, user behaviour and involvement is very much necessary. The users should wear, charge, and synch their devices regularly to be used effectively, otherwise they may have cases of gaps in data, and the accuracy of data may be compromised. To enhance adherence, prolonged interaction by users poses challenges that require the development of interfaces that are easy to use and comprehensive training. Also, it is possible to stimulate user engagement with the help of motivational rewards and pedagogical strategies.

New Directions of Future Intelligent Wearable Sensors: The future of wearable sensors will drive demand-based design, state-of-the-art technology, and edge-AI connectivity, which will allow the wearable sensors to work in a manner that is clinically reliable, affordable, and safe. The customized and sustainable and privacy-sensitive solutions supported by the enhanced user engagement and administrative alignment can achieve scalable deployment of proactive and precise healthcare monitoring [14]. Figure 14.1 illustrates the major challenges of wearable sensors including data integrity, energy efficiency, sensor functionality, system coordination, user acceptance, and security concerns that affect the implementation of healthcare systems.

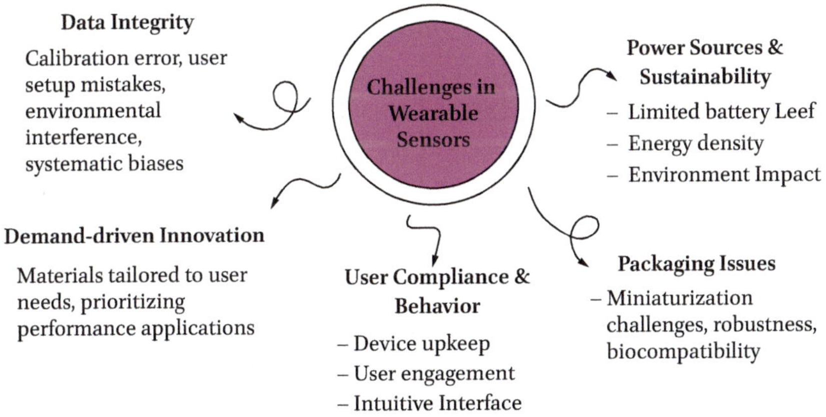

FIGURE 14.1 Key challenges in wearable sensor technologies. (Source: Own elaboration)

14.6 Future Pathways of Wearable Biosensors in Intelligent Healthcare

Customized medicine, prevention of illness and even genome integration may lead to a transformational wearable biosensor. The healthcare as biosensors is wearable will take time to develop and be integrated with the revolutionary technologies. In the health monitoring will be preventive, tailor-made and deeply entrenched in the individual journey toward health. With constant tracking of the physiological information and the customization of treatment to personal response, wearable biosensors offer real-time individualized healthcare. The combination of genetic susceptibility and metabolic profile therapies has the potential to improve treatment outcomes. Portable sensor data with genetic data facilitate precision medicine:

- Continuous preventive care and early identification of the risk of diseases become possible surveillance with ML analytics.
- By providing the feedback of exercise programs, sleep patterns, and diet to avoid chronic illnesses, wearable bio sensors are brought to assist in the optimization of lifestyles.
- Applications in drug discovery are informed using real-time indicators and trends in genetic reactions, drug choice, and dosing.
- Customized insights result in better participation and autonomy of patients, which is motivating, Adherence to treatment and health.
- Remote and cooperative is improved by real-time data exchange between patients and medical personnel care models and enhance clinical outcomes.
- Wearable data that has been collected and classified contribute to the progress of medical and the health of the population.
- End to end encryption, collaborative learning, secure cloud-edge architecture, and user-defined.
- Consent mechanisms are instances of privacy-preserving systems developed in the future.
- Guarantee safe storage of medical and genomic information and maintaining ethical utilization of regulatory compliance, trust and wearable biosensor methods [15].

14.7 Conclusion

The importance of the portable sensor-based systems, state-of-the-art AI, and RPM in this was well researched on the management of chronic diseases and smart healthcare study. Wearable sensors make continuous and noninvasive data collection possible and help preemptive care, personalized treatment, and early diagnosis of the disease as well as reduced hospitalization. Dependency and healthcare costs, patient safety, real-time decision-making, and senior care are enhanced even more by the addition of intelligent materials, ML, edge-cloud architectures, and electronic fencing. In spite of these advantages, problems related to data accuracy, sensor efficiency, system sustainability, system adoption, user compliance, privacy, system ethics, and system regulations remain some of the key barriers to mass adoption. Challenges related to demand oriented sensor design, robust clinical assessment, energy-efficient design, security conscious system architecture, and user friendly systems can be addressed through an appropriate design approach.

References

[1] Zhao S, et al. Elements of chronic disease management service system: an empirical study from large hospitals in China. Scientific Reports. 2022;12(1):5693. https://doi.org/10.1038/s41598-022-09784-8.

[2] Adepoju VA, et al. Wearable technology in the management of chronic diseases: a growing concern. Chronic Diseases and Translational Medicine. 2025;11(02):117–21. https://doi.org/10.1002/cdt3.156.

[3] Lewczak Z, Mitchell M, Mitchell MG. Wearable technology and chronic illness: balancing justice and care ethics. Cureus. 2024;16(11). https://doi.org/10.7759/cureus.73686.

[4] Yadav SP, et al. An improved deep learning-based optimal object detection system from images. Multimedia Tools and Applications. 2024;83(10):30045–72. https://doi.org/10.1007/s11042-023-16736-5.

[5] Rani P, et al. Simulation of the lightweight blockchain technique based on privacy and security for healthcare data for the cloud system. International Journal of E-Health and Medical Communications (IJEHMC). 2022;13(4):1–15. https://doi.org/10.4018/IJEHMC.309436.

[6] Zainab H, et al. Integration of AI and wearable devices for continuous cardiac health monitoring. International Journal of Multidisciplinary Sciences and Arts. 2024;3(4):123–39. https://doi.org/10.47709/ijmdsa.

[7] Ali U. The impact of wearable devices on the management and treatment of chronic medical conditions. Archives of Clinical Psychiatry. 2022;49(2). https://doi.org/10.15761/0101-60830000000415.

[8] Adeyinka A, et al. Machine learning efforts that enhance personalized patient care and chronic disease management. International Journal of Scientific Research in Science and Technology. 2022;9(2):647–71. https://doi.org/10.32628/IJSRST2302551.

[9] Wang W-H, Hsu W-S. Integrating artificial intelligence and wearable IoT system in long-term care environments. Sensors. 2023;23(13):5913. https://doi.org/10.3390/s23135913.

[10] Nazir A, et al. Wearable technology and its potential role in cardiovascular health monitoring and disease management. Health Science Reports. 2025;8(11):e71486. https://doi.org/10.1002/hsr2.71486.

[11] Tsvetanov F. Integrating AI technologies into remote monitoring patient systems. Engineering Proceedings. 2024;70(1):54. https://doi.org/10.3390/engproc2024070054.

[12] Arun S, Sykes ER, Tanbeer S. RemoteHealthConnect: Innovating patient monitoring with wearable technology and custom visualization. Digital Health. 2024;20552076241300748:10. https://doi.org/10.1177/20552076241300748.

[13] Qadrie Z, et al. Wearable technology in healthcare: opportunities, challenges, and future directions. Smart Wearable Technology. 2025;1:A12–A12. https://doi.org/10.47852/bonviewSWT52025578.

[14] Vo D-K, Trinh KTL. Advances in wearable biosensors for healthcare: current trends, applications, and future perspectives. Biosensors. 2024;14(11):560. https://doi.org/10.3390/bios14110560.

[15] Chen S, et al. Transforming healthcare: intelligent wearable sensors empowered by smart materials and artificial intelligence. Advanced Materials. 2025;37:2500412. https://doi.org/10.1002/adma.202500412.

CHAPTER 15

Smart Wearables for Real-Time Monitoring and Rehabilitation of Musculoskeletal Disorders

Logashanmugam Edeswaran, Anitha Josephine, Vishakha Khambhati

Abstract Musculoskeletal disorders (MSDs) represent a major global health challenge due to long-term functional impairment, high rehabilitation demands, and the need for continuous clinical supervision. Conventional assessment and rehabilitation approaches remain limited by episodic evaluation, subjective observation, and restricted accessibility outside clinical environments. Advances in smart wearable technologies, combined with the intelligent analytics and Internet of Medical Things (IoMT), enable continuous and objective monitoring of biomechanical and physiological signals. This chapter focused on the purpose of smart wearable systems in real-time monitoring and rehabilitation support for MSD within connected medical ecosystems. This chapter discusses wearable sensor architectures, including inertial, electromyographic, and pressure-based systems, integrated through IoMT frameworks for continuous data transmission. Intelligent analytics, including adaptive and generative models, enable movement pattern interpretation, functional assessment, and personalized rehabilitation guidance. Clinical and remote rehabilitation applications demonstrate improved tracking of mobility, posture, muscle activity, and therapy adherence. System-level challenges related to interoperability, data security, accuracy, and user compliance are also examined. Smart wearable technologies integrated with IoMT and intelligent analytics provide a scalable and effective paradigm for continuous musculoskeletal health monitoring and rehabilitation, supporting personalized care, remote supervision, and data-driven clinical decision-making.

Keywords Smart wearables – musculoskeletal disorders – Internet of Medical Things – real-time monitoring – rehabilitation systems – intelligent analytics

15.1 Introduction

Musculoskeletal disorders (MSDs) are a permanent body injury or pains that touch on bones, joints, spinal discs, muscles, ligaments, and nerves. Functional disability and immobility are common symptoms of MSD, which are accompanied by muscular pain, numbness, tingling, and stiffness of joints. The diseases are a significant cost in terms of higher expenditure on healthcare, decreased productivity, and employee involvement [1]. MSDs are severe and common health problems that

 | HTTPS://DOI.ORG/10.1515/9783112237083-015

plague vast majority of individuals regardless of their age, gender, and socioeconomic considerations. Self-reported musculoskeletal discomfort in adolescents predisposes adolescents to chronic musculoskeletal morbidity in adulthood [2].

MSDs are a number of illnesses, such as rheumatoid arthritis (RA), osteoarthritis (OA), osteopenia, osteoporosis, fractures, and sarcopenia. These conditions are treated in multidisciplinary approach, which includes a large number of medical personnel and an enormous number of interventions, starting with pharmacological treatment and finishing with invasive treatment and noninvasive rehabilitation measures [3]. Musculoskeletal rehabilitation is an important factor that reduces occurrence of disability and promotes functional recovery. New developments have introduced novel rehabilitation methods that promote the effectiveness of the treatment with the least invasiveness [4].

The use of telemedicine to enhance access to chronic musculoskeletal pain rehabilitation services has been adopted. Effective and adaptable patient monitoring and treatment administration are made possible by telerehabilitation. The patients would be able to receive rehabilitation services faster and without the need to spend many hours waiting in hospitals and be provided with tailored care as per their unique needs. Recent developments of smart wearable devices and the integration of Internet of Medical Things (IoMT) have allowed the transformation of the paradigm to continuous and musculoskeletal surveillance of health. Wearable systems equipped with miniaturized and low-power sensors can continually gather biomechanical and physiological information while performing everyday activities [5, 6].

Physical therapy is non-pharmacological management of MSD widely applied, and it encompasses numerous frontline therapies addressing chronic pain as suggested by clinical practice guidelines [7]. Development of wearable sensors (WSs) specifically used inertial measurement units (IMUs) to aid the capture of human motions in an efficient and effective manner. IMU sensors find extensive application in the commercial health monitoring wearable products [8].

Physiotherapy is among treatment processes for several MSDs and this aspect assists the patients to maintain their muscular capacity. The physiotherapy process assists in enhancing the fineness of the muscle function besides minimizing the muscle pain. Manual muscle testing is very tedious in assessing muscle function, and this compromises the effectiveness of the physiotherapy process. MSDs are closely associated with occupational disorders in modern workplaces [9, 10].

15.1.1 Problem Statement

MSD requires prolonged rehabilitation and continuous functional assessment, yet existing clinical practices rely largely on periodic, clinic-based evaluations that are subjective, resource-intensive, and poorly suited for long-term monitoring.

The absence of objective and real-time data constrains the timely identification of functional deterioration, the individualization of treatment and evaluation of the adherence of patients beyond the supervised environment. Despite the potential solutions of WSs and integrated healthcare systems, there are still requirements to combine dissimilar sensor data, provide valid analytics, and provide safe and scalable on-distance monitoring. These gaps need to be addressed to ensure that rehabilitation remains continuous and data-driven to improve the clinical outcomes of persons with MSD. This chapter presents intelligent wearables and connected healthcare models, which address these problems through enhancing continuous, secure, and data-driven rehabilitation of musculoskeletal conditions.

15.2 Overview of Musculoskeletal Disorders (MSDs)

This chapter gives an introduction to MSDs. MSDs consist of a group of issues of the tendons, muscles, joints, bones,, and related connective tissues that result in pain, loss of motion, and functional loss. The disorders are the major causes of disability in the global society and pose a great impact on the quality of life, healthcare systems, and work productivity. MSDs are caused by acute traumas, repeated load, and improper posture, degeneration of the body in old age, and chronic inflammatory activity, and in many cases, they need long-term treatment and rehabilitation.

The most prevalent forms of MSDs are low back pain, neck and shoulder disorders, OA, RA, tendinopathies, carpal tunnel syndrome, and sports-related injuries. At the workplace and in day-to-day life settings, the occupations and lifestyles of the individuals are marked by extended stasis, repetitive activities, heavy physical tasks and poor ergonomic status leading to the development and accumulation of these conditions. The increasing age population also adds to the burden of MSDs globally because degenerative alterations in musculoskeletal structures expose the body to chronic pains and mobility restrictions.

The clinical presentation of MSDs can be diverse, and it has mild pains up to severe disability, in most cases, with muscle weakness, joint stiffness, movement abnormalities, and impaired functioning. The management is usually effective when it is based on clinical assessment, physical therapy, rehabilitation exercises, and long-term monitoring. Since many MSDs are chronic and progressive, ongoing evaluation and individualized treatment is crucial, which necessitates high-technology solutions that would allow objective real-time monitoring and provide long-term rehabilitation assistance [11].

15.3 Technical Foundations of Smart Wearable Systems for Real-Time Monitoring and Rehabilitation of Musculoskeletal Disorders

The technical basis of smart wearable systems is the key reason that can be used to determine the effectiveness of these systems in the process of real-time monitoring and rehabilitation of MSD. These are highly developed wearable sensing platforms, trusted communication protocols, and powerful IoMT frameworks, which together can allow uninterrupted data gathering, transfer, and smart analysis in interdependent healthcare systems. WS technologies, IoMT architecture, and intelligent clinical applications for MSD assessment and rehabilitation are illustrated in Figure 15.1.

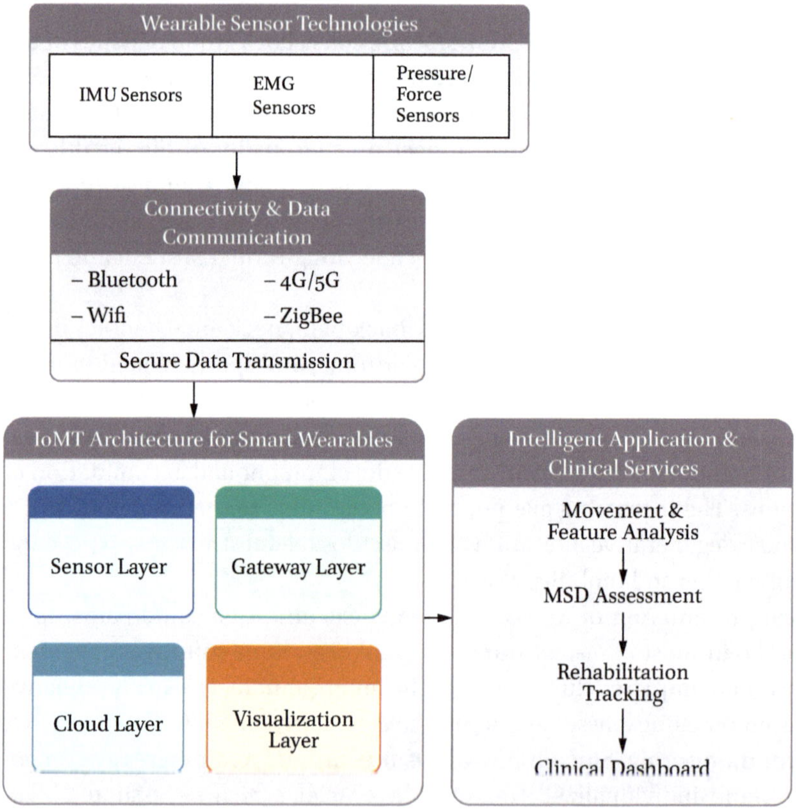

FIGURE 15.1 Architecture of smart wearable-based IoMT system for musculoskeletal disorder monitoring and rehabilitation. (Source: Own elaboration)

15.3.1 Wearable Sensor Technologies for Musculoskeletal Monitoring

WSs form the core of smart wearable systems by capturing biomechanical and physiological signals relevant to musculoskeletal health. The most widely adopted sensor modalities include IMUs, electromyography (EMG), and pressure-based sensors, each contributing complementary information for comprehensive functional assessment. IMUs are essential to satellite technology because they allow for accurate assessment of object orientation and movement. EMG is a diagnostic test that evaluates the health of muscles and the motor neurons controlling them by recording their electrical activity. Pressure-based sensors are devices that measure the force exerted by liquids or gases (pressure) and convert it into a proportional electrical signal for monitoring, recording, or control, working by detecting the deformation of a sensitive element like a diaphragm and translating it via technologies like piezoresistive, capacitive, or piezoelectric effects. These sensors are crucial in diverse fields, from automotive systems to medical devices.

IMUs, which are complexes of magnetometers, accelerators, and gyroscopes, find broad application in the measurement of joint kinematics, limb orientation, gait patterns, and posture dynamics. IMUs facilitate constant tracking of movement quality, activity levels, and range of motion in the clinical and free-living setting, which makes them especially useful in assessing the MSD in the long term and tracking rehabilitation progress [12].

EMG sensors record the signals depicting muscle activation, which allows to obtain an idea of neuromuscular activity, muscle exhaustion, and coordination patterns [13]. Wearable surface EMG systems are becoming more commonly incorporated into rehabilitation systems to measure abnormal muscular recruitment, train and measure therapy progress, and provide biofeedback-based interventions such as low back pain and poststroke musculoskeletal deficits:

$$x[m] = \sum_{i=1}^{k} \mathrm{IMF}_i[m] + z_k[m] \tag{15.1}$$

where $x[m]$ represents the original signal of discrete time and sample index, k is the total number of intrinsic mode functions (IMFs) that are extracted in the signal, $\mathrm{IMF}_i[m]$ represents the i th IMF that is used to capture oscillatory components at a variety of frequency scales, m is the time or sample index, and $z_k[m]$ is the remaining signal (residue or trend) that has not been extracted. The IMFs represent oscillatory components of x.

Load distribution, weight-bearing symmetry, and contact forces are measured using pressure sensors, such as force-sensitive resistors and plantar pressure insoles. The sensors are particularly useful in gait processing, posture stability, and post-lower limb trauma or joint replacement surgeries rehabilitation [14].

15.3.2 Connectivity and Data Communication Technologies

High-quality and energy-saving connectivity is critical to facilitate real-time data transfer of wearable gadgets to local or remote processing units. The wearable systems are usually based on the short-range wireless communication protocols that are optimized in terms of low power consumption and constant operation [15].

Wearable healthcare applications that rely on Bluetooth Low Energy (BLE) technology are widely used because of its low power consumption, adequate data throughput, and compatibility with a wide range of smartphones and gateway devices. BLE allows a smooth synchronization of sensor data to provide real-time feedback and analytics to a cloud.

Where much higher bandwidth is needed, e.g., in multi-sensor systems or in constant streaming of high-resolution signals, Wi-Fi connectivity is used at the cost of increased power consumption. The currently developing 5G technologies provide the capability of low-latency deployment of large numbers of wearable systems in remote rehabilitation and monitoring of the population.

15.3.3 IoMT Architecture for Smart Wearable Systems

The IoMT architecture offers an organized model of incorporating the wearable devices into the connected healthcare systems allowing scalable, secure, and interoperable solutions to musculoskeletal monitoring.

Sensor Layer: This layer includes miniaturized wearable or implantable sensors, which receive patient-specific biomechanical and physiological information. Low-power communication devices such as MedRadio spectrum, Bluetooth, and Wi-Fi are used to transmit the collected signals wirelessly to the next system layer to provide the ability to monitor them, in real time, continuously, and reliably.

Gateway Layer: Poor computation and storage ability of IoMT-enabled WSs, raw data are sent to the gateway layer without massive processing. The layer typically comprises the smartphone of the patient or a dedicated access point with enhanced processing and storage capabilities. The gateway devices execute initial functions such as data validation, short-term buffering, and preprocessing with the assistance of artificial intelligence (AI). The refined sensor data are then sent safely to cloud-based system through the Internet to perform advanced analytics, long term storage and clinical integration.

Cloud Layer: This is also charged with the responsibility of storing gateway layer data to be safely stored, in large scale and under control. The sophisticated use of data processing and smart analytics is used to detect the changes in the musculoskeletal health of the patient and to provide useful information to clinicians and patients. The layer also includes key management services that are in charge of creation and management of unique identifiers and cryptographic keys to system components. Also, the cloud layer allows remote configuration of WSs, access control, and management, to enhance scalable and secure rehabilitation monitoring.

Visualization Layer: Interpreted data and rehabilitation data are visualized to clinicians and patients using user-friendly interfaces, allowing for the continuous monitoring of musculoskeletal health continuously. Depending on the functional status and progress of the patient, clinicians are in a position to prescribe and change intervention strategies, such as rehabilitation plans, intensity of exercises, and prescriptions or changes in dosage. The layer provides customized treatment planning, distant overseeing, and knowledgeable clinical decision-making.

15.4 Generative and Adaptive Models for Intelligent Musculoskeletal Monitoring

The implementation of AI in the framework of smart wearable and IoMT systems has contributed to the power of MSD monitoring and rehabilitation systems to a substantial degree [16]. The analytics based on AI process raw biomechanical and physiological signals into clinically meaningful insights, which can be used to assess a person in real time, provide rehabilitation, and make decisions independently. Specifically, adaptive learning model and generative AI solutions have a central role in developing intelligent analysis and rehabilitation guidance. Figure 15.2 represents adaptive and generative models.

15.4.1 Adaptive Learning-Based Treatment Planning

The model allows to conduct individual musculoskeletal rehabilitation, examining personal data of the patient through the use of smart wearables, such as movement patterns, muscle activity, and functional indicators. The system suggests and constantly revises the best treatment strategies in accordance with the past performance and the current condition of the patient. The data-driven and adaptive approach improves the effectiveness of therapy, allows remote monitoring, and allows patient-centered and personal rehabilitation [17].

15.4.2 Dynamic Treatment Adjustments

Smart wearable devices continuously record the responses of patients to musculoskeletal rehabilitation, and real-time biomechanical, physiological, and functional data are recorded. These measures include movement quality, fatigue, and compliance, enabling the dynamic adaptation of exercise intensity, duration, and progression. The real-time adaptation provides optimized and personalized therapy,

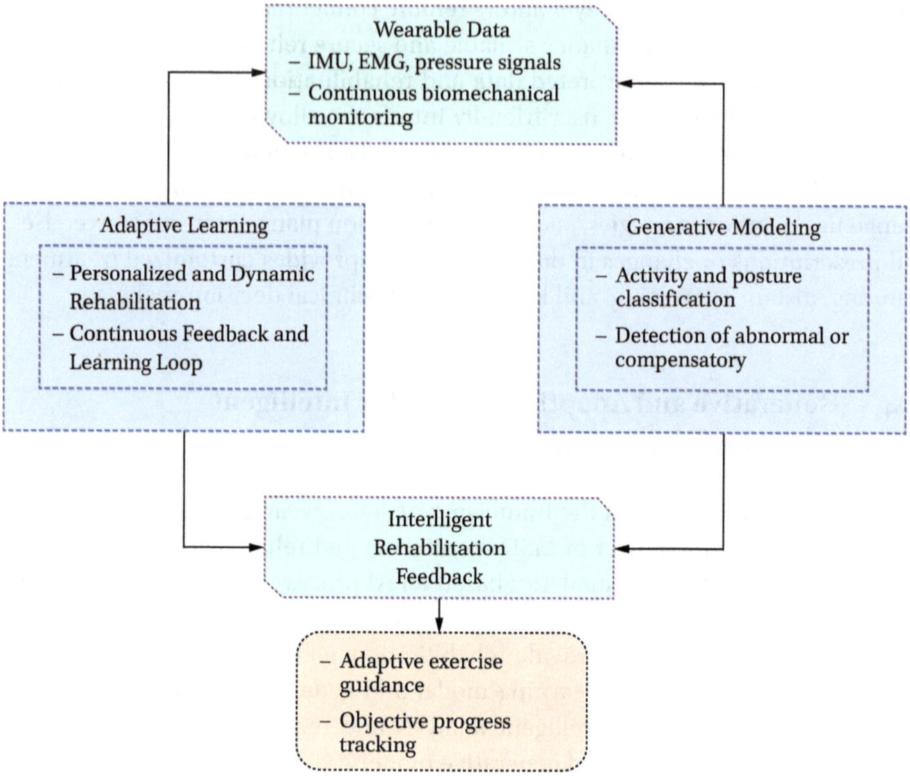

FIGURE 15.2 Generative and adaptive intelligence for musculoskeletal health monitoring. (Source: Own elaboration)

which improves the recovery and makes remote supervision in connected healthcare eco-systems possible.

15.4.3 Multimodal Data Integration

The system incorporates real-time information of smart wearables, clinical evaluation, and medical imaging to track musculoskeletal capability. The advantage of this holistic approach is the ability to create an adaptive and personalized rehabilitation plan to maximize the benefits of therapy and facilitate ongoing and patient-oriented care.

15.4.4 Continuous Feedback Loop for System Refinement

The rehabilitation systems with smart wearable would use a continuous feedback loop to promote efficacy of treatment. Monitoring of patient responses in real time will enable such deviations to be intervened on time. The outcome data collected is processed into retraining and refining adaptive models to ensure that the diagnostic and treatment recommendations are constantly improved and adjusted to the changing needs of patients.

15.4.5 Learning Loop

In musculoskeletal rehabilitation, predictive models are updated with feedback obtained using smart wearables. This lifelong learning upholds system intelligence, which facilitates adaptive and personalized treatment and enhanced capacity of the system to handle new situations:

$$\hat{x} = \text{softmax}\left(W^{(n)} \star f\left(W^{(m-1)} \star f\left(\ldots W^{(1)} \star x + b^{(1)} \ldots\right) + b^{(n-1)}\right) + b^{(n)}\right) \tag{15.2}$$

Equation (15.2) represents a multi-layer neural network where input features $x$from wearables are transformed through weighted layers $W^{(i)}$ with biases $b^{(i)}$ and an activation function$f$, producing class probabilities $\hat{x}$ via softmax.

15.4.6 Generative Models for Movement Synthesis and Predictive Analysis

Generative AI models are beyond pattern recognition and learn the underlying distribution of musculoskeletal movement information. Variational autoencoders, the generative adversarial networks, and diffusion-based models are techniques that allow one to generate natural motions and muscle activity patterns. The capabilities facilitate the process of predictive modeling of movement outcomes and rehabilitation progress:

$$Y \sim M(\mu(x), \sigma(x)^2 \tag{15.3}$$

where Y denotes the output random variable (e.g., a biomechanical or clinical outcome), $\mathcal{M}$ represents the assumed probabilistic model or distribution, x denotes the input feature vector (such as WS measurements), $\mu(x)$ is the conditional mean function, and $\sigma^2(x)$ is the conditional variance function that captures the uncertainty or variability of Y as a function of x.

The generative models can be used in the rehabilitation setting to propose optimal movement patterns and compare them to the motions performed by the patient, as recorded by WSs. The error of generated patterns of reference and actual performance gives objective measures of functional impairment and recovery. Moreover, generative models allow to predict the future state of affairs and estimate the possibility of an injury recurrence or slow healing following the biomechanical trends [18].

15.4.7 Intelligent Feedback and Virtual Rehabilitation Assistance

Intelligent feedback processes in smart wearables would be possible due to the fusion of adaptive and generative models. Real-time inference enables the systems to provide context-sensitive feedback using haptic, visual, or mobile health application. Virtual coaching systems use AI-based information to assist patients in performing corrective exercises, adjust their posture, and use movement optimization techniques.

AI-based virtual rehabilitation aides support patients round-the-clock at home and remotely, where it is not possible to have clinical sessions. These systems will automatically identify improper exercise performance, suggest change, and adjust rehabilitation programs to patient improvement or exhaustion [19].

15.4.8 Integration within IoMT Ecosystems

AI models are deployed across edge, fog, and cloud layers within IoMT architectures to balance latency, energy efficiency, and computational complexity. Lightweight adaptive inference is often performed at the wearable or edge device level for immediate feedback, while computationally intensive generative modeling and longitudinal analysis are executed in cloud environments. Secure data exchange and interoperability standards ensure seamless integration of AI analytics with electronic medical data and clinical decision-support systems [20].

15.5 Clinical Applications

Smart wearable systems integrated with the IoMT and intelligent analytics have enabled a wide range of clinical applications for real-time monitoring, rehabilitation, and patient engagement in MSD. Clinical application of smart wearable systems is shown in Figure 15.3.

15.5.1 Continuous Musculoskeletal Monitoring

WSs enable longitudinal monitoring of biomechanical and physiological parameters associated with MSDs. EMG, IMUs, force, and pressure sensors capture movement kinematics, muscle activation patterns, joint loading, and posture during daily activities.

In clinical practice, such continuous monitoring supports mobility and gait assessment in conditions such as OA, low back pain, poststroke musculoskeletal impairment, postural analysis for spinal disorders, workplace ergonomics, rehabil-

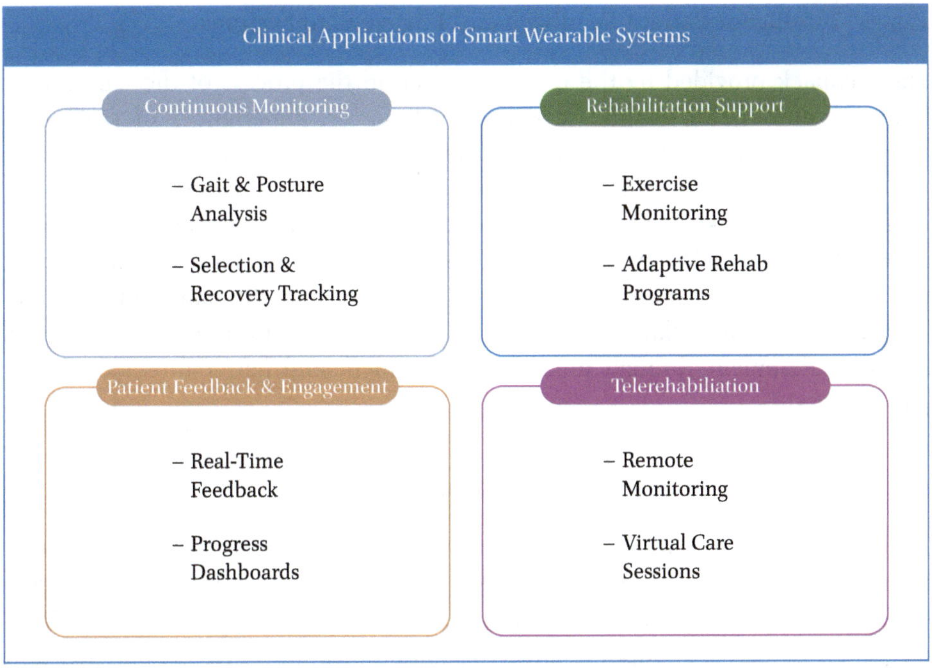

FIGURE 15.3 Clinical applications of smart wearable systems in musculoskeletal care. (Source: Own elaboration)

itation adherence, and muscle fatigue and activation tracking in neuromuscular and overuse injuries.

Data transmitted through IoMT platforms enable clinicians to classify deviations from normal movement patterns, detect early signs of deterioration, and assess recovery trends objectively over time, reducing reliance on episodic clinical evaluations [21].

15.5.2 Wearable-Assisted Rehabilitation Support

The wearables with sensors assist in both clinic-based and home-based rehabilitation programs that enable monitoring of exercises in nonconventional healthcare settings. The main areas of rehabilitation use are monitoring of exercise execution, correct joint angles, range of motion, and muscle activity of therapeutic exercises. The adaptive rehabilitation protocols are ML models that tailor the exercise intensity, repetition, and progression depending on the patient performance and recovery measures.Wearables are particularly valuable during rehabilitation in the postoperative period, especially following orthopedic surgeries such as knee replacement or rotator cuff repair, where they can objectively track the recovery process [22].

15.5.3 Intelligent Patient Feedback and Engagement Systems

The feedback provided by the users is crucial in the process of the continuous improvement of the healthcare delivery platforms. Experiences of real-world users crucially depict the real aspects of usage and situational circumstances which guide user experiences. Patients and healthcare providers can exchange useful, and even essential, insights into system usability, effectiveness and reliability. This feedback can be used to determine which parts are working, which make the system easier and more efficient, and which are difficult to understand, contain errors, are lacking, or not of much use, thus informing the process of systems refinement through iterative development and better interaction with patients [23].

15.5.4 The Clinical Use of Telerehabilitation

The concept of telerehabilitation in clinical practice of musculoskeletal conditions is focused on safety due to exercise instructions, real-time warnings, and access due to the ability to provide rehabilitation at home and community settings. Digital platforms serve to support effective communication between patients and the clinicians and allow timely feedback and adherence monitoring. WSs allow the objective analysis of movement, posture, and functional performance, and assist with individual and sustained musculoskeletal care through remote assessment [24].

15.6 Challenges

Due to the presence of smart wearable technologies and IoMT-enabled rehabilitation systems, several aspects will require fixing to be able to rely on their application in large-scale and reliable management regarding MSD management. Meanwhile, new areas of research show a lot of potential evidence-based and personalized musculoskeletal care.

Body posture and movement: Sensors deployed in on-body medical devices, as well as auxiliary sensors are usually deployed in groups. The patients are very mobile and thus the movements of the patients are dynamic and inconsistent, which causes sensor orientation, signal quality and connectivity to vary. This means that transmission of data relating to the monitoring of the posture and the body movements needs to be optimized dynamically by considering the effects of movements on the signal quality to facilitate its reliable and correct monitoring [25]. The IoMT-based systems are usually characterized by heterogeneous devices and sensors designed by various manufacturers that use different hardware configurations and communication protocols and system architectures.

15.6.1 Data Security and Privacy

Information security is an important issue in IoMT healthcare given that patient health data is sensitive. Strong privacy-protective features such as protection of authentication, encryption, and access control are necessary at all the layers of the IoMT. Spaces of secure communication help to secure sensitive information that is being transferred among wireless systems without unauthorized access, data breaches, and adversarial threats in the e-healthcare systems. A lot of medical institutions still use old medical equipment that does not have embedded security capabilities, and the process of updating or refitting the equipment to have current security-related applications is usually intricate, expensive, and technical in nature [26].

15.7 Conclusion

This chapter has highlighted the significant potential of smart wearable technologies integrated with the IoMT for real-time monitoring and rehabilitation of MSD. WSs such as IMUs, EMGs, and pressure sensors enable continuous, objective assessment of movement, posture, muscle activity, and joint loading in both clinical and real-world environments. These capabilities overcome the limitations of conventional musculoskeletal care, which often relies on episodic, subjective, and clinic-centered evaluation. The incorporation of intelligent analytics, particularly adaptive and generative models, enhances the clinical relevance of wearable data. Adaptive learning models support personalized assessment and rehabilitation by accounting for individual variability and recovery dynamics, while generative models enable movement synthesis, predictive analysis, and data augmentation for robust decision support. Together, these approaches facilitate tailored rehabilitation programs and proactive management of musculoskeletal conditions. Clinical applications including continuous monitoring, wearable-assisted rehabilitation, intelligent feedback systems, and telerehabilitation demonstrate improved therapy adherence, patient engagement, and functional outcomes. Despite difficulties associated with data security, interoperability, and user compliance, smart wearable systems integrated with IoMT and AI offer a scalable, patient-centered framework for personalized rehabilitation and continuous musculoskeletal health management.

15.7.1 Future Directions

Future developments should focus on ultra-low-power, multimodal WSs with improved comfort and long-term reliability. Advanced federated and explainable AI models can enhance personalization while preserving data privacy. Integration with digital twins and predictive analytics enables proactive intervention and injury prevention. Standardized IoMT frameworks, improved interoperability, and robust cybersecurity mechanisms will support large-scale clinical adoption. Long-term clinical validation and user-centered design will be essential to translate smart wearable systems into routine musculoskeletal care.

References

[1] Pacheco MP, et al. Prevalence of postural changes and musculoskeletal disorders in young adults. International Journal of Environmental Research and Public Health. 2023;20(24): 7191. https://doi.org/10.3390/ijerph20247191.

[2] Huang H, et al. LncRNAs as a new regulator of chronic musculoskeletal disorder. Cell Proliferation. 2021;54(10): e13113. https://doi.org/10.1111/cpr.13113.

[3] De Sire A, Invernizzi M. Musculoskeletal rehabilitation: State-of-the-art. Applied Sciences. 2021;11(14):6243. https://doi.org/10.3390/app11146243.

[4] Perna A, Proietti L. Editorial on: Musculoskeletal rehabilitation: Current challenges and new perspectives. Journal of Clinical Medicine. 2023;12(12):3981. https://doi.org/10.3390/jcm12123981.

[5] Rani P, et al. Simulation of the lightweight blockchain technique based on privacy and security for healthcare data for the cloud system. International Journal of E-Health and Medical Communications (IJEHMC). 2022;13(4): 1–15. https://doi.org/10.4018/IJEHMC.309436.

[6] Awasthi S, et al eds. AI and IoT-based Intelligent Health Care & Sanitation. Bentham Science Publishers. vol. 1. 2023. https://doi.org/10.2174/97898151365311230101.

[7] Bhavsar NA, et al. Prevalence and predictors of no-shows to physical therapy for musculoskeletal conditions. PLoS One. 2021;16(5): e0251336. https://doi.org/10.1371/journal.pone.0251336.

[8] Paloschi D, et al. Validation and assessment of a posture measurement system with magneto-inertial measurement units. Sensors. 2021;21(19): 6610. https://doi.org/10.3390/s21196610.

[9] Kim Y-M, Cho S-I. Work-related musculoskeletal disorders and digitalization: Past adoption, current utilization, and future concerns. BMC Public Health. 2025;25(1):2336. https://doi.org/10.1186/s12889-025-23466-w.

[10] Lal S, et al. Lipid metabolism in age-related musculoskeletal disorders: Insights into sarcopenia and osteoporosis. BMC Biology. 2025;23(1): 265. https://doi.org/10.1186/s12915-025-02383-9.

[11] Marzocchi I, et al. Stability and fluctuations in musculoskeletal disorder symptoms: The roles of chronic and acute job demands and resources. Safety Science. 2026;-196:107100. https://doi.org/10.1016/j.ssci.2025.107100.

[12] Zhao J, Obonyo E, Bilén SG. Wearable inertial measurement unit sensing system for musculoskeletal disorders prevention in construction. Sensors. 2021;21(4):1324. https://doi.org/10.3390/s21041324.

[13] Sankaran V, et al. An Internet of Medical Things-Based Smart Electromyogram Device for Monitoring of Musculoskeletal Disorders. Engineering Proceedings. 2024;82(1): 108. https://doi.org/10.3390/ecsa-11-20351.

[14] Manickam P, et al. Artificial intelligence (AI) and internet of medical things (IoMT) assisted biomedical systems for intelligent healthcare. Biosensors. 2022;12(8): 562. https://doi.org/10.3390/bios12080562.

[15] Oztoprak O, Ryu J-C. Development of a Mobile Health Monitoring and Alert Application for Agricultural Workers. Applied System Innovation. 2025;8(5):133. https://doi.org/10.3390/asi8050133.

[16] Ghubaish A, et al. Recent advances in the internet-of-medical-things (IoMT) systems security. IEEE Internet of Things Journal. 2020;8(11): 8707–18. https://doi.org/10.1109/JIOT.2020.3045653.

[17] Maheshwari RU, et al. Intelligent HealthTech: Building an Adaptive Learning Ecosystem for Optimized Patient Care. CompSci& AI Advances. 2024;1(1): 51–63. https://doi.org/10.69626/csi.2024.0051.

[18] Dindorf C, et al. Enhancing biomechanical machine learning with limited data: Generating realistic synthetic posture data using generative artificial intelligence. Frontiers in Bioengineering and Biotechnology. 2024;12:1350135. https://doi.org/10.3389/fbioe.2024.1350135.

[19] Islam U, et al. A hybrid fog-edge computing architecture for real-time health monitoring in IoMT systems with optimized latency and threat resilience. Scientific Reports. 2025;15(1): 25655. https://doi.org/10.1038/s41598-025-09696-3.

[20] Khalid U, et al. Impact of AI-powered solutions in rehabilitation process: Recent improvements and future trends. International Journal of General Medicine. 2024;-17:943–69. https://doi.org/10.2147/IJGM.S453903.

[21] Kumar R, et al. Emerging diagnostic approaches for musculoskeletal disorders: advances in imaging, biomarkers, and clinical assessment. Diagnostics. 2025;15(13): 1648. https://doi.org/10.3390/diagnostics15131648.

[22] Dubey A, et al. Role of artificial intelligence in musculoskeletal interventions. Cancers. 2025;17(10): 1615. https://doi.org/10.3390/cancers17101615.

[23] Batra P, Dave DM. Revolutionizing healthcare platforms: The impact of AI on patient engagement and treatment efficacy. International Journal of Science and Research (IJSR). 21275 2024;13(10):613–24. https://dx.doi.org/10.21275/SR24201070211.

[24] Baroni MP, et al. The state of the art in telerehabilitation for musculoskeletal conditions. Archives of Physiotherapy. 2023;13(1): 1. https://doi.org/10.1186/s40945-022-00155-0.

[25] Bhushan B, et al. Towards a secure and sustainable internet of medical things (IoMT): Requirements, design challenges, security techniques, and future trends. Sustainability. 2023;15(7): 6177. https://doi.org/10.3390/su15076177.

[26] Pradyumna GR, et al. Empowering healthcare with IoMT: Evolution, machine learning integration, security, and interoperability challenges. IEEE Access. 2024;12:20603–23. https://doi.org/10.1109/ACCESS.2024.3362239.

CHAPTER 16

Wearable Sensors for Remote Monitoring and Management of Chronic Renal Diseases

Dimpal Khambhati, G. Sundari, Manivannan Karunakaran

Abstract Chronic renal diseases (CRDs) require continuous monitoring and long-term management to slow progression and prevent complications. Traditional clinical follow-ups provide episodic snapshots, often missing daily physiological fluctuations. Wearable sensor technologies offer real-time, remote monitoring, capturing vital parameters such as exercise, blood pressure, pulse variability, balance of fluids, and the quality of sleep. Advanced biosensors enable noninvasive detection of renal biomarkers in sweat or interstitial fluid, supporting early diagnosis and personalized treatment. Integration with mobile health platforms and cloud-based analytics facilitates continuous data transmission, trend analysis, and clinical decision support. Machine learning algorithms applied to wearable data achieve high predictive performance for CRD-related events, with accuracy up to 99.0%, F_1-score, true positive rate of 99.2%, and false negative rate as low as 0.8%, enabling timely interventions and individualized therapy. Remote monitoring enhances patient engagement, improving self-management, treatment adherence, and lifestyle modification, which reduces hospital readmissions. Despite these benefits, challenges such as sensor accuracy, data reliability, interoperability, and privacy protection remain. Overall, wearable sensor-based monitoring, combined with artificial intelligence-driven analysis, represents a patient-centered approach that can improve early detection, optimize treatment, and enhance long-term outcomes in CRD management.

Keywords Chronic renal diseases – machine learning – wearable sensor technologies – clinical adoption

16.1 Introduction

Chronic renal diseases (CRDs) affecting thousands of individuals around with significant implications for morbidity, mortality, and costs. The important thing is that early diagnosis and periodic observation would help to avoid severe complications such as end-stage renal disease and heart disease, since they often develop without any symptoms and gradually worsen the functioning of the kidneys over time [1]. Regular hospital appointments and lab tests are typical elements of traditional clinical treatment, which, as such, only provide infrequent snapshots of a

 | HTTPS://DOI.ORG/10.1515/9783112237083-016

patient's physiological state and can fail to detect some major daily changes [2]. The types of intermittent observing may lead to a decrease in the effectiveness of intervention methods and delay the detection of adverse outcomes. The use of wearable sensor technology as a disruptive innovation in healthcare in recent years has enabled noninvasive, continuous monitoring of both behavioral and physiological predictors, which in the real-world environment is possible in non-disruptive healthcare [3]. These wearable devices could maintain updates on indicators of health including blood pressure, heart rate, fluid balance, and physical activity, and give real-time data concerning the health of a person without considering clinical environments [4]. Along with assisting in personalized health insights and active involvement in self-management methods, continuous monitoring simplifies the early prevention of aberrant trends. Wearable sensors could contribute to reducing the burden on healthcare systems [5] by lowering the overall cost and increasing the number of patients to access care in underserved or remote communities. The widely recognized use of wearable technology as an essential component of modern healthcare practices to manage chronic diseases is expanding with the advances in sensor design, connectivity, and data analytics. For patients with long-term chronic renal disorders, their integration into the long-term care paradigm could be considered a promising measure to improve the quality of life, and adopting health monitoring that is more patient-centered and responsive [6]. The primary objective of this research is to develop a wearable sensor-based system for monitoring patients with chronic renal diseases (CRDs), using artificial intelligence (AI) and machine learning (ML) to support personalized treatment and clinical decision-making.

16.2 Role of Wearable Sensor Technologies

Recent developments through miniaturization of artificial hearts and ventricular assist devices could be utilized in renal replacement devices, which present new prospects of continuous renal therapies without necessarily necessitating the patient to be immobilized or the quality of life to be impaired. The wearable peritoneal dialysis devices, including Vicenza wearable artificial kidney and automated wearable artificial kidney, permit continuous or tidal dialysis recirculation with sorbent-based regeneration, reducing the fluid and glucose load, limiting the degradation of the peritoneal membrane, and lowering the risk of infection [7]. A system comprises wearable hemodialysis gadgets, which are constructed on battery-powered pumps and sorbent cartridges and are employed in succession to effectively eliminate uremic toxins and electrolytes. Integrated with biosensors, fluid status, electrolytes, and ultrafiltration could be continuously monitored, so that real-time and patient-focused care is presented. These systems address the learning objective of exploring wearable technologies to provide a way of remote monitoring, personalize the treatment due to individual differences, enhance patient self-management of CRDs, and bridge the gap between efficacy of treatment and patient control and clinical outcomes in the long term.

16.3 Advancements in Wearable Sensor Technologies for Kidney Health

To create and implement cutting-edge wearable sensor platforms that would allow for continuous, noninvasive monitoring of chronic kidney disease (CKD) biomarkers in sweat and interstitial fluid (ISF) creatinine, urea, and cystatin C. To offer accurate measurements of early diagnosis and disease treatment, the project aims at incorporating colorimetric, electrochemical, and microneedle-based sensors into wearable devices with the ability to detect in real time and quantitatively. Critical technologies encompass pumping out ISF using reverse iontophoresis, secretion of sweat at rest using iontophoresis (sweat-producing chemicals), and control of the flow of sweat using microfluidic channels to prevent the evaporation, mixing, and other sample effects. To detect trace biomarkers such as cystatin C, high-sensitivity amplifying transducers such as organic electrochemical transistors might be used. The simultaneous measurement of sweat rate and pH would be able to manage changes in biomarker concentrations caused by fluid dynamics and biofluid properties to ensure proper measurements. The wearable platform aims to reduce skin irritation, pain, and power usage in addition to providing real-time and continuous monitoring, similar to the continuous glucose monitoring systems [8]. It is alleged that the final system will facilitate an improved clinical decision-making process, patient involvement, treatment adherence, and overall quality of life, as it would allow early CKD detection, real-time biomarker monitoring, and customized treatment changes. It demonstrates how high-tech wearable technology might be utilized to take proper control of the health of the kidneys.

16.4 Patient Engagement and Self-Management

The patient activation issue is highly crucial to the successful self-management of CRDs, as it reflects their knowledge, skills, and belief in their ability to control their illness. Remote monitoring devices that have wearable sensors could improve patient activation by offering personalized feedback on the parameters of such physiological measurements as blood pressure, physical activity, sleep quality, and fluid status in real time [9]. Improved awareness of the disease, making informed decisions, and adherence to the treatment plan and lifestyle change could be enhanced due to the availability of health data in real time. The association between the increased activation and the subsequent higher self-management practices, enhanced management of symptoms along with quality of life in connection with a positive effect in healthcare. Reduction of hospital readmission rates can also be achieved by wearing a sensor to prevent late clinical deterioration diagnosis and its subsequent cause of early intervention. Therefore, wearable technology was a variant of a patient-centered strategy to improve long-term outcomes of self-management and involvement in CRD care.

Patients with CRD would be able to use mobile health solutions that combine wearable, Internet of Things, and AI to obtain more self-management. Even though the cloud-based analytics and mobile health apps allow delivering personal feedback, warnings, and treatment options, they offer real-time checking of vital signs with the help of smartwatches, health trackers, and other wearable devices [10]. The sharing of data between clinicians and patients could be facilitated by such systems to encourage medication and lifestyle changes and regular follow-ups that enhance health outcomes and quality of life. The existing evidence-based clinical research suggests that it is possible to promote disease awareness, interaction with patients, and self-management in different chronic diseases with the help of smart digital applications. These CRD care technologies provide patient-centered care in early detection, early intervention, and reduced hospitalizations.

16.5 Artificial Intelligence for Personalized CKD Management

In CKD management, the early and correct predictions of renal and cardiovascular risks are essential. The use of AI methods could be applied to supplement conventional regulatory approaches, like estimated glomerular filtration rate and proteinuria, and identify trends that humans cannot observe by viewing large and complex auras. These wearable sensor data, when coupled with the AI algorithms, allow one to offer real-time monitoring, early detection, and customized therapy of the CKD and cardiovascular risks linked to it. The application of clinical, genetic, and imaging data, along with continuous physiological data, such as blood pressure, heart rate variability, hydration condition, activity, and the quality of sleep, boosts risk assessment individually. Deep learning techniques can be used to refine predictions over time and be applied to timely offers, more effective treatments, and better results in patients [11].

16.5.1 Random Forest (RF)

This chapter proposes the continuous monitoring and management of vital signs and renal biomarkers using wearable sensors for the early detection and prediction of chronic renal diseases (CRDs). Complex multidimensional data and identification of the major features were analyzed using ML, especially random forest (RF) algorithm. The ensemble of decision trees offered by RF offers high predictive accuracy, missing data, and overfitting reduction. The methodology encourages earlier diagnosis, individualized treatment, and better patient interaction, minimizing hospital admission readmission [12]. Traditional models (support vector machines (SVM), linear discriminant analysis (LDA), quadratic discriminant analysis (QDA), Naïve Bayes, ridge, and decision tree), ensemble models (voting and stacking), and advanced models (logistic regression (LR), K-nearest neighbors (KNNs), AdaBoost, decision tree classifier (DTC), RF, multilayer perceptron (MLP), MKR stacking, and MKR voting), and the results presented below indicate that ensemble and advanced methods are superior to classical ones:

$$\mathrm{D}\,(p) = \max_p F_O \sum\nolimits_i^k (c_i(W) = R) \tag{16.1}$$

$$q\;(r) = 1 - j_1^2 - j_0^2 \tag{16.2}$$

$$j_k = \frac{gd}{g} \tag{16.3}$$

In eqs. (16.1)–(16.3)$D(p)$was a decision or performance function that captures the maximum expected output as indicated by the intended outcome function and depends on the probability p. F_O is the phrase $\sum_{i-1}^{k}(c_i(W) = R)$ sums over k components, each of which is c_i of the weight vector W satisfies the reference requirement R, hence expressing the total score or contribution of these elements. A supplementary metric, like error or uncertainty, is expressed as a function of r by the function $q(r)$, which is calculated using coefficients j_1^2 and j_0^2, which measure certain system contributions or normalized probability. The normalized contribution of the k-th element is denoted by j_k. It is calculated as the ratio of the partial measure gd to the overall measure g.

$$\delta n\;(s) = r(h) - j_n m(g_s) - j_w n(g_x) \tag{16.4}$$

$$Kl\,(\theta) = \sum\nolimits_s \sum\nolimits_h \delta z_\theta(a, A) \tag{16.5}$$

In eqs. (16.4) and (16.5), $\delta n\;(s))$ is the difference for the nth element at state s. It is computed as the difference between the weighted contributions of two components and $r(h)$, a response or measured value based on h, $(j_n\, m\,(g_s))$ signifies a particular measure's normalized contribution gs with weight j_n and j-$j_w\, n\,(gx)$ signifies another measure's weighted contribution gx with weight $jw.Kl(\theta)$ is an aggregated metric for parameter θ, which is generated by adding the values of $r(h)$ over states s and $h.(a, A)$, which records variations or discrepancies related to parameters a and A for the specified θ. In this case, s and h denote state or condition indices, whereas n and w denote component indices. j_n and j_w. represent the normalized weights corresponding to the respective measures.

16.6 Integration with Mobile Health and Cloud Platforms

Two such digital therapies are mobile health platforms and wearable sensors, which could play a major role in helping patients with CKD make the desired lifestyle changes, improve their health literacy, and self-management. Wearable computers can relay real-time information to patients and medical practitioners that could continuously monitor physiological variables such as cardiac rate, blood pressure, water intake, and activity [13]. It enables one to treat and care about people in a timely manner. The combination of blood pressure measurement and physical exercise monitoring on smartphones has shown potential and enhanced patient knowledge about CKD. Although the time constraint might be the determining factor of their effectiveness, digital health interventions that include interpersonal support, informational recordings, and physical activity are all accessible and advantageous for patients to participate. Remote monitoring and telemedicine contribute to better CKD management since they enable the interdisciplinary team care to address comorbidities, including hypertension, and offer less expensive care, especially to individuals in remote areas. The purpose of the facilitation of constant and distance monitoring and personalized management of CKD that would improve patient outcomes and life quality appears to be aligned with the addition of wearable sensors to these digital platforms.

16.7 Results and Conclusion

Wearable sensors allow for the use of models generated by ML to support the implementation of CRD management and continuous surveillance. In addition to delivering prediction data on tailored modifications of treatment, the desire to increase previous awareness of physiological alterations such as fluid congestion, hypertension, and heart strain is also present. The comparison of these models was carried out, i.e., the LR, DTC, ridge, RF, SVMs, Naive Bayes, KNNs, AdaBoost, MLP, and the ensemble learning models (voting and stacking). The models were made to compete with each other in terms of standard measures, accuracy, precision, F_1-score, sensitivity, specificity, false positive rate (FPR), false negative rate (FNR), error rate (ER), true positive rate (TPR), and time of execution.

16.7.1 AI-Driven Monitoring of Chronic Renal Disease

Wearable sensors have not been fully utilized by the management of CRD yet because of the strength of the current AI techniques in medical applications, due to the continuous and real-time information that is offered by the wearable sensors. The existing models, such as LR, SVM, RF, DTC, LDA, ridge, KNN, AdaBoost, MLP, naive Bayes, ensemble models, and QDA, such as voting and MKR stacking/voting, do not provide an all-encompassing and real-time decision-support system with vital signs and activity patterns and renal biomarkers [14, 15]. To facilitate timely treatments and improve patient-centered care of CRD, this section describes the proposed methodology based on the integration of wearable sensor data with the abovementioned ML models to provide timely treatments and early detection.

16.7.2 Quantitative Performance Metrics and Model Comparison in CRD Monitoring

- **Accuracy:** The model's overall reliability in predicting CRD-related events, such as high blood pressure, excessive fluid retention, and cardiovascular stress, is measured by the percentage of accurate predictions.
- **Recall:** The proportion of the number of times CRD was observed to be true to the model.
- **Sensitivity:** The capacity of the model to identify the real positive CRD incidents, which meet the requirements of identifying the important situations of the patient in time.
- **Precision:** It is the values of the correct positive predictions that minimizes the false alarms and maximizes the utilization of medical resources to act in time.
- **F_1-Score**: The model's performance might be accurately estimated using recall and accuracy in the harmonic mean, especially because the distribution of CRD events is skewed.
- **TPR:** It is the true positive CRD instances that the computational model effectively recognized.
- **FNR:** The number of instances that comprise real CRD incidents that architecture can appropriately capture.
- **FPR:** It describes the amounts of the real affirmative CRD that the ML manages to recognize.
- **TNR:** A percentage of non-CRD events, which were falsely modeled as CRD events.
- **ER:** The percentage of incorrect predictions, and this value is obtained to determine the risk of a misclassification of an entire model.

- **Execution Time:** It is the time that the model requires to calculate, and to produce predictions; it can be used in real-time wearable sensor monitoring.
- **RMSE:** The extent of model prediction of real patient outcomes is quantified, implying the mean value of the error in continuous physiological outcomes prediction.
- R^2 **Score:** It is utilized to illustrate the overall fit and power of the prediction, which illustrates the percentage of the data variation (e.g., blood pressure changes and fluid balance) clarified by the model. Standard performance metrics and equations are utilized for assessing the classification and regression techniques' performance.

Comparison of Model Performance: RF model [15] and MKR stacking [15] are the most successful, having a prediction accuracy of 99.00%. MKR voting [15] was rated at 98.00%, in which AdaBoost [15] was rated slightly lower at 98.75%. The classical models that had the accuracy of 87.89–90.53% consisted of SVM [14], QDA [14], and stacking ensemble [14]. The lowest score of 84.74% was on decision tree [14]. It demonstrates that the more complex and advanced ensemble ML models are more effective than straightforward conventional classifiers for predicting CRD, as shown in Table 16.1 and Figure 16.1.

Precision and F_1-Score Comparison: AdaBoost [15] scored the highest with 100%, then 99.20% and 99.17% with MKR stacking [15] and MLP, respectively. The similar tendencies could be observed in F_1-scores as well because RF [15] and MKR stacking [15] are reported to have an identical score, which is 99.20% that was outstanding in terms of recollection and accuracy. Naive Bayes [14] and QDA [14] were reported to have an F_1-score of more than 91%, and decision tree [14] had the lowest F_1-score of 87.71%.

Comparison of Sensitivity and Specificity: The highest sensitivity of 94.98% was SVM [14], with the capability to improve CRD events. The decision tree [14] was the least sensitive (86.61%) and, as a result, it was unresponsive to any real cases of CRD, but QDA [14] was responsive as well, being at 94.14%. Ridge [14] was the most specific with 84.40%, meaning that this modelis more reliable to predict the occurrence and non-occurrence of the events compared to SVM [14], QDA [14], and decision tree [14], which had a lower specificity of 81.56–81.56%. In Table 16.2, the ensemble techniques and the more sophisticated techniques, such as stacking ensemble and ridge regression, are more likely to provide a more sensible sensitivity and specificity ratio in regards to CRD monitoring and are shown in Figure 16.2.

16.7.3 CRD Model Performance Comparison

The MKR stacking model [15] achieved the best performance, with a true positive rate (TPR) of 99.20%, a false negative rate (FNR) of 0.80%, and a true negative rate (TNR) of 98.67%, demonstrating excellent recognition of both CRD events and

TABLE 16.1 Accuracy, precision, F_1-score, and performance comparison of ML models for CRD monitoring [14, 15].

Model	Accuracy (%)	Precision (%)	F_1-score (%)
SVM [14]	90.00	89.72	92.28
LDA [14]	87.89	90.38	90.38
QDA [14]	89.47	89.64	91.84
Naïve Bayes [14]	89.21	90.57	91.51
Ridge [14]	89.47	90.98	91.93
Decision tree [14]	84.74	88.84	87.71
Voting ensemble [14]	89.74	89.96	91.80
Stacking ensemble [14]	90.53	90.44	92.65
LR [15]	97.25	98.38	97.79
KNN [15]	94.75	96.36	95.77
AdaBoost [15]	98.75	100	98.99
DTC [15]	97.25	96.14	97.84
RF [15]	99.00	98.81	99.20
MLP [15]	96.50	99.17	97.14
MKR stacking [15]	99.00	99.20	99.20
MKR voting [15]	98.00	99.19	98.39

non-event situations. Although RF and DTC had the highest TPR of 99.60% and FPR of 6.67%indicated that they still give a certain part of false alarms. AdaBoost [15] achieved the lowest error rate (1.25%) and the highest specificity, with a true negative rate (TNR) of 100%, indicating its suitability for real-time applications. In contrast, FNRs (92.80% and 4.80%) and better error rates (2.75 and 5.25) corresponded to the fact that the classical methods, such as LR [15] and KNN [15], slightly underperformed in the detection. The MLP and MKR voting [15] fared on average since their TPRs were more than 95%, and FPRs were extremely minimal (1.33%). But the MLP (0.50 s) and MKR stacking (4.66 s) would take more time and would imply that there would be a trade-off between the accuracy and the computational efficiency. In general, one can say that the ensemble models, including the MKR stacking, RF, and AdaBoost, are the most efficient and the most successful at predicting CRD events, as presented in Table 16.3 and Figure 16.3.

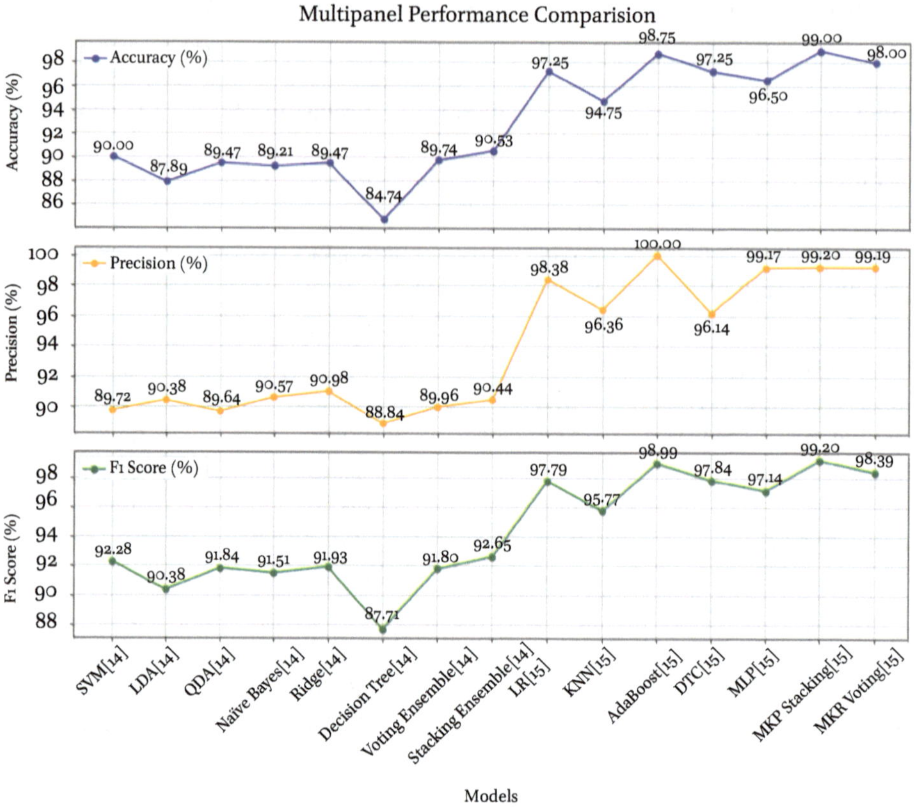

FIGURE 16.1 Comparative analysis of predictive performance metrics across ML models for accurate and reliable wearable-based CRD monitoring. (Source: Own elaboration)

TABLE 16.2 Sensitivity and specificity comparison of ML models for CRD monitoring [14].

Model	Sensitivity (%)	Specificity (%)
SVM [14]	94.98	81.56
LDA [14]	90.38	83.69
QDA [14]	94.14	81.56
Naïve Bayes [14]	92.47	83.69
Ridge [14]	92.89	84.40
Decision tree [14]	86.61	81.56
Voting ensemble [14]	93.72	82.27
Stacking ensemble [14]	94.98	82.98

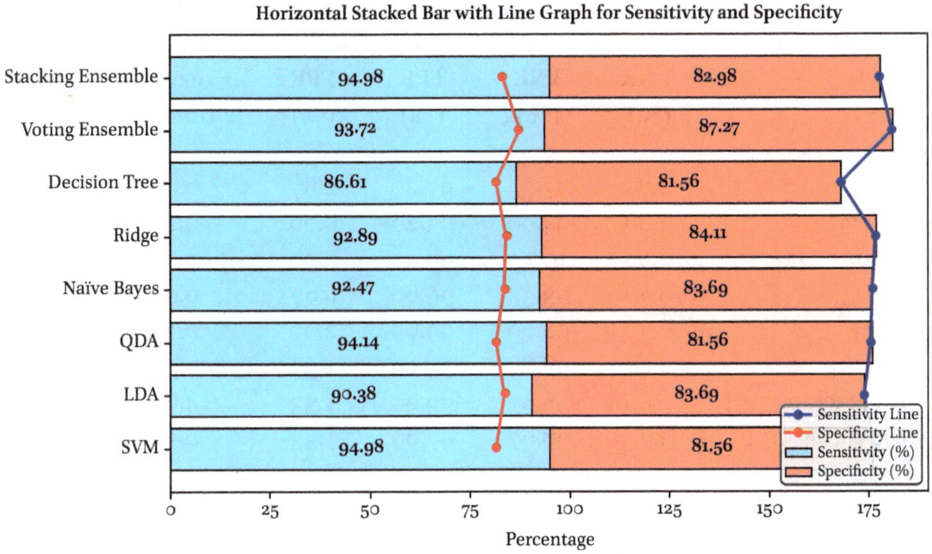

FIGURE 16.2 Sensitivity and specificity across ML models for wearable-based CRD monitoring. (Source: Own elaboration)

16.8 Future Trends and Challenges

A wearable sensor development in the CRD management is founded on the advancement of multimodal sensing, smart analytics of data, and a reduction in the size of sensors. The wearable technology that will be accessible in the nearest future will provide the opportunity to monitor the well-being of kidneys in a holistic, continuous manner since all three types of sensors will be integrated within one platform and turned into biochemical, physiological, and behavioral. The development of noninvasive biosensors is one such area of investigation because it could determine the renal distinct pointers in the form of urea, creatinine, and electrolytes against the collections of the sweat or ISF. These sensors and federated ML use computational edge technology to predict risk in real time and avoid infringement of patient privacy. The fact that the AI-support decisions in medicine systems will forecast the development of diseases, fluid imbalance, and cardiovascular diseases at the most initial level will enhance individual treatment. The introduction of the electronic health records and telemedicine systems integration and the ability to conduct timely interventions will help to improve the effectiveness of the clinical workflow. Nevertheless, there are many problems that should be addressed to accept it on a large scale. Of the technical problems, the long-term precision of the sensors, the same behaviour of the calibration, and the dependence of the working situation are essential. The information security, the privacy

TABLE 16.3 Comparison of FPR, TPR,TNR, ER, FNR, and execution time for CRD models [15].

Models	FNR (%)	TNR (%)	TPR (%)	FPR (%)	Execution time(s)	Error rate
LR [15]	2.80	97.33	97.20	2.67	0.09	2.75
KNN [15]	4.80	94.00	95.20	6.00	0.17	5.25
AdaBoost [15]	2.00	100	98.00	0.00	0.05	1.25
DTC [15]	0.40	93.33	99.60	6.67	0.08	2.75
RF [15]	0.40	98.00	99.60	2.00	0.13	1.00
MLP [15]	4.80	98.67	95.20	1.33	0.50	3.50
MKR stacking [15]	0.80	98.67	99.20	1.33	4.66	1.00
MKR voting [15]	2.40	98.67	97.60	1.33	0.59	2.00

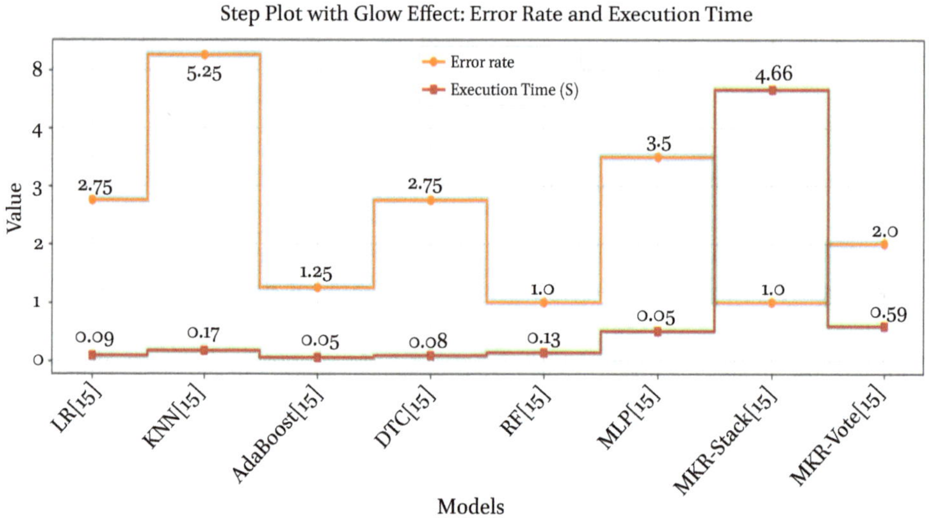

FIGURE 16.3 Comparative evaluation of ML models for wearable sensor-based CRD remote monitoring. (Source: Own elaboration)

protection, and the adherence to the regulations are at stake, specifically, the cloud-based surveillance system. The other barrier to the seamless movement of data is the impossibility of integrating heterogeneous devices and healthcare systems. Patient adherence, battery duration, and comfort of the user also influence the efficacy of wearable solutions. The technology, ethical, and clinical issues of this nature ought to be mitigated to ensure that the CRD wearable sensor-based remote monitoring is maximally utilized.

16.9 Conclusion

Mobile health platforms like AI and wearable sensors will offer an innovative method of treatment and management of CKDs in their continuous form. Such types of technologies will enable the performance of early diagnosis, treatment, and proactive intervention due to the possibility of tracking vital rates, kidney biomarkers, and predictors of a particular lifestyle in real time. The defined features of the ML models are the high accuracy of the predictions, in this case, the sophisticated ensemble algorithms, such as the MKR stacking or AdaBoost, which enhance the speed of the decisions made and minimize the number of undesired results. In addition, wearable technology leads to patient self-management, patient engagement, and adherence to therapy, which lead to enhanced improvements in health and decreased hospitalization. The prospective solution to these moral and technical problems of sensor dependability, information protection, and interoperability is the wearable sensor integration into CRD care, a possibility, and a patient-centered model of managing chronic diseases.

References

[1] Caturano, Alfredo et al. The dual burden: Exploring cardiovascular complications in chronic kidney disease. Biomolecules. 2024;14(11):1393. https://doi.org/10.1007/s10653-022-01373-y.

[2] Salama, Ramiz et al. Benefits of Internet of Things (IoT) Applications in health care-an overview. 2023 International Conference on Computational Intelligence, Communication Technology and Networking (CICTN). 2023;1(1):778–84. https://doi.org/10.1109/CICTN57981.2023.10141452.

[3] Khosravi, Arezoo et al. Mxene-based wearable contact lenses: Integrating smart technology into vision care. Nano-Micro Letters. 2026;18(1):20. https://doi.org/10.1007/s40820-025-01863-5.

[4] Pujahari, Rakesh Mohan, Rijwan Khan, and Satya Prakash Yadav. Federated autonomous deep learning for distributed healthcare system. Federated Intelligent System for Healthcare: A Practical Guide. 2025;1(1):25–59. https://doi.org/10.1002/9781394271382.ch2.

[5] Cheng, Jianqun et al. Recent progress in flexible wearable sensors for real-time health monitoring: materials, devices, and system integration. Micromachines. 2025;16(10):1124. https://doi.org/10.3390/mi16101124.

[6] Ali, Jawad, Manzoor Hussain, and Trisiani Dewi Hendrawati. A comprehensive remote monitoring system for automated diabetes risk assessment and control through smart wearables and personal health devices. Engineering Proceedings. 2025;107(1):91. https://doi.org/10.3390/engproc2025107091.

[7] Groth, Thomas et al. Wearable and implantable artificial kidney devices for end-stage kidney disease treatment: Current status and review. Artificial Organs. 2023;47 (4):649–66. https://doi.org/10.1111/aor.14396.

[8] Han, Sanggil et al. Wearable sensors for monitoring chronic kidney disease. Communications Materials. 2024;5(1):153. https://doi.org/10.1038/s43246-024-00606-0.

[9] Lightfoot, Courtney J. et al. Patient activation: The cornerstone of effective self-management in chronic kidney disease?. Kidney and Dialysis. 2022;2(1):91–105. https://doi.org/10.3390/kidneydial2010012.

[10] Liu, Yu, and Boyuan Wang. Advanced applications in chronic disease monitoring using IoT mobile sensing device data, machine learning algorithms and frame theory: A systematic review. Frontiers in Public Health. 2025;1510456:13. https://doi.org/10.3389/fpubh.2025.1510456.

[11] Simeri, Andrea et al. Artificial intelligence in chronic kidney diseases: Methodology and potential applications. International Urology and Nephrology. 2025;57(1):159–68. https://doi.org/10.1007/s11255-024-04165-8.

[12] Faysal, Jabed Al et al. XGB-RF: A hybrid machine learning approach for IoT intrusion detection. Telecom. 2022;3(1):52–69. https://doi.org/10.3390/telecom3010003.

[13] Shrestha, Sanjivani et al. Care of adults with advanced chronic kidney disease. Journal of Clinical Medicine. 2024;13(15):4378. https://doi.org/10.3390/jcm13154378.

[14] Iftikhar, Hasnain et al. Clinical application of machine learning models for early-stage chronic kidney disease detection. Diagnostics. 2025;15(20):2610. https://doi.org/10.3390/diagnostics15202610.

[15] Bijoy, Md Hasan Imam et al. A robot process automation based mobile application for early prediction of chronic kidney disease using machine learning. Discover Applied Sciences. 2025;7(6):528. https://doi.org/10.1007/s42452-025-06980-9.

CHAPTER 17

Machine Learning Algorithms for Predicting Disease Progression and Personalized Treatment in Chronic Diseases

Ashutosh Roy, Swapnil M. Parikh, Joany R. M.

Abstract Chronic diseases like diabetes, cardiovascular disorders, chronic kidney disease, and autoimmune conditions require continuous monitoring and individualized treatment due to their long-term and heterogeneous progression patterns. Conventional clinical decision-making approaches often fail to capture complex relationships among clinical, and lifestyle factors. This chapter summarizes the request of machine learning (ML) methods for predicting disease progression, supporting personalized treatment in chronic disease management. ML models use longitudinal electronic health records, laboratory results, imaging-derived indicators, and patient lifestyle data, to temporal patterns and early risk markers associated with disease worsening or recovery. Methods with random forests, *K*-nearest neighbors, logistic regression, gradient boosting, and support vector machines effectively model nonlinear interactions and high-dimensional data while retaining clinical interpretability. Feature engineering based on summaries of time-series data and clinically meaningful variables further improves predictive accuracy. ML-based frameworks enable early risk stratification, disease-stage prediction, and treatment response estimation, allowing clinicians to personalize medication plans, and lifestyle interventions. Evidence from recent studies indicates that ML approaches outperform traditional methods in prediction accuracy, robustness across various chronic diseases. Overall, ML offers a scalable and data-driven foundation for precision medicine in chronic disease care, supporting proactive decision-making and improved patient outcomes.

Keywords Machine learning (ML) – chronic diseases – clinical decision-making – disease-stage prediction – progression patterns

17.1 Introduction

Diseases are disturbance of the normal physiological functioning due to genetic, metabolic, environmental, infectious, and degenerative causes. Generalized classifications of diseases are acute, chronic, and degenerative diseases, with each having a different onset and clinical course. Acute diseases are normally characterized by short term presentation and a fast solution, where chronic and degenerative

 | HTTPS://DOI.ORG/10.1515/9783112237083-017

diseases occur over a long duration of time and with accumulation of biological effects [1].

Chronic diseases form a significant portion of the world health issues because of the inability to eliminate the symptoms and permanent harm to the organs and inability to stop treatment. These disorders tend to be intertwined by biological mechanisms in terms of fundamentals, lifestyle factors, and prolonged exposure to therapy. The course of disease in chronic diseases has nonuniform pathways, and patients have unpredictable response patterns and levels of clinical deterioration [2].

Machine learning (ML) algorithms represent a new data-centric paradigm of improving the disease progression prediction and customized treatment plans in chronic illnesses. Predictive intelligence based on the algorithmic learning assists in the early stratification of risk, continuous monitoring of the disease, and the therapeutic alignment of individuals [3]. It provides adaptive clinical behavior in response to patient-specific data profiles instead of population averages.

Although there is a lot of potential, there are a number of obstacles limiting the successful implementation of ML systems into chronic disease management. Data quality variability, incomplete longitudinal data and heterogeneous clinical records decrease model robustness [4]. The absence of interpretability of algorithmic predictions is a barrier to clinical confidence and adoption of decision support.

Purpose: Build an ML-based model of predicting chronic disease evolution and treatment success based on longitudinal clinical and lifestyle data, allowing the opportunity to stratify risks early, make more personalized interventions and explainable decision support to achieve better patient results in the long run.

17.2 Overview of Chronic Disease Progression

Chronic disease course is an irreversible and progressive change of physiological and clinical conditions during prolonged time periods. An example of progression patterns is heterogeneity at the population, as well as within people over time [5]. Longitudinal analysis of disease progression will also be necessary to accurately describe disease progression, as opposed to individual clinical observations. ML frameworks promote systematic modeling of progression dynamics, allow a finer insight into temporal transitions, and recovery phases needed to manage chronic diseases with precision.

17.2.1 Characteristics of Chronic Disease Trajectories

The course of chronic diseases has nonlinear courses with progressive worsening, non-recurring exacerbations, and frail phases of temporal stability. The development of the disease is characterized by the occurrence of clinically distinct stages, which depend on the metabolic dysfunction, damage to specific organs and treatment. Subclinical alterations are often present prior to the manifestation of the symptoms, which will delay clinical identification until late phases [6]. The fluctuation in the rate and degree of progression indicates past history of cumulative biological stress and treatment response. ML models use temporal relationships and rate-of-change patterns on longitudinal data, modeling subtle disease dynamics over fixed limits.

17.2.2 Patient Specific and Population-Level Variability

It is caused by genetic diversity, comorbidity profiles, and socioeconomic factors that affect disease progression. Different progression and treatment rates are often observed among patients with similar diagnoses. Intra-patient variability indicates time compliance, stress, and interactions of diseases [7]. Conventional population-based approaches cannot take into account this stratified heterogeneity. ML algorithms use patient-related characteristics and time history to train personalized progression signatures.

17.2.3 Role of Early Risk Identification and Continuous Monitoring

The timely detection of the risks is essential to avoid irreversible complications of the diseases and functional impairment. Laboratory, vital, and behavioral changes tend to evolve gradually long before a patient worsens [8]. ML procedures work with high-frequency data streams in order to identify deviation patterns related to upcoming exacerbations. Continuous surveillance systems enhance surveillance accuracies, and encourage dynamic disease management plans that are consistent with the changing health conditions of patients.

17.2.4 Clinical Decision Challenges in Chronic Care Settings

The ongoing treatment of chronic diseases is a complex clinical decision that relies on changing patients, and late therapeutic responses. There is a significant cognitive load involved in the synthesis of the heterogeneous sources of data over a long period of time by clinicians [9]. Decision support systems developed based on MLs combine multidimensional data to create context-sensitive insights that are consistent with patient-specific trajectories. The solution to these challenges improves the reliability of decisions and promotes sustainable, data-driven chronic disease management.

17.3 Methodology

In this part, the data-driven ML frameworks are combined to forecast the development of chronic diseases and tailor treatment. The longitudinal clinical, laboratory, imaging, and lifestyle data are preprocessed, and cleaned, normalized, and aggregated based on time. Feature engineering associates clinically meaningful variables, followed by dimensionality reduction to optimize method performance. Methods such as support vector machine (SVM), random forest (RF), logistic regression (LR), gradient boosting (GB), and *K*-nearest neighbors (KNNs) capture nonlinear patterns and patient-specific variability, as shown in Figure 17.1.

17.3.1 Dataset

This section presents various multimodal healthcare datasets integrating clinical, laboratory, imaging, wearable, and lifestyle data that were utilized to enable ML-based illness prediction. Several multimodal healthcare datasets were available on Kaggle platform.

The chronic kidney disease (CKD) dataset contains laboratory and clinical test results and is designed to predict the presence of CKD. It includes 25 features that provide comprehensive biomedical indicators for ML-based risk prediction.

The dataset (https://www.kaggle.com/datasets/uciml/human-activity-recognition-with-smartphones) contains smartphone accelerometer and gyroscope signals capturing daily physical activities, representing wearable-based lifestyle data useful for activity monitoring, and adherence assessment in health-focused ML studies.

The cardiovascular dataset (https://www.kaggle.com/datasets/sulianova/cardiovascular-disease-dataset) contains longitudinal, multimodal clinical records, including physical activity, smoking, weight, age, glucose, gender, blood pressure, and labels for risk prediction.

17.3.2 Data Preprocessing and Feature Engineering

It is the foundational stage of ML pipelines for chronic disease prediction. Preprocessing transforms such data into structured and analytically suitable representations. The feature engineering also boosts the predictive power because it orchestrates informative patterns of clinical variables, temporal variables, and behavioral variables.

Data Cleaning and Missing Value Handling: Internal consistency is ensured through data cleaning, which corrects any anomaly and eliminates any inconsistency and implausible values that distort the learning outcomes. The systematic

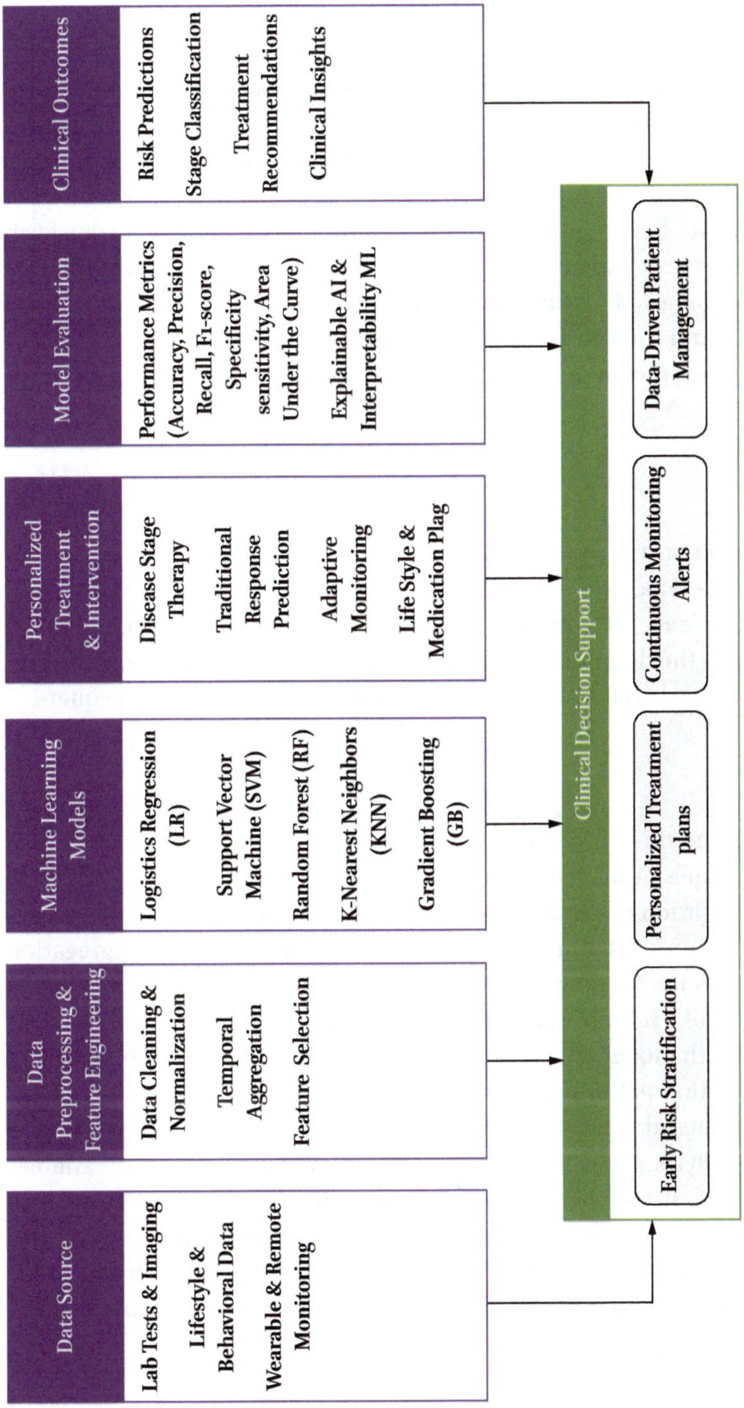

FIGURE 17.1 ML-based framework for chronic disease prediction and personalized care. (Source: Own elaboration)

treatment of the missing data maintains the integrity of the dataset and prevents biased predictions [10].

Normalization and Scaling Techniques: Normalization and scaling standardize the features in datasets of clinical data that have variables with varying numbers. Predictive disease progression and individual treatment in chronic diseases have a number of normalization and scaling approaches to predict the disease. The most common ones include z-score normalization and min-max scaling that transform variables into similar scales and maintain significant variation. Normalization by z-score converts the values of features to a standardized scale using reference data that is distributed, and thus, it makes sure that the contribution of all features is uniform when training the method [11]. The value of normalization is calculated as follows:

$$z = \frac{x - \mu}{\sigma} \tag{17.1}$$

In this case, x represents the practical feature rate, μ is the feature means, and σ is the dispersion of the data. The min-max scaling puts the feature standards into a fixed numerical sequence, often in the range 0 to 1 [12]. The approach maintains the original shape of the distribution and at the same time, it sets the same magnitude of features among variables through calculation using the following equation:

$$x' = \frac{x - x_{\min}}{x_{\max} - x_{\min}} \tag{17.2}$$

where x' denotes the original values, and $x_{\min}$ and $x_{\max}$ are the lowest and the highest observed values of the feature, respectively.

Temporal Aggregation of Longitudinal Data: The datasets on chronic diseases are longitudinal with repeated measurements over a long duration. Aggregation methods are trend summarizing techniques based on mean, slope, and rate of change with clinically relevant windows [13]. Temporal features construction enhances disease path modeling and assists with the early detection of progression patterns or stabilization patterns in chronic disease groups.

Clinically Meaningful Feature Construction: The feature construction is a combination of analysis and clinical knowledge to create informative variables that are indicative of disease behavior. The feature characterizes the central tendency and the variability of time, and provided quantitative summaries of health status of the patients. Clinical meaningful features are a domain knowledge coded in the form of disease-stage indicators, risk scores, and markers of treatment adherence [14].

17.3.3 Dimensionality Reduction and Feature Selection Strategies

Healthcare data in high dimensions create redundancy, noise and the risk of overfitting, which require dimensionality reduction and feature selection. The process of feature selection uses informative variables identified by a method that is relevant, correlated, or has a model-driven importance [15].

17.4 Machine Learning (ML) Algorithms for Disease Progression Prediction

ML algorithms have been able to offer robust computational paradigms to understand the progression of chronic diseases by training complex associations between clinical, biological, and lifestyle factors. The nonlinear dynamics and interdependence over time and variability of disease trajectories are patient-specific and defy the traditional methods to a question [16].

17.4.1 Logistic Regression (LR) for Baseline Risk Modeling

LR is a generalizable ML method of modeling baseline risks in diseases because the algorithm is interpretable and produces probabilistic predictions. During this step, LR will be optimized to maximize predictive performance by taking into consideration the clinical properties of the CKD data in the form of the following equation [17]:

$$P\ (y=1/Z)=\frac{1}{1+f^{-x}} \tag{17.3}$$

where $P\ (y=1/Z)$ denotes the forecasted likelihood that a particular person has CKD when input features Z, y denotes the binary predictable impact, Z denotes the column of input health features utilized in prediction, and e^{-x} denotes the natural log base used in the sigmoid transformation.

17.4.2 Support Vector Machines (SVMs) for Margin-Based Classification

SVMs perform disease progression prediction by identifying optimal decision boundaries that maximize class separation margins [18]. SVMs demonstrate strong robustness in biomedical applications involving heterogeneous risk factors. Commonly used kernels include the linear kernel (dot), the polynomial kernel, and the radial basis function (RBF) kernel using the following equations:

$$K(y,x) = y^S x + c = \tag{17.4}$$

$$K(y,x) = \left(\propto y^S x + c\right)^d = \tag{17.5}$$

$$K(y,x) = \exp(-\gamma \|y - x\|^2) \tag{17.6}$$

where x_i, x_j denote distinct samples drawn from the training dataset, $K(y,x)$ specifies the kernel-based similarity measure, $y^S x$ expresses the vector inner product, c denotes the kernel bias, $\propto$ represents the dot product in the polynomial kernel, d denotes the complexity of the decision boundary, γ denotes the influence range of each training sample, and $x - y^2$ denotes the squared Euclidean distance between two feature vectors.

17.4.3 *K*-Nearest Neighbors (KNNs) for Similarity-Based Prediction

KNN predicts disease outcomes based on similarity relationships among patient records within feature space [19]. The KNN algorithm is a simple yet effective instance-based ML technique that classifies data points based on their similarity to labeled training samples, expressed as follows:

$$D = \sqrt{(x_1 - x_2)^2 + (y_1 - y_2)^2} \tag{17.7}$$

where D denotes the Euclidean distance that quantifies the similarity between two data instances in the feature space, $x_2,\ y_2$ represent the corresponding values of the same feature for a neighboring data point, $x_1,\ y_1$denote the charge of the first feature for the query data point, and $\sqrt{(x_1 - x_2)^2 + (y_1 - y_2)^2}$ denotes the squared difference along the first feature dimension.

17.4.4 Random Forest (RF) for Nonlinear Pattern Discovery

RF is the term used to describe the ensemble learning techniques that combine the prediction of a series of decision trees to exploit complex nonlinear interaction relationships [19]. RF performs consistently across diverse chronic disease prediction tasks involving heterogeneous clinical variables, and is defined as follows:

$$\text{Ent}\,(D) = -\sum_{i-1}^{n} p_i \log_2 p_i \tag{17.8}$$

$$\text{Gain}\,(D,a) = \text{Ent}\,(D) - \sum_{v=1}^{V} \frac{|D_v|}{|D|}\,\text{Ent}\,(D_v) \tag{17.9}$$

where D denotes the complete training sample set, Ent (D) represents the information entropy measuring the impurity of the dataset D, p_i denotes the probability of samples, n signifies the entire amount of class labels in the dataset, Gain $(D,\ a)$ represents the information gain obtained by splitting dataset D using feature a, $|D_v|$

denotes the count of samples in the subset D_v, $|D|$ indicates the total sample size D, and V denotes the total number of distinct values of feature a.

17.4.5 Gradient Boosting (GB) Methods for Enhanced Predictive Performance

GB methods construct sequential ensembles where each method corrects the prediction errors of preceding methods. Boosting algorithms train loss functions directly, and are directly aligned with the clinical prediction needs [20]. The algorithm provides reduction of a loss function that has been set through the ensemble method iteratively at the function space, which is stated as follows:

$$L(y) = \min_F \sum_{i=1}^{N} l\ (y_i,\ F\ (x_i)) \tag{17.10}$$

$L(y)$ is the total loss, which is to be minimized on the training set, $F\ (x_i)$ is the predicted label of the ensemble mechanism of input x_i, N is the total number of training methods, and y_i is the actual target label of the ith training sample.

17.4.6 Personalized Treatment and Intervention Modeling

Individualized treatment and intervention modeling is one of the most important applications of ML in the management of chronic diseases. The nature of chronic conditions is heterogeneous in their progression and unpredictable in response to treatment and thus do not respond to a standardized set of clinical protocols but instead tailored therapeutic methods [21]. Individualized intervention models better long-term disease administration by evidence-based decision support as shown in Figure 17.2.

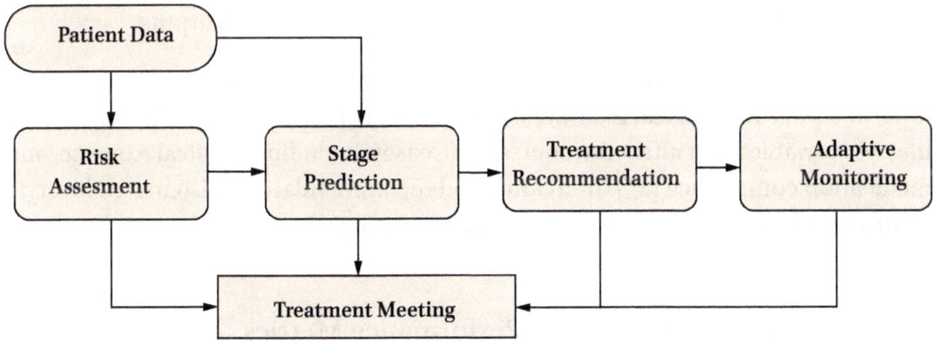

FIGURE 17.2 ML clinical decision workflow.(Source: Own elaboration)

17.4.7 Treatment Response Prediction

Treatment response prediction deals with the estimation of the strength of response of individual patients to certain treatment interventions. The ML techniques examine past treatment data, and compliance history to detect response patterns [22]. The integration of prediction of response in clinical processes aids informed decision-making and enhances consistency in the outcomes of chronic disease management across groups of patients.

17.4.8 Disease-Stage-Specific Therapy Recommendation

Treatment approaches are aligned to the present stage of severity and progression of chronic disease through disease-stage-specific therapy recommendation. ML technologies identify stages of diseases based on longitudinal clinical indicators and stages of progression [23]. Model-based recommendations give more importance on the interventions that have been tested to be effective at a particular stage of the disease.

17.4.9 Adaptive Monitoring Schedule Generation

Adaptive monitoring schedule generation maximizes the frequency of follow-up and diagnostic tests according to risk assessment on an individual basis. ML techniques are projection short-term progression risk with the help of current clinical patterns, pattern of treatment response, and lifestyle signs [24]. Monitoring plans based on data help to identify the lack of improvement promptly and act in time to reinforce long-term disease monitoring of the chronic care system.

17.4.10 Lifestyle and Medication Personalization Using ML Outputs

Lifestyle and medication personalization is a combination of predictive, behavioral, and pharmacological planning. The products of ML recognize the changeable lifestyle variables that affect the course of diseases, including physical exercise, and medication compliance [25]. Individualized recommendations also match lifestyle changes with personal health profiles and treatment objectives.

17.5 Method Evaluation and Performance Metrics

The method evaluation will be used to examine the reliability, validity, and clinical applicability of ML methods to predict chronic diseases. Performance measures transform computational output to clinically relevant measures. Strict assessment is imperative to validate the reliability of the integration of predictive systems into clinical practice to be safe:

- **Accuracy** represents the percentage of the correct predictions in comparison with the total number of cases, which is a broad measure of model performance across all classes.
- **Precision** is an expression of the proportion of positive cases that are accurately predicted, which is a statement of how well the method predicts occurrences of diseases.
- **Recall** is a measure of the rate of successful identification of the actual number of positive cases that the method identifies correctly, indicating the effectiveness of the method in detecting the presence of the disease.
- ***F*1-score**, the harmonic mean of the measures of precision and recall, provides a balanced performance measure, which is especially effective when the classes are not distributed evenly.
- **Area under the curve (AUC)** is a measure of the overall capacity of the method to differentiate positive and negative cases at all possible thresholds, and it is a measure of the discriminatory power of the method.
- **Specificity** is a measure of the correct percentage of false negatives that are actually being identified by the method, which shows how well this method avoids false positives.
- **Sensitivity** is used to measure the rate of true positive cases that are correctly measured, where a method is emphasized in detecting the actual incidences of disease.

TABLE 17.1 Comparative performance evaluation of ML methods using standard classification metrics.

Methods	Accuracy (%)	Precision (%)	Recall (%)	F1-Score (%)	Specificity (%)	Sensitivity (%)
LR [17]	86	84	88	82	–	–
KNN [19]	65.80	51	–	53.80	56.58	56.79
RF [19]	76	51	–	60.02	65.6	55.6

Source: Adapted from [17, 19]

Table 17.1 compares LR [17], which has the best accuracy (86%) and recall (88%). The KNN [19] techniques register lower accuracy (65.80%), RF [19] has an accuracy of 76%, and moderate sensitivity (55.6%).

Table 17.2 overviews the fact that the highest accuracy of 98.50 with an ideal AUC of 1.000 was obtained using the dot-product SVM. Poly-kernel SVM has good

TABLE 17.2 Comparison of performance of SVM classifiers with dissimilar kernel functions.

Methods	AUC	Accuracy (%)
SVM (dot) [18]	1.000	98.50
SVM (polynomial) [18]	0.986	97.25
SVM (RBF) [18]	0.999	92.25

Source: Adapted from [18]

discrimination properties with AUC of 0.986 and an accuracy of 97.25%. RBF-kernel SVM achieves a high AUC of 0.999 and a lower accuracy of 92.25%.

TABLE 17.3 The measure of prediction quality and computation time.

Methods	Accuracy (%)	Misclassification rate (%)	Runtime (s)
GB [20]	97.46	2.54	2.752

Source: Adapted from [20]

Table 17.3 shows that the GB [20] approach has an accuracy of 97.46% and a small misclassification rate of 2.54%, and the evaluation time is 2.752 s, demonstrating a low-run time, which effectively used in practical chronic disease prediction problems.

17.6 Explainable and Interpretable Machine Learning (ML)

Elicitable and understandable ML models meet the address transparency criterion that is a must in clinical adoption models. The high-stakes decisions that are meant in chronic disease management require clarity in terms of approach and rationale. Explainable ML methods offer detailed information on how to feature importance, and their decision journey.

17.6.1 Need for Transparency in Clinical ML Systems

Clinical ML systems are applied in the setting that involves accountability, safety, and responsibility. Transparency makes reasoning in clinics clear and helps clinicians to evaluate the drivers of prediction and gauge dependability. Absence of clarity interferes with adoption and prevents clinical confidence [26]. Openness will be core in the approach to integrate ML in the diagnostic and prognostic processes.

17.6.2 Feature Importance and Rule-Based Explanations

The methods of feature importance measure the effect that a single variable has on prediction of a method. Ranking of features on the basis of contribution provides outstanding clinical and behavioral determinants that influence disease progression. At the feature level, understanding of the methodology improves credibility, evidence-based interpretation [27]. The joint application of feature importance and rule-based approaches increase the transparency and interpretability of healthcare-oriented ML systems.

17.6.3 Building Clinician Trust through Explainable ML

A precondition to long-term ML tool adoption is clinician trust. Easy to understand approaches validate trust with the explanation of the reasoning behind prediction and uncertainty. Trust in ML recommendations facilitates proper decision-making. The trust-centered design encourages integration over the long term and also boosts the acceptance of intelligent systems within the chronic disease organizations.

17.7 Discussion

ML has been more actively investigated to assist individual therapy and chronic disease management in the long run with the analysis of complex clinical and lifestyle data. Previous methods focus on the treatment planning and outcome improvement but make use of the individualized features of patients, which are quite static or aggregated and cannot consider disease progress over time, limiting their capacity to express clinical interpretability [28]. Basing it on retrospective databases, insufficient external validation, possible balance of data, and weaker externalizable between various CKD samples [29]. The approach in this approach overcomes these limitations by combining longitudinal feature modeling, the robust learning algorithms that have the strength to handle heterogeneous inputs, and explainable prediction mechanisms that facilitate clear clinical choices.

17.8 Conclusion

The purpose of this chapter was to witness how well ML algorithms predict chronic disease progression and allow designing personalized treatment plans. The linked longitudinal clinical, laboratory, and lifestyle data were applied in the form of a comprehensive ML-based framework with preprocessing, feature engineering, and a number of predictive algorithms. The process in the methodology was data normalization, time aggregation, feature selection, as well as comparative analysis of classification procedures. This chapter demonstrated the benefits of the advanced ML algorithms on accuracy and robustness and explainable ML on clinical interpretability. This chapter, however, had a limitation on quality of data, nature of the longitudinal records, and possible bias due to a heterogeneous dataset, which influences generalizability. The future will be oriented to integrating real time inputs of data, improving the interpretation of methods, ensuring that the issue of privacy of data is solved, and ensuring that the chapter method is validated among various groups of people.

References

[1] Yogi KK, Yadav SP. Chicken Diseases Detection and Classification Based on Fecal Images Using EfficientNetB7 Model. 2024;314–30. https://doi.org/10.5109/7172288.

[2] Upadhyay P, Tomar P, Yadav SP. ADNet: Deep learning based model for Alzheimer's disease classification. Iran Journal of Computer Science. 2025;1–13. https://doi.org/10.1007/s42044-025-00243-x.

[3] Islam R, Sultana A, Islam MR. A comprehensive review for chronic disease prediction using machine learning algorithms. Journal of Electrical Systems and Information Technology. 2024;11(1):27. https://doi.org/10.1186/s43067-024-00150-4.

[4] Tian YE, et al. Heterogeneous aging across multiple organ systems and prediction of chronic disease and mortality. Nature Medicine. 2023;29(5):1221–31. https://doi.org/10.1038/s41591-023-02296-6.

[5] Hao Y, et al. Vanin1 (VNN1) in chronic diseases: Future directions for targeted therapy. European Journal of Pharmacology. 2024;962:176220. https://doi.org/10.1016/j.ejphar.2023.176220.

[6] Scott AJ, et al. Depression and anxiety trajectories in chronic disease: A systematic review and meta-analysis. Psychotherapy and Psychosomatics. 2023;92(4):227–42. https://doi.org/10.1159/000533263.

[7] Park Y, et al. Interactive Impact of Tacrolimus Inter-Patient Variability and Intra-Patient Variability in Allograft Outcomes in Kidney Transplantation: FR-PO782. Journal of the American Society of Nephrology. 2022;33(11S):537. https://doi.org/10.1681/ASN.20223311S1537b.

[8] Moon D, Krahel JP. Continuous risk monitoring and assessment: New component of continuous assurance. Journal of Emerging Technologies in Accounting. 2020;17(2):173–200. https://doi.org/10.2308/JETA-18-01-09-1.

[9] Lapp L, et al. Decision support tools in adult long-term care facilities: Scoping review. Journal of Medical Internet Research. 2022;24(9):e39681. https://doi.org/10.2196/39681.

[10] Sharifnia AM, et al. A primer of data cleaning in quantitative research: handling missing values and outliers. Journal of Advanced Nursing. 2025;81(2):345–58. https://doi.org/10.1111/jan.16908.

[11] Peng L, et al. Dual-Structure elements morphological filtering and local z-score normalization for infrared small target detection against heavy clouds. Remote Sensing. 2024;16(13):2343. https://doi.org/10.3390/rs16132343.

[12] Ali PJM. Investigating the Impact of min-max data normalization on the regression performance of K-nearest neighbor with different similarity measurements. ARO-The Scientific Journal of Koya University. 2022;10(1):85–91. http://dx.doi.org/10.14500/aro.10955.

[13] Alarcon Falconi TM, et al. Effects of data aggregation on time series analysis of seasonal infections. International Journal of Environmental Research and Public Health. 2020;17(16):5887. https://doi.org/10.3390/ijerph17165887.

[14] Desai N, Albrecht E. Minimal clinically important difference: Bridging the gap between statistical significance and clinical meaningfulness. Journal of Clinical Anesthesia. 2024;111366:94. https://doi.org/10.1016/j.jclinane.2023.111366.

[15] Alhassan AM, Zainon WMNW. Review of feature selection, dimensionality reduction, and classification for chronic disease diagnosis. IEEE Access. 2021;87310–17:9. https://doi.org/10.1109/ACCESS.2021.3088613.

[16] Lei N, et al. Machine learning algorithms' accuracy in predicting kidney disease progression: A systematic review and meta-analysis. BMC Medical Informatics and Decision Making. 2022;22(1):205. https://doi.org/10.1186/s12911-022-01951-1.

[17] Alanazi R. Identification and prediction of chronic diseases using a machine learning approach. Journal of Healthcare Engineering. 2022;2022(1):2826127. https://doi.org/10.1155/2022/2826127.

[18] Purwaningsih E. Improving the performance of support vector machine with forward selection for prediction of chronic kidney disease. Jitk (Jurnal Ilmu Pengetahuan Dan Teknologi Komputer. 2022;8(1):18–24. https://doi.org/10.33480/jitk.v8i1.3327.

[19] Aldhyani THH, Alshebami AS, Alzahrani MY. Soft clustering for enhancing the diagnosis of chronic diseases over machine learning algorithms. Journal of Healthcare Engineering. 2020;2020(1):4984967. https://doi.org/10.1155/2020/4984967.

[20] Ganie SM, et al. Chronic kidney disease prediction using boosting techniques based on clinical parameters. Plos One. 2023;18(12):e0295234. https://doi.org/10.1371/journal.pone.0295234.

[21] Mohsin SN, et al. The role of artificial intelligence in prediction, risk stratification, and personalized treatment planning for congenital heart diseases. Cureus. 2023;15(8). https://doi.org/10.7759/cureus.44374.

[22] Jankowsky K, et al. Predicting treatment response using machine learning: A registered report. British Journal of Clinical Psychology. 2024;63(2):137–55. https://doi.org/10.1111/bjc.12452.

[23] Argirò A, et al. Stage-specific therapy for hypertrophic cardiomyopathy. European Heart Journal Supplements. 2023;25(Supplement_C):C155–C161. https://doi.org/10.1093/eurheartjsupp/suad042.

[24] Alam MA, et al. Big data and chronic disease management through patient monitoring and treatment with data analytics. Academic Journal on Artificial Intelligence, Machine Learning, Data Science and Management Information Systems. 2024;1(01):77–94. https://doi.org/10.69593/ajaimldsmis.v1i01.133.

[25] Shang F, Yu L. Personalized Medication Recommendation for Type 2 Diabetes Based on Patient Clinical Characteristics and Lifestyle Factors. Journal of Advanced Computing Systems. 2025;5(4):1–16. https://doi.org/10.69987/.

[26] Marcinkevičs R, Vogt JE. Interpretable and explainable machine learning: A methods-centric overview with concrete examples. Wiley Interdisciplinary Reviews: Data Mining and Knowledge Discovery. 2023;13(3):e1493. https://doi.org/10.1002/widm.1493.

[27] Winter PD, Carusi A. (De) troubling transparency: Artificial intelligence (AI) for clinical applications. Medical Humanities. 2023;49(1):17–26. https://doi.org/10.1136/medhum-2021-012318.

[28] Adeyinka A, et al. "Machine learning efforts that enhance personalized patient care and chronic disease management." International Journal of Scientific Research in Science and Technology (2022): 647–71.https://doi.org/10.32628/IJSRST2302551.

[29] Bai Q, et al. Machine learning to predict end stage kidney disease in chronic kidney disease. Scientific Reports. 2022;12(1):8377. https://doi.org/10.1038/s41598-022-12316-z.

CHAPTER 18

Wearable Sensors for Real-Time Monitoring of Mental Health Disorders

S. Lalithakumari, Divya Paikaray, Suman Sau

Abstract Wearable sensors have emerged as a promising technology for real-time monitoring and management of mental health disorders by enabling continuous, objective, and noninvasive assessment of physiological and behavioral signals. These devices integrate multimodal sensors such as electrodermal activity, heart rate and heart rate variability, skin temperature, accelerometry, and sleep patterns to capture sign of stress, anxiety, depression, and mood dysregulation. By collecting high-frequency data in naturalistic settings, wearable systems overcome limitations of traditional self-reports and intermittent clinical assessments, which are often subjective and prone to recall bias. Advanced signal processing and machine learning techniques are employed to extract meaningful features, detect anomalies, and model temporal patterns linked to mental health states. Real-time analytics enable early warning systems and personalized interventions, supporting timely clinical decision-making and adaptive treatment strategies. Integration of wearable data with mobile health platforms facilitates remote monitoring, patient engagement, and longitudinal mental health tracking. Despite challenges related to data privacy, sensor reliability, user adherence, and clinical validation, advancements in low-power electronics and intelligent algorithms have improved feasibility and accuracy. Overall, wearable sensor-based monitoring represents a scalable and transformative approach for proactive mental healthcare, enabling continuous assessment, early detection of symptom changes, and support for individuals.

Keywords Machine learning (ML) – wearable sensors – real-time monitoring – signal processing – mental health assessment

18.1 Introduction

Wearable sensors are no longer a negligible technological trend in the contemporary healthcare, especially within the area of mental health. These are small, wearable gadgets that are capable of constant recording of physiological and behavioral cues as individuals continue with their daily activities. Examples of mental health disorders include anxiety, depression, stress-related illnesses, and mood disorders, which tend to be complicated to notice at an early stage due to nature of their

 | HTTPS://DOI.ORG/10.1515/9783112237083-018

symptoms, which vary and people may not necessarily notice or report the change in their emotions. Conventional method of assessment is highly dependent on use of clinical visits, questionnaires, or self-reports, which only reveal short periods of time and can be biased by the memory or individualism [1]. Wearable sensors, on the contrary, give a chance to observe real-world patterns which can be related to underlying emotional or psychological states. With the growth of digital health technologies, wearable systems are becoming more and more perceived as supportive tools to supplement clinical judgment and self-awareness in patients. Their increased application points to wider trend toward active, ongoing, and individualistic mental health management [2]. The fundamental tenet of wearable sensor-based mental health monitoring is the fact that the mental health is able to control the physiological and behavioral responses of the body. Wearables can detect the signal of heart rate (HR), skin reactions, body movement and sleep, and temperature. The change of these signs may sometimes speak of emotional distress, mood swings, or mental stress. By capturing such information at present, wearable devices can be used to create a stream of visualization of mental health of an individual, rather than snapshot. Such systems are practical in real-life contexts, and this means that the data are real-life processes and not controlled or clinical settings [3, 4].

It is not aimed at replacing mental health professionals but rather to contribute to the increase of awareness and monitor and detect potential areas of concern in the early stages. Changes in mood or mind could be demonstrated by the patterns in the received information over time. It is the concept that is used as a manifestation of more broad-based digital health goals that are geared toward prevention, continuous care, and longitudinal follow-up to learn more about the relationship between everyday life experience and psychological well-being [5]. Wearable sensors permit real-life, real-time monitoring of signs associated with mental conditions and assists in recognizing the existence of emotional or behavioral changes timely. They may be applied to complement the conventional care, since they provide objective observations that are not reliant on memory and self-reporting. Wearable devices also promote self-reflection, patient engagement, and long-term tracking, which may supplement more proactive and personal mental health support [6]. Wearable systems also create privacy concerns, need cooperation on the part of the user, and are incomplete and unreliable in some instances. They are also able to make the complex psychological conditions simple and can be misjudged without the assistance of a professional [7]. Main constraints are the risk to data privacy, inaccuracy between devices, and user adherence in the long term. The ways of reduced through them include safe data management, better sensor design, user-friendly equipment, ethics, and effective cooperation between the developers of technologies, clinicians, and users.

18.2 Wearable Sensor Deployment

The implementation of wearable sensors constitutes the initial action in real-time mental health surveillance, since the precision and dependability of gathered physiological and conductive data intensely rely relies on the nature and area of devices. These wearable systems are multimodal and incorporate the sensors that are aimed at measuring the stressful indicators, anxiety, depression, and mood swings. Typical sensors are electrodermal activity (EDA) sensors to measure skin conductance, HR, and heart rate variability (HRV) sensors to measure the cardiovascular dynamics, skin temperature sensors, accelerometers to trace physical activity and movement patterns, and sleep-monitoring sensors to trace the changes in the circadian rhythm.

The choice of sensors is informed by the mental health metrics such as interest, signal fidelity, and of user comfortable. As an example, wrist-worn devices can concurrently log HR, HRV, EDA, and accelerometry with very little user interaction compared with the chest straps, which can be more accurate on HRV, but are less convenient in extended monitoring. Placement is thus made to give compromise between data accuracy and wearability, and adherence, where persons can wear the device all the time without experiencing inconvenience or interference with their everyday lives.

Moreover, wearable deployment also takes into account low-power electronics, wireless connectivity, and synchronization among sensor modules, which makes it possible to collect high-frequency data smoothly. Calibration, safe attachment, as well as baseline measurement before improvement of the signal quality and reduction of artifacts are performed. By the wise selection and use of multimodal sensors, the system ensures that one gets a good acquisition of physiological and behavioral signals, which provide a good foundation to continue with the analysis, in relation to machine learning (ML) and real-time monitoring, and personal intervention plans. Lastly, the best application of wearable mental health monitoring systems is the implementation of sensors strategically to allow their appropriate application at large scale and to ensure proactive and continuous assessment in real-world applications [8]. The important points about the deployment of wearable sensors are outlined in Table 18.1, focusing on the selection of sensors, where and how to wear them, and how comfortable they are, how reliable they are, and what are their real-life applications in mental health monitoring.

18.3 Machine Learning-Based Analysis

The main analysis characteristic of mental health monitoring systems with wearables is ML that would enable the interpretation of high-frequency physiological

TABLE 18.1 Wearable sensor deployment for mental health monitoring.

Aspects	Description	Purpose in mental health monitoring
Sensor types	Multimodal devices, such as EDA, HR, HRV, skin temperature, accelerators, and sleep sensors	Record physiological and behavioral pointers of stress, anxiety, depression, and mood fluctuations
Sensor placement	Devices, which are worn on the wrist and straps, are placed on the chest chosen on accuracy and comfort	Balances indicate loyalty and long-term user-friendliness and wearability
User comfort	Lightweight, nonintrusive, and wearable designs	Ensures continuous usage without interfering with daily activities
Power efficiency	Low-power electronics integrated into wearable devices	Enables long-term, uninterrupted real-time monitoring
Wireless connectivity	Secure wireless data transmission and synchronization between sensors	Supports seamless, high-frequency data collection
Calibration and baseline	Initial calibration and baseline physiological measurement	Improves signal quality and reduces noise and artifacts
Data quality assurance	Secure attachment and proper deployment of sensors	Ensures reliable and consistent physiological data capture
Real-world deployment	Continuous monitoring in naturalistic environments	Enables proactive and ongoing assessment of mental health states

and behavioral data into useful information about the mental state of an individual. Wearable sensors generate multimodal data, which is usually of large size, and is continuously changing (HRV, EDA, skin temperature, physical activity, and sleep patterns). These are time-dependent nonlinear signals, and they exhibit a time-varying stress, anxiety, depression and mood. These temporal dynamics, and subtle variations, and abnormalities of normal behavioral or physiological default are then approximated by using ML algorithms.

Recurrent neural networks, long short-term memory networks, and temporal convolutional networks are especially well-suited to capturing the temporal dependencies and enable the system to discover how the current sensor readings are correlated with the past tendencies. Statistical, frequency domain, and entropy-based features are used in machine learning models to detect stress, mood changes, and early signs of depression. Isolation forests or autoencoders, which are anomaly detectors, alert to abnormal physiological or behavioral patterns which might need intervention and offer a vital kind of proactive monitoring [9].

In addition to anomaly detection, predictive modeling can assist the system in the prediction of mental health-related states, which facilitates early interventions. To classify sensor data as a mental health condition (e.g., low, moderate, or high stress), classification models (e.g., support vector machines, random forests, or deep neural networks) are used, whereas regression-based models provide continuous estimates of the intensity of arousal, anxiety, or mood. Personalized feedback and adaptive interventions are based on such predictions, and by using personalized interventions, clinicians or automated systems can react proactively to the development of mental health risks. ML-based analysis transforms raw wearable sensor data into actionable insights, enabling continuous and objective real-time mental health monitoring through anomaly detection, predictive analytics, and temporal modeling. Such a layer of analysis enables wearable systems to not only detect the existing symptoms, but also anticipate potential mental health challenges and, therefore, contribute to providing early, personalized, and evidence-based changes to reality-based conditions.

18.4 Real-Time Analytics and Early Warning System

Real-time analytics is a significant aspect of wearable sensor-based mental health monitoring, as the system can process real-time physiological and behavioral data and convert it into useful information in real time. Upon the collection and preprocessing of data obtained with the use of multimodal sensors, HR, HRV, EDA, skin temperature, accelerometry, and sleep patterns, the data is transmitted to analytics engines that would be able to recognize patterns and anomalies in real time. Such engines use predictive models and adjustive algorithms to identify the onset of stress, anxiety, depression, or mood dysregulation when they happen. The early warning system operates by continuously comparing incoming data against personalized baselines or population-wide thresholds to detect anomalies that may indicate an elevated risk of mental health issues. Methods such as threshold-based alerts, ML classifiers, or anomaly detection models (e.g., autoencoders and isolation forests) are used to indicate physiological or behavioral patterns that are predictive of an impending stress episode or sleep disturbance, or emotional dysregulation. The system can identify risk situations as they occur and therefore take timely measures to stop situations before they get out of hand, especially in proactive mind management [10].

When a risk event has occurred, the system might send real-time notifications to users, carers, or clinicians either via mobile device, wearable interface, or connected health platform. Alerts can either provide individual recommendations, coping mechanisms, or clinical consultation suggestions. The communication of these alerts can be improved with integration into mobile health (mHealth) platforms, which provide the ability to monitor remotely and react to intervention strategies. Real-time analytics can also be used to promote dynamic adaptation, where predictive models are continuously updated on new incoming information to adjust the alert thresholds and increase sensitivity and specificity as time goes on. This makes sure that notifications are highly customized, relevant to clinical users, and the least invasive to help users adhere to them and trust them. In general, the real-time analytics and early warning system turn raw wearable sensor data into a proactive and ongoing mental health assessment, allowing the detection of risks in real time and intervention to prevent adverse outcomes and improve clinical decision-making in naturalistic and daily settings. Figure 18.1 illustrates a real-time wearable sensor-based mental health monitoring framework integrating multimodal data, ML, early warning alerts, and clinical support systems.

18.5 Personalized Intervention and Feedback Mechanisms

Individualized intervention and feedback systems cannot be ignored in mental health monitoring through the use of wearable sensors since they convert real-time analytical information to useful user support. The system provides context-sensitive advice to assist individuals in stress, anxiety, mood swings, emotional fluctuation, among other mental states and reducing risks once they are identified. Unlike offering generic information, the interventions are customized to the physiological base, behavioral trends, the trend in history, and the current mental state of an individual. Feedback can be in the form of real-time stress-reduction prompts, guided breathing or relaxation practices, cognitive reframing messages, sleep hygiene tips, and activity encouragement, which can be administered via mobile applications, wearable interfaces, or clinician dashboards and does not disrupt daily routines. The system improves with time and learns the most effective interventions for each user to make the system more relevant and effective. Collaborative care is also supported by personalized feedback because it allows clinicians to assess summarized trends and alert histories in order to make timely changes to therapeutic plans. Overall, these systems lead to the increase of user engagement, trust, and proactive mental health self-management [11].

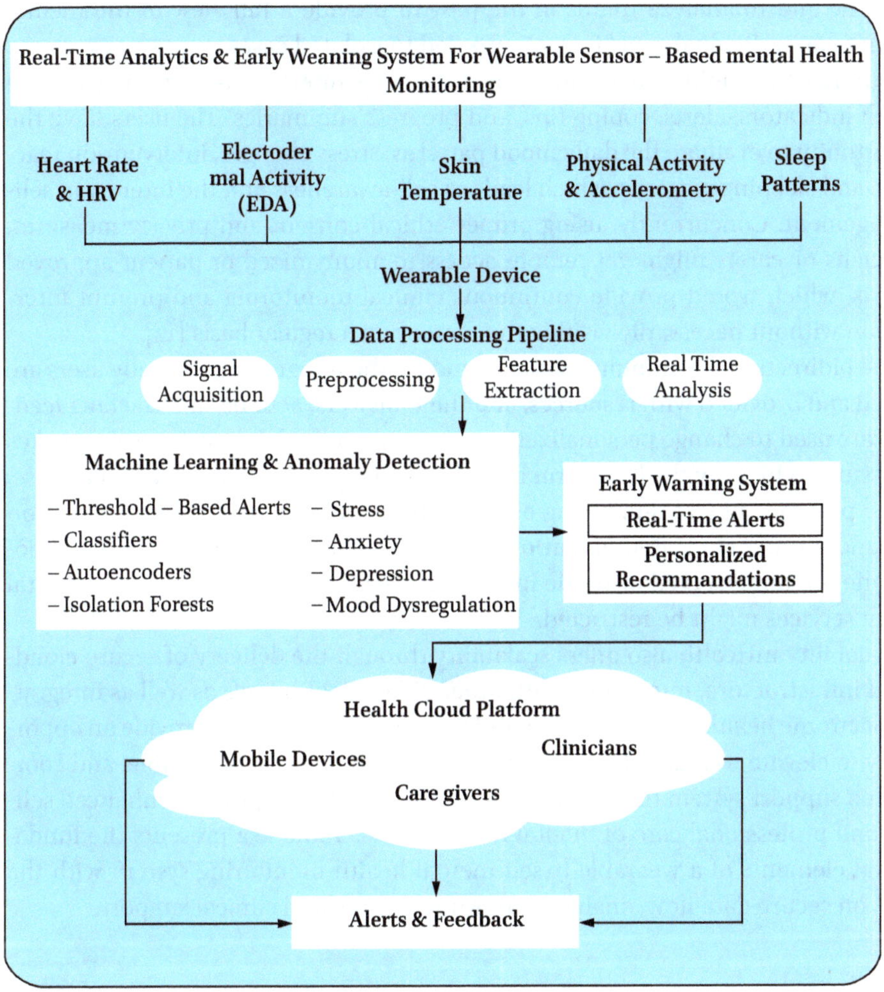

FIGURE 18.1 Real-time wearable mental health monitoring framework. (Source: Own elaboration)

18.6 Integration with Mobile Health Platforms

Combination with an mHealth solution is a key factor in making the wearable sensor-powered mental health monitoring something more than an independent data collection system and a detached ecosystem that can be long-term tracked, remotely managed, and collaborate between a patient and a clinician. Some of the data collected in the wearable devices (HR, EDA, temperature, movement, and sleep) are safely sent to mHealth applications, where these are aggregated, visualized, and aligned with cloud-based analytics platforms. This allows it to monitor in real time and to analyze trends in the past to provide a full view of the mental health of an individual over time, as opposed to isolated cases.

Through the mobile platforms, a user-friendly interface is accessing mental health indicators, alerts, coping tips, and progress summaries. The users have the opportunity to evaluate the daily mood patterns, stress changes, intervention reactions, and sleeping patterns, which leads to self-awareness and the interest in self-management. Concurrently, using proper ethical consent and privacy measures, clinicians or carers might get remote access to anonymized or patient-approved reports, which would provide continuous clinical monitoring and prompt intervention without necessarily visiting the patient on a regular basis [12].

The bidirectional communication also aids in the integration, whereby users are guided and provided with resources, and their answers, actions, and tracking feedback are used to change personalization. The longitudinal data storage also makes it possible to identify the long-term behavioral changes, the occurrence of relapse, or the presence of stress-inducing events, which are necessary in the prevention care and prompt clinical intervention. The features are particularly handy to people who are underserved or reside in remote places, where their access to mental health services might be restricted.

Scalability mHealth also offers scalability through the delivery of secure cloud-based infrastructure, integration with other digital health tools as well as integrating electronic health records where required. mHealth platforms provide an opportunity to elevate wearable-based information on an ongoing, accessible, and coordinated support system through mental health analytics to provide enhanced self-care and professional care of mental health issues. Table 18.2 presents the fundamental elements of a wearable-based mental health monitoring system with the focus on secure data flow, analytics, user interaction, and clinical support.

18.7 User Adherence and Engagement Monitoring

A very important part of a wearable sensor-based mental health system is user adherence and engagement monitoring because the accuracy, reliability, and clinical utility of continuous monitoring are directly related to the regular use of the device and active participation in the platform. The best sensing and analytical technologies are useless, even when the user does not put on the devices periodically, alert the device, and access suggested interventions. Thus, the monitoring of adherence is aimed at monitoring the levels of frequency of use, connectivity of the sensor, data completeness, and sensitivity to system prompts to guarantee the ability of the system to be effective in the real world. Engagement monitoring also looks at the interaction between users with mHealth apps, such as seeing insights about mental health, responding to alerts, self-reporting and engaging in coping or

TABLE 18.2 System components for wearable mental health monitoring.

Components	Description	Role in mental health monitoring
Wearable data sources	The overall physiological and behavioral data (HR, EDA, temperature, movement, and sleep) are gathered by wearable sensors	Gives ongoing and objective contribution in mental health assessment
Secure data transmission	Automated wireless transmission of sensor data to mobile applications and cloud servers	Guarantees real-time accessibility and preservation of data privacy and security
Mobile health application	Data visualization, alerts, and feedback platform exposed to the user	Increases user interaction, self-reflection, and everyday self-assessment
Cloud-based analytics	Long-term wearable information storage and processing are centralized	Allows real-time analysis and long-term trends
Visualization and feedback	Mood and stress level, sleep quality, and intervention results display	Helps in self-directed behavior and management
Remote clinical access	Access of summarized reports by clinicians or caregivers	Remote monitoring and timely clinical intervention
Bidirectional communication	Sharing of alerts, recommendations and user response.	Facilitates customized personalization and ongoing and coordinated care
Longitudinal tracking	Archiving of historical data over a long period of time.	Promotes timely relapse pattern and preventive measures

Source: Own elaboration

behavioral exercises suggested by the app. Through these behavior patterns, the system is able to detect a decrease in engagement or possible obstacles, including discomfort, alert fatigue, privacy issues, or perceived low usefulness. The personalized reminders, inspirational messages, gamification methods, or simplified user interfaces might then be used as adaptive strategies to keep the user engaged without overwhelming them.

Notably, contracting and monitoring adherence and engagement also permit personalized support channels. As an example, in the situation when the system identifies that the system is not used over a certain time, it can initiate a supportive check-in or a notification to the clinician (with consent) to facilitate continuity of care. Engagement data also aid in the perfecting of the personalization models with time, so that the tone, timing, and frequency of interaction can match the user preference and tolerance levels [13]. The observation of compliance is also a technical goal but a behavioral and ethical focus to strengthen trust, empowerment, and joint decision-making. Open communication about the use of data, adjustable settings, and privacy controls by the user contributes even more to the desire to participate constantly. In general, user compliance and adherence tracking can make wearable mental health systems reliable, user-centered, and clinically significant, facilitating consistent data collection, interventions in due time, and self-management of mental health in the daily setting.

18.8 Ethical Considerations and Data Privacy

The design and implementation of wearable sensor-based mental health monitoring systems are based on ethical concerns and data confidentiality due to the sensitivity of the services that these systems gather in terms of physiological, behavioral, and emotional data. Responsible data handling process is necessary not only to comply with laws and regulations but also to guarantee user trust and promote continued involvement. First in ethical design is informed consent, where the users are well aware of what information is being gathered, how it is going to be processed, the persons who can access it, and what it is that they can use it. Consent procedures must be open and voluntary, and they should be revisited to be able to make changes regarding permissions.

Effective data privacy and data security are guaranteed to make sure that the data related to information on mental health is not accessed or misused by unauthorized parties. These may include encryption of the transmission and storage, pseudonymization or anonymization of identifiable data, secure authentication and role-based access controls, particularly when the clinician or caregiver is part of the scenario. Validity and security of the system can be improved by following the pertinent health information policy and moral principles. Besides technical protection, one more aspect of ethical practice is avoiding the algorithmic bias, discrimination, and unfair treatment. Models are required to be applied on multiple and different populations to ensure that the predictability of model is comparable across the different populations regardless of their age, sex, ethnicity, or socioeconomic status. Information provided by users should be under their control and permit them to make a selection of withdrawal, sharing or deleting the information because it encourages autonomy and respect to acknowledge personal dignity [14].

Clear governance structures like accountability, transparency reporting, and auditing must be used to define clearly the data stewardship responsibilities. The other feature of ethical deployment is avoidance of psychological harm including lowering false alarms, eradicating excessive surveillance and providing notifying warning in a supportive and non-infectious manner. Lastly, high ethical and privacy protection will allow establishing an atmosphere of trust in which individuals will not be afraid of sharing confidential information regarding mental health problems. The monitoring systems based on wearable sensors will be able to help in proactive mental healthcare by observing high levels of ethical responsibility. Confidentiality, fairness, and user autonomy are priorities to them in order to make them socially acceptable and trustworthy. Table 18.3 demonstrates some major ethical principles of wearable mental health monitoring that focus on privacy, fairness, user autonomy, data protection, and responsibly governed systems.

18.9 Future Directions and Clinical Implications

The wearable-based mental health monitoring systems of the next generation will be able to be more intelligent, personalized, and clinically applicable in early diagnosis, treatment, and long-term management. Multimodal data fusion is the way to precisely evaluate the states of emotions, whereas adaptive learning provides individualized management. Surgery and telemental health systems are applications that enhance proactive intervention, remote monitoring, and clinically valid and ethical implementation [15].

18.10 Conclusion

Wearable sensors are an important evolution of continuous monitoring and management of mental illnesses because they can provide real-time, objective, and noninvasive measurement of physiological and behavioral indicators. The main purpose of the study is to emphasize how wearable technologies together with ML can be used to assist in the early identification, individualized care, and proactive management of mental health. The results suggest that multimodal wearable sensing has the potential to offer valuable information on stress, anxiety, depression, and mood changes, which can overcome the limitations of the traditional method of self-reports, as well as intermittent clinical measurements. Real-time analytics also aid in effective timely clinical decision-making and dynamic intervention. Nevertheless, the issues, including the problem of data privacy, sensor

TABLE 18.3: Ethical considerations in wearable mental health monitoring.

Ethical aspects	Description	Significance in mental health monitoring
Informed consent	Clear and free consent procedures describing data gathering, usage and access privileges	Provides awareness to the users, autonomy, and trust in the monitoring system
Data security measures	Data transmission and data storage encryption, secure authentication, and access controls	Secures the confidentiality of mental health information against unauthorized users or losses
Data anonymization	Eradication or concealment of personal data	Minimizes the risk of loss of privacy and makes data analyzable and researchable
Regulatory compliance	Compliance with codes of ethics and health data security laws	Makes the system legal and ethically legitimate
Algorithmic fairness	Testing of the ML models among the various population demographics	Avoids prejudices, discrimination, and unequal mental health forecasts
User data control	Alternatives to view, limit, remove, or erase personal information	Enhances the sense of autonomy and respect of individual dignity
Governance and accountability	Specified roles of data stewards, auditing and transparency	Ensures good data practices and accountability of the system
Psychological safety	Reducing the false alarms and alerts that are intrusive or distressing	Mitigates damage and promotes moral interaction with users

Source: Own elaboration

reliability, the compliance of the user, and the necessity of a strong clinical validation, are also significant limitations. Further studies are needed to enhance the security of data, precision of sensors, user interaction, and a smooth combination with clinical processes to enable its mass adoption and successful implementation in mental healthcare.

References

[1] Almalawi A, et al. Managing security of healthcare data for a modern healthcare system. Sensors. 2023;23(7):3612. https://doi.org/10.3390/s23073612.

[2] Xing Y, et al. Intelligent sensing devices and systems for personalized mental health. Med-X. 2025;3(1):10. https://doi.org/10.1007/s44258-025-00057-3.

[3] Robinson T, et al. Self-management of subclinical common mental health disorders (anxiety, depression and sleep disorders) using wearable devices. International Journal of Environmental Research and Public Health. 2023;20(3):2636. https://doi.org/10.3390/ijerph20032636.

[4] Rani P, et al. Simulation of the lightweight blockchain technique based on privacy and security for healthcare data for the cloud system. International Journal of E-Health and Medical Communications (IJEHMC). 2022;13(4):1–15. https://doi.org/10.4018/IJEHMC.309436.

[5] Pujahari RM, Khan R, Yadav SP. Integration of quantum artificial intelligence in healthcare system. Quantum Computing for Healthcare Data. Academic Press. 2025;139–66. https://doi.org/10.1016/B978-0-443-29297-2.00005-8.

[6] Wu J-Y, et al. Emerging wearable biosensor technologies for stress monitoring and their real-world applications. Biosensors. 2022;12(12):1097. https://doi.org/10.3390/bios12121097.

[7] Torku A, et al. Wearable sensing and mining of the informativeness of older adults' physiological, behavioral, and cognitive responses to detect demanding environmental conditions. Environment and Behavior. 2022;54(6):1005–57. https://doi.org/10.1177/00139165221114894.

[8] Winslow BD, et al. Automated stress detection using mobile application and wearable sensors improves symptoms of mental health disorders in military personnel. Frontiers in Digital Health. 2022;4:919626. https://doi.org/10.3389/fdgth.2022.919626.

[9] Gedam S, Dutta S, Jha R. Analyzing mental stress in Indian students through advanced machine learning and wearable technologies. Scientific Reports. 2025;15(1):20610. https://doi.org/10.1038/s41598-025-06918-6.

[10] Iqbal T, et al. Stress monitoring using wearable sensors: A pilot study and stress-predict dataset. Sensors. 2022;22(21):8135. https://doi.org/10.3390/s22218135.

[11] Saito T, Suzuki H, Kishi A. Predictive modeling of mental illness onset using wearable devices and medical examination data: Machine learning approach. Frontiers in Digital Health. 2022;4:861808. https://doi.org/10.3389/fdgth.2022.861808.

[12] Koinis L, et al. A commentary on the potential of smartphones and other wearable devices to be used in the identification and monitoring of mental illness. Annals of Translational Medicine. 2022;10(24):1420. https://doi.org/10.21037/atm-21-6016.

[13] Liberman JN, et al. Mental health care provider's perspectives toward adopting a novel technology to improve medication adherence. Psychiatric Research and Clinical Practice. 2022;4(2):61–70. https://doi.org/10.1176/appi.prcp.20210021.

[14] Solaiman B, Malik A, Ghuloum S. Monitoring Mental health: Legal and ethical considerations of using artificial intelligence in psychiatric wards. American Journal of Law & Medicine. 2023;49(2-3):250–66. https://doi.org/10.1017/amj.2023.30.
[15] Adler DA, et al. Beyond detection: Towards actionable sensing research in clinical mental healthcare. Proceedings of the ACM on Interactive, Mobile, Wearable and Ubiquitous Technologies. 2024;8(4):1–33. https://doi.org/10.1145/3699755.

CHAPTER 19

Real-Time Monitoring of Chronic Gastrointestinal Diseases with Wearable Sensors

A R Chayapathi, K. V. Karthikeyan, Rishali

Abstract Chronic gastrointestinal diseases (CGDs), such as adaptive motility problems, irritable bowel syndrome, and inflammatory bowel disease, require continuous monitoring due to fluctuating symptoms and unpredictable exacerbations. Traditional approaches relying on episodic clinical evaluations, invasive diagnostics, and subjective symptom reporting limit timely intervention and personalized management. Wearable sensor technologies enable noninvasive, real-time monitoring of gastrointestinal health. This chapter explores wearable modalities, such as electrogastrography, bioimpedance, acoustic sensing, and motion-based indicators, integrated within Internet of Medical Things frameworks for continuous data transmission and remote access. Advanced signal processing and intelligent analytics detect abnormal motility patterns, physiological changes, and early signs of disease exacerbation. The models demonstrate high precision in gastrointestinal anomaly detection: single-shot detector (90.42%), You Only Look Once (version 4) (93.02%), YOLO (version 5) (99.071%), and a novel feature fusion network (97%). Applications show improved symptom tracking, personalized feedback, and proactive intervention. Considerations of patient comfort, long-term adherence, signal variability, data privacy, and clinical integration are discussed. Wearable sensor-based real-time monitoring offers a promising pathway for personalized, continuous, and proactive management of CGD, supporting data-driven clinical decisions and enhanced patient-centered care.

Keywords Wearable sensors – gastrointestinal diseases – real-time monitoring – Internet of Medical Things – digital health – chronic disease management

19.1 Introduction

Chronic gastrointestinal disease (CGD), such as inflammatory bowel disease (IBD), irritable bowel syndrome (IBS), and other disorders of functional motility in millions of people throughout the world, are associated with a high burden of symptoms, as well as decreased quality of life [1]. These disorders are normally identified by some typical symptoms, which include belly pain, bloating, abnormal bowel movement, and stomach upsets, and might vary dramatically over a long period and might be contingent on physiological and lifestyle conditions like stress, sleep,

 | HTTPS://DOI.ORG/10.1515/9783112237083-019

and activity levels [2]. That kind of variability makes the continuous observation of a patient's health a necessary step enabling timelier assessment, a deeper understanding of the symptoms, and a constructive contribution to the process of managing the disease.

Conventional ways of monitoring the conditions of gastrointestinal rest on periodic visits to the clinic, self-registered monitoring of the symptoms, and invasive diagnostics. Though these methods are useful in providing pointers, they are not effective in most cases to capture the dynamic changes in the disease activity as they could occur outside the clinical environment [3]. Clinical assessments are episodic and thus cannot detect the exacerbations of the symptoms in time, and thus fail to intervene early and respond to the symptoms. Self-reported symptoms are also unreliable and subjective, and thus, it is hard to assess the disease accurately [4].

The digestive system of the human organism proves that food gets into the body and out of it. It begins at the mouth, where the digestion occurs mechanically and chemically, and then passes to the esophagus, through which the stomach is reached, where further digestion of the food occurs. The small intestine was secreted with enzymes and bile to aid in digesting the nutrients and absorption into the liver, pancreas, and gallbladder. The small intestine absorbs nutrients into the bloodstream, while the large intestine absorbs water, forming feces.

One of the possible solutions to these issues should be wearable sensors that offer a real-time monitor of a person. An immediate tracking of physiological and behavioral biomarkers with wearable sensors, such as heart rate variability, activity, and biochemical biomarkers, could provide important insights into the symptom pattern, daily variation, and overall gastrointestinal health that cannot be easily observed through periodic clinical measurements [5]. Active and individual approaches to care might be based on such technologies because it is possible to track the significant changes in the activity of the disease earlier. One way to significantly change the patient care process is by integrating constant monitoring into the treatment of treating chronic gastrointestinal illnesses. By introducing objective and continuous data collection to conventional care, it would be capable of improving disease progression knowledge and patient engagement in self-management and ultimately result in better health outcomes and quality of life of individuals with chronic gastrointestinal disorders [6]. The development of a wearable sensor-based, noninvasive system for ongoing, real-time monitoring of chronic gastrointestinal disorders was the objective of this project. To provide dependable data transfer and remote clinician access, the investigation combines multimodal wearable sensing with an Internet of Medical Things (IoMT) infrastructure. Deep learning models and sophisticated signal processing are used to accurately identify early illness exacerbations and aberrant gastrointestinal patterns. The suggested strategy enhances patient comfort, adherence, and clinical judgment while promoting proactive, individualized illness care.

19.2 Wearable Sensors in Gastrointestinal Health

The use of wearable and ingestible sensors in real-time detection and management of CGD. The investigation focused on electrogastrogram devices, which chart the contraction of the smooth muscle. It achieves this through the noninvasive recording of the electrical activity of the stomach. The information was quite significant in the therapy of disorders such as IBS, functional dyspepsia, and gastroparesis. It also discusses acoustic gastrointestinal biosensors, which are a type of bowel sounds measurement employing embedded abdominal microphones to provide an indication of delayed stomach emptying, surgical ileus, and personalized dietary maturation [7]. Moreover, the digestible sensor tech to detect gut microbiota metabolites to enhance diagnostics and nutrition is also being examined to constantly test the pH, pressure, temperature, and luminal contents. It was strategy that the purpose of the study was to support continuous data exchange, remote patient monitoring, intelligent analytics, proactive clinical interventions, and enhancements in symptom monitoring, patient-specific feedback, patient comfort, and whole-patient-centered care by linking these technologies in interconnected healthcare systems.

The goal of this research is to explore the possibility of using wearable sensor technologies to monitor the maintenance and acute conditions in real time, and continuously, through the gastrointestinal tract (GIT). Chronic gastrointestinal diseases (CGDs), such as functional mobility disorders, IBS, and IBD, are associated with different symptoms and unpredictable exacerbations. They need continuous and accurate monitoring that could make it possible to manage them individually. On the same note, gastrointestinal crises such as necrotizing enterocolitis, obstruction or stenosis of the digestive tract, and gastrointestinal bleeding are equally dangerous and should therefore be detected early to prevent severe adverse outcomes. The traditional ways of keeping track that use episodic clinical evaluation, invasive diagnostics, and application of subjective symptoms are likely to delay intervention and limit successful control [8]. In systems that use IoMT, objective, noninvasive, and continuous gastrointestinal activity monitoring could be achieved with the aid of wearable and connected gadgets like the electrogastrography, bioimpedance, acoustic sensing, and motion-based physiological signals. It is combined with advanced signal processing and intelligent analytics to identify the abnormal motility patterns, assess the enteral nutrition tolerance, and anticipate the worsening of a disease. The strategy assists in maintaining proactive and evidence-based, patient-centered care, clinical decision-making, prognosis, and timely intervention in CGDs and acute gastrointestinal crises.

This chapter proposes an all-encompassing system of real-time, patient-centered CGD management through wearable sensors, artificial intelligence (AI), and IoMT systems. The CGDs that necessitate constant monitoring include functional motility difficulties, such as IBS and IBD, as the symptoms are constantly changing and the disease could trigger an exacerbation at any time. Wearable sensors make it possible to continuously monitor important physiological variables such as gastrointestinal motility, heart rate, skin temperature, and physical activity in a noninvasive manner to offer real-time data on the health of the patients [9]. The system could be used to identify abnormal patterns and predict exacerbations as well as provide more individualized interventions to enhance patient outcomes and compliance by combining these sensors with edge computing and AI-driven analytics. In patients with comorbid conditions, like diabetes, the framework also includes a smart transdermal drug delivery system that could dynamically change the level of medication in real time depending on the values of vital signs collected with wearable sensors and connected continuous glucose monitoring devices, which improves the treatment efficiency and security. Moreover, the investigation examines the AI-aided automatic detection and localization of colorectal cancer lesions on an image, decreasing the number of examinations required of the clinician, enhancing the accuracy of lesion detection, and facilitating prompt treatment planning. Through continuous physiological monitoring, smart therapeutic delivery, and automated diagnostics, this investigation could develop an all-encompassing, data-driven approach to proactive disease management, personalized care, better symptom tracking, and better clinical decision-making in connected healthcare settings.

19.3 Integration within Internet of Medical Things (IoMT)

Gastrointestinal disorders, diabetes mellitus, and cancer are the types of chronic conditions requiring a permanent assessment and active treatment due to the alteration of symptoms, complex physiological patterns, and potential critical complications. The conventional surveillance systems may be termed as timely, intrusive, or having low predictive power, and, hence, there is a need to have real-time data-driven solutions. All these conditions can be monitored, predicted, and personalized interventions could be introduced with the help of the current wearable sensors and AI-driven technologies. The electrogastrography, bioimpedance, acoustic sensing, and motion-based physiological indicators that are incorporated into the IoMT frameworks, in the example of gastrointestinal diseases, could be employed to measure the abnormal motility patterns and symptom exacerbation at early stages [10]. It has been discovered that the application of AI and machine-learning approaches including random forest methods, trees of decisions, and artificial neural networks (ANNs) can enhance the accuracy of continuous glucose monitoring, predict glycemic patterns, and give personalized nutrition and treatment plans in managing diabetes. In cancer, solutions using AI to interpret imaging, clinical, pathology, and genetic data can optimize the detection of cancer at an early stage, cancer subtypes, prognosis, and individual treatment plans. By integrating wearable sensors, sophisticated analytics, and intelligent decision-support systems, AI-based models could offer a comprehensive, real-time, patient-centered solution to monitoring and management of chronic disease, improve clinical outcomes, and take proactive measures.

19.4 Machine Learning Models for Gastrointestinal Symptom Assessment Using Wearable Sensors

To predict physiological or gastrointestinal disorders in real time, wearable devices generate and forecast physiological markers using machine learning and deep learning technologies. In order to facilitate active control and customized administration, these models analyze complicated sensor data to identify noteworthy patterns and trends. A detailed analysis of each model's performance is provided in the "Results" section.

19.4.1 Bidirectional LSTM (Bi-LSTM)

Bi-LSTM (Long Short-Term Memory) networks are utilized to compare past and future data sequences to predict physiological measurements in the case of wearable gastrointestinal sensors. The ability of the Bi-LSTM to identify aberrant patterns of motility, physiological changes associated with symptoms, and early signs of further worsening of the disease are true since it recognizes temporal dependencies in both directions [11]. This chapter explores how wearable monitoring of gastrointestinal health facilitates proactive intervention and personalized patient feedback. The strategy would be able to enhance patient-centered care and guarantee timely CGD management through data-driven clinical decision-making, enhanced symptom monitoring, and enhanced variability and real-time signal interpretation issues, as shown in Figure 19.1:

$$b_g^s = \sum_{l=1}^{K} w_k^s x_{kg} + \sum_{g',s>0}^{G} a_{g'}^{s-1} x_{g'g} \tag{19.1}$$

$$b_g^s = \theta_g\left(b_g^s\right) \tag{19.2}$$

In eqs. (19.1) and (19.2), b_g^s represents the node g pre-activation aggregated feature at layer s, which aggregates contributions from its input features and nearby nodes. The input feature x_{kg} indicates the kth property of node g, and w_k^s is the associated weight for that feature at layer s, where K is the total number of input features. The initial sum, $\sum_{l=1}^{K} w_k^s \, x_{kg}$ determines the weighted sum of all input characteristics for the node g. The second sum, $\sum_{g',s>0}^{G} a_{g'}^{s-1} x_{g'g}$ collects data from nearby nodes g' throughout the preceding layer $s-1$, where $a_{g'}^{s-1}$is the weight or adjacency coefficient that shows how node g influences node $x_{g'g'}$, which represents the edge characteristic or interaction between nodes g' and g, and G indicates the total number of nodes in the graph. Only contributions from earlier layers, excluding the input layer, are included, thanks to the requirement $s > 0$. Lastly, the activation function $\theta g\ (\cdot)$ is applied to b_g^s generating the post-activation value, which adds nonlinearity and produces the node g' finals updated representation at the layer s.

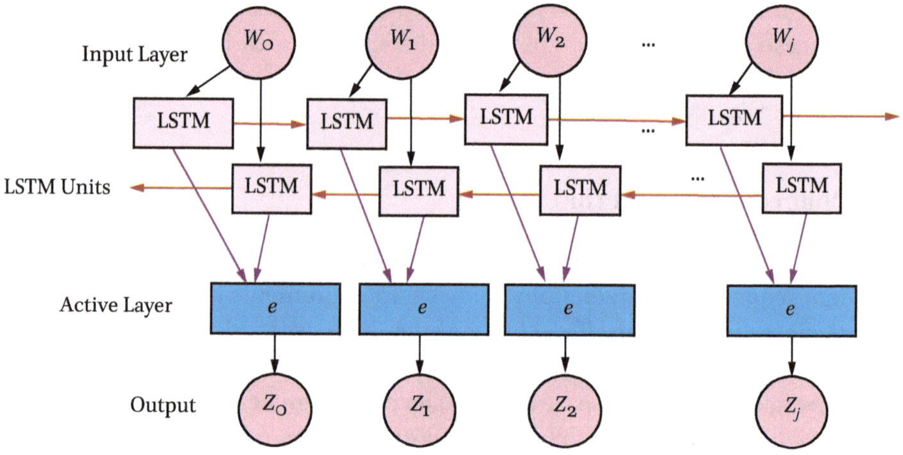

FIGURE 19.1 Bidirectional LSTM model for real-time gastrointestinal disease monitoring using wearable sensor data. (Source: Own elaboration)

19.5 Clinical Applications and Self-Management

To connected healthcare systems could transform raw physiological signals into clinical insights with the combination of ingestible and wearable sensor data with cognitive analytics. The pH, pressure, and temperature monitoring, motility patterns, gas production, and luminal contents would be monitored constantly to enable individual gastrointestinal profiles to be reflective of normal functioning and pathophysiology of an individual. Machine-learnt analytics might be applied to sensor-based biomarkers and symptoms, along with dietary measurements, medication intake, and lifestyle, to identify the initial signs of a disease outbreak or abnormal functioning. This evidence-based practice could assist in ensuring the timeliness of the clinical intervention, according to the personal dietary prescriptions, and simplify the treatment plan. In addition, the feedback given in mobile health applications on time helps the patient to manage independently and improves compliance and long-term outcomes [12]. It was possible to monitor remotely, supervise clinicians, and analyze the population level through interconnectivity with electronic health records and IoMT platforms. The convergence of ingestible sensors, wearable technologies, and intelligent analytics represents a significant step toward active, noninvasive, patient-centered management of chronic gastrointestinal ailments in modern connected healthcare settings, despite the difficulties associated with sensor miniaturization, data dependability, personal data security, and regulatory approval.

19.6 Results and Conclusion

Continuous monitoring and early diagnosis are offered by techniques that utilize AI, and conventional episodic techniques possess poor real-time capacity to monitor chronic gastrointestinal disorders. A new architecture with the advanced deep learning models, including single-shot detector (SSD), You Only Look Once (version 4) (YOLOv4), YOLO (version 5) (YOLOv5), ANN, and feed-forward neural network (FFNN) models with GoogLeNet and AlexNet as the additional characteristics of LBP, was adopted to achieve, high classification with the wearable sensor data of patients, yet the multi-sensor integration and the detection of abnormal motility with the help of the bachelor-student and novel feature fusion models. Such measures as accuracy, precision, sensitivity, specificity, F_1-score, mean average precision (mAP), and area under the curve (AUC) were used to measure such models. The analysis of sensor data using AI was effective in real time to provide customized feedback and make thoughtful decisions in the healthcare industry, particularly in preventing interventions.

19.6.1 AI-Driven Monitoring of Chronic Gastrointestinal Diseases (CGDs)

CGD datasets are yet to be tackled, and AI has greatest performance in healthcare. The multimodal physiological data, which are measured with wearable sensors, such as electrogastrography, bioimpedance, auditory indicators, and motion-based indicators, were used in this work. ANN [13], FFNN [13], SSD [14], YOLOv4

[14], YOLOv5 [14], and novel feature fusion models [15] were categorized as aberrant motility detection models. These models could be combined to provide proactive monitoring, tailored feedback, and timely intervention of the multi-sensor feature analysis and real-time sensor data. The AI-based technology could lead to an accurate, efficient, and receptive procedure for treating chronic GIT disorders by covering the weaknesses of the traditional episodic and intrusive tracking strategies:

- **Accuracy**: The term "accuracy" refers to the general accuracy of the models to detect CGD-based physiological abnormalities and symptom patterns. The correlation use of the AlexNet features to generate predictions of reliability by implementing the aforementioned features to models with the use of FFNN indicated that the latter provided a higher overall accuracy, which, in turn, suggests that the latter provided a higher reliability in the prediction of reliability.
- **Precision:** It is the percentage of the correct predictions that are actually correct of what is being predicted. The results of ANN and FFNN deep feature extraction models were great, and false alarms were minimized, thereby enhancing the plausibility of the detected events.
- **Sensitivity (Recall):** The capacity to detect correctly true positive CGD events of the model. The sensitivity with features of AlexNet was introduced into the FFNN models, identifying the most important patterns of abnormalities to intervene in a timely manner.
- **Specificity**: Competence in accurately identifying the non-anomalous physiological events, but not false alerts. The advanced features integration models were virtually impeccable in their specificity to make sure that the unnecessary interventions are minimized to the utmost.
- **Area under the Curve (AUC):** It is a measure that was applied to give an idea of the cumulative discriminative force, the strength of the models in distinguishing between normative and abnormal physiological situations. This is because the neural structure that had the rich inputs in terms of the features produced improved AUC, which represents superior classification capability.
- **Mean Average Precision (mAP):** By making it possible to use wearable sensors monitor chronic gastrointestinal disorders in real time. To guarantee accurate identification of aberrant gastrointestinal patterns, one of the most important objectives was to assess detection models using measures such as precision and mAP. The strategy intends to improve prompt intervention, tailored feedback, and accurate condition monitoring.

The performance was evaluated using the ANN and FFNN models with a hybrid feature extraction method. Whereas fuzzy color histogram, gray-level co-occurrence matrix, local binary pattern (LBP), and first-order statistical features were used to get the texture and the statistical feature, deep features were obtained in the GoogLeNet and AlexNet architectures. The efficacy of the models was assessed with the help of standard evaluation criteria, including sensitivity, precision, specificity, accuracy, and AUC. These steps provide a complete analysis of strength, discrimination of classes, consistency of prediction, and general classification dependability. The results demonstrate that handmade and deep attributes significantly boost the two neural models. The FFNN-based structure has a superior generalization and a higher discrimination performance in the evaluated configurations that show the suitability to do accurate and reliable classification of tasks, as shown in Figure 19.2 and Table 19.1.

TABLE 19.1 Performance comparison of ANN and FFNN models using deep features for gastrointestinal anomaly detection [13].

Metrics	**ANN**		**FFNN**	
	GoogLeNet feature (%)	**AlexNet feature (%)**	**GoogLeNet feature (%)**	**AlexNet feature (%)**
Accuracy (%) [13]	98	99.1	98.5	99.3
Precision (%) [13]	98.25	99	98.6	99.2
Sensitivity (%) [13]	97.8	98.8	98.2	99
Specificity (%) [13]	99.4	99.8	99.75	100
AUC (%) [13]	98.69	99.76	98.83	99.87

In Table 19.2, to measure detection accuracy and dependability, a sample image dataset was evaluated to determine the performance of object detection models, including SSD, YOLOv4, and YOLOv5. Precision, mAP, and recall are standard assessment measures used to evaluate the accuracy of the projected bounding boxes, the overall performance of the detectors across all the classes, and their ability to correctly identify relevant items. YOLO-based architectures demonstrated improved results in the models, with YOLOv5 showing the most consistency among the models and the most reliable detection ability. The results show that a combination of advanced YOLO models with well-annotated image samples leads to highly accurate and reliable object detection and creates a very good balance between recall and precision, and ensures successful classification of classes.

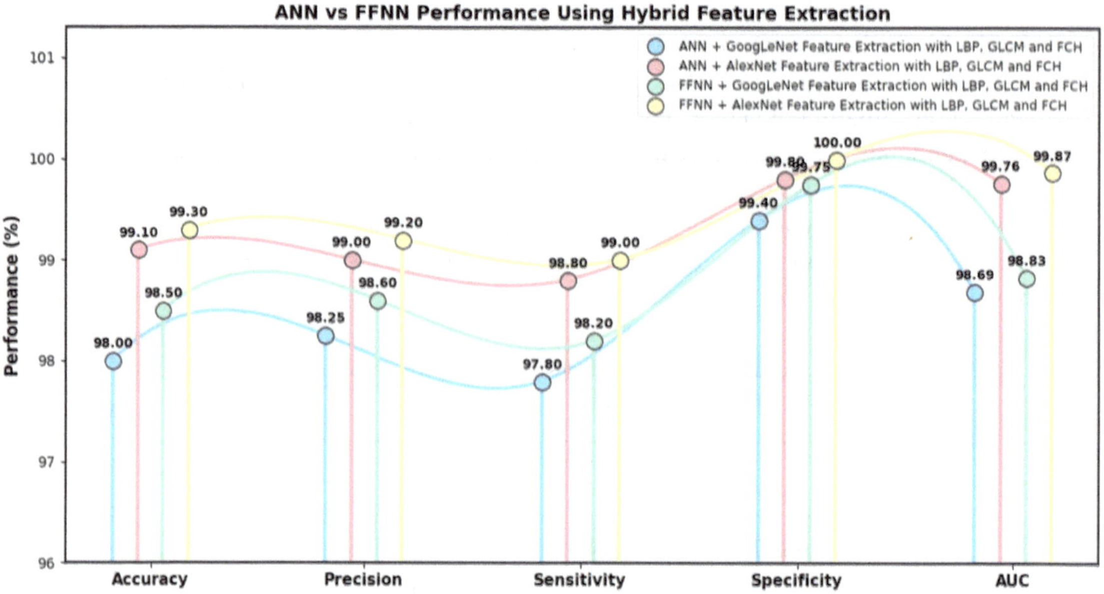

FIGURE 19.2 Comparison of ANN and FFNN performance using hybrid feature extraction for gastrointestinal disease monitoring. (Source: Own elaboration)

TABLE 19.2 Performance comparison of SSD, YOLOv4, and YOLOv5 models for gastrointestinal anomaly detection metrics [14].

Methods	Precision (%)	mAP (%)	Recall (%)
SSD [14]	90.42	87.5	91.32
YOLOv4 [14]	93.02	92.7	92.83
YOLOv5 [14]	99.071	98.8	98.064

In Table 19.3, integrating multiple representations of discriminative features to enhance the classification effectiveness, the efficacy of the novel feature fusion network [15] such as EfficientNet B0, Effimix network was evaluated with the help of a common dataset. In order to assess the overall classification accuracy, positive prediction reliability, the ability to identify the relevant cases, and the precision/recall tradeoff, standard performance measures, such as precision, accuracy, recall, and F_1-score, were employed. Based on the experimental findings, the model has good learning ability in the given dataset and also attains very good classification accuracy. Moreover, the recall of the model confirms that it is capable of detecting most of the relevant samples, and its high accuracy indicates that the model has a low false positive rate. The obtained F_1-score demonstrates the novel feature fusion network's stability and balanced performance for the necessary dataset.

TABLE 19.3 Performance metrics of AI models applied to wearable sensor data for GI diseases [15].

Methods	Accuracy (%)	Precision (%)	F_1-score (%)	Recall (%)
Novel feature fusion network [15]	98	97	97	98

19.7 Future Trends and Challenges

Future technological innovations in the fields of cognitive analytics, integrated healthcare systems, and multimodal wearable device sensors will drive the real-time monitoring of chronic gastrointestinal disorders. The next generation of wearables shall be incorporate many sensing modalities on small, stretchable, and skin-capable platforms to enhance physiological coverage and comfort of patients, such as electrogastrography, bioimpedance, audio, and motion signals. Individualized background modeling, an early indication of symptom flare-up, predictive analysis of the possibility of disease exacerbation, etc., will become possible with the combination of AI and machine learning models. The privacy of the patients will be ensured, and low-latency analytics will be supported with the assistance of edge computing and federated learning approaches. It will assist in increasing the engagement of physicians and the implementation of the proactive and data-driven interventions by including the clinical decision-support systems and electronic health records. In addition, personalized feedback, behavioral feedback, and self-management feedback will still be the focus of the patient-facing applications; however, a number of obstacles exist. Variability in signals, as a result of individual variances, motion artifact, and ambient noise, influences data dependability and clinical interpretability. The factors of sustained monitoring are long-term wearability, battery efficiency, and compliance with the users. The key issues include data security, protection of privacy, and regulatory compliance, especially in a large-scale IoMT ecosystem. Further, it had not been widely adopted, because of the absence of clinical recommendations and set validation procedures. Interdisciplinary interaction, widespread clinical study, and common frameworks should solve these issues to institute the use of wearable sensor-based gastrointestinal monitoring in the medical field for typical clinical applications.

19.8 Conclusion

Wearable devices, AI, and IoMT systems can revolutionize the workflow of CGDs. These medical conditions are dynamic; hence they cannot be effectively measured with self-reported supervision and traditional episodic clinical studies that usually postpone instant therapies. Continuous, noninvasive measurements of physiological and behavioral biomarkers, such as electro-gastrography, bioimpedance, acoustic, and motion sensors, became possible due to wearable technology. These methods were very accurate, sensitive, specific, and precise in determining the aberrant patterns of motor movements in combination with modern machine learning algorithms such as ANN, FFNN, SSD, YOLOv4, YOLOv5, and new feature fusion networks. Proactive and personal care could be supported by AI-based analytics because it could notify in real-time and offer customized interventions and more involvement in patients. In addition, the interface to the IoMT platforms makes the transfer of data, remote monitoring, and the foundation of clinical decision-making seamless. This chapter discusses potential paradigm changes and solutions to change the current state. Wearable sensor-based monitoring was a potential approach to patient-centered, predictive, and proactive management of gastrointestinal problems, but the potential risks include signal variability, long-term compliance, data privacy, and regulatory regulations. Such intelligent healthcare systems should continue to get the best out of interdisciplinary investigations, clinical verification, and technical developments.

References

[1] Severo JS, et al. Physical exercise as a therapeutic approach in Gastrointestinal diseases. Journal of Clinical Medicine. 2025;14(5):1708. https://doi.org/10.3390/jcm14051708.

[2] Bhardwaj J, et al. A review in wavelet transforms based medical image fusion. In Evolving Role of AI and IoMT in the Healthcare Market 2022 (vol. 1, pp. 199–214). Cham: Springer International Publishing. https://doi.org/10.1007/978-3-030-82079-4_9.

[3] Koppelman LJM, et al. Modifiable factors influencing disease flares in inflammatory bowel disease: A literature overview of lifestyle, psychological, and environmental risk factors. Journal of Clinical Medicine. 2025;14(7):2296. https://doi.org/10.3390/jcm14072296.

[4] Thomas Abraham JV, et al. A deep-learning approach for identifying and classifying digestive diseases. Symmetry. 2023;15(2):379. https://doi.org/10.3390/sym15020379.

[5] Yadav SP, Yadav S. Image fusion using hybrid methods in multimodality medical images. Medical & Biological Engineering & Computing. 2020;58(4):669–87. https://doi.org/10.1007/s11517-020-02136-6.

[6] Forood K, Mohammad A, et al. Development and Validation of a Multimodal Wearable Belt for Abdominal Biosignal Monitoring with Application to Irritable Bowel Syndrome. Micromachines. 2025;16(11):1255. https://doi.org/10.3390/mi16111255.

[7] Kim S, et al. Advances in Digital Health for Gastrointestinal Nutrition. Current Treatment Options in Gastroenterology. 2024;22(4):267–78. https://doi.org/10.1007/s11938-024-00458-0.

[8] Mao Y, Yang Y, Xiong T. Gastrointestinal emergencies and monitoring in preterm infants: A review. Annals of Medicine. 2025;57(1):2525397. https://doi.org/10.1080/07853890.2025.2525397.

[9] Sarkar M, Lee T-H, Sahoo PK. Smart healthcare: Exploring the internet of medical things with ambient intelligence. Electronics. 2024;13(12):2309. https://doi.org/10.3390/electronics13122309.

[10] Manickam P, et al. Artificial intelligence (AI) and internet of medical things (IoMT) assisted biomedical systems for intelligent healthcare. Biosensors. 2022;12(8):562. https://doi.org/10.3390/bios12080562.

[11] Hamayel MJ, Owda AY. A novel cryptocurrency price prediction model using GRU, LSTM and bi-LSTM machine learning algorithms. Ai. 2021;2(4):477–96. https://doi.org/10.3390/ai2040030.

[12] Chong KPL, Woo BKP. Emerging wearable technology applications in gastroenterology: A review of the literature. World Journal of Gastroenterology. 2021;27(12):1149. https://dx.doi.org/10.3748/wjg.v27.i12.1149.

[13] Fati SM, Senan EM, Azar AT. Hybrid and deep learning approach for early diagnosis of lower gastrointestinal diseases. Sensors. 2022;22(11):4079. https://doi.org/10.3390/s22114079.

[14] Aliyi S, Dese K, Raj H. Detection of gastrointestinal tract disorders using deep learning methods from colonoscopy images and videos. Scientific African. 2023;20:e01628. https://doi.org/10.1016/j.sciaf.2023. e01628.

[15] Ramamurthy K, et al. A novel multi-feature fusion method for classification of gastrointestinal diseases using endoscopy images. Diagnostics. 2022;12(10):2316. https://doi.org/10.3390/diagnostics12102316.

CHAPTER 20

Human-Centered Design of Wearable Sensor Technologies for Chronic Disease Management

Raj Kumar Dhakad, Manivannan Karunakaran, Rekha Chakravarthi

Abstract Chronic disease management increasingly relies on wearable sensor technologies to support continuous monitoring, early intervention, and long-term self-management. Despite technological advancements, many wearable systems face limited adoption due to discomfort, usability challenges, cognitive burden, and lack of alignment with user needs. In this chapter, the authors intend to discuss the ways in which the principles of human-centered design (HCD) can be used to make wearable sensor technologies more effective, usable, and acceptable in chronic disease management. This chapter addresses the principles of HCD and its use in wearable healthcare systems. The essential ones are its wearability, ergonomics, as well as optimization of the user experience in the long-term use. The acceptance of the user, the trust, and the perceived value across the chronic care settings are explored together with the inclusive and accessible interface design. The issues of ethical considerations, privacy protection, and the problem of long-term engagement are also discussed, focusing on the aspect of a balance between intelligent automation and human control. HCD is very important in ensuring that wearable sensor technologies efficiently contribute to the management of chronic diseases by enhancing comfort, usability, trust, and long-term involvement, among others, to deliver credible patient-oriented digital health solutions.

Keywords Human-centered design – wearable sensors – chronic disease management – user experience – digital health – Internet of Medical Things (IoMT)

20.1 Introduction

The current paradigms in modern medicine emphasize the maintenance of wellness and dealing with chronic diseases to enhance the health of the population and cut costs. The world has major causes of morbidity and mortality and a large proportion of healthcare costs is attributed to cardiovascular disease, diabetes, tumors, among other chronic illnesses. The techniques of coordinated care are applied in the treatment of chronic illnesses in order to manage the already existing medical conditions. The chronic disease management optimizes the treatment of the patient through premeditated course of treatment, whether it is through

 | HTTPS://DOI.ORG/10.1515/9783112237083-020

adherence to medication and frequent assessment [1]. Wearable devices in the treatment of chronic illnesses have become a significant and growing concern of the medical field. In case of patients with chronic illnesses, these devices, including trainers, smartwatches, and other devices with sensors can provide 24/7 tracking and real-time data collection. It is possible to collect symptoms, the activity level, sleeping patterns and other data, which will provide valuable data about the state of a person. Wearable devices make individuals stay in touch and safe, as well as from prevention and treatment of dangerous illnesses [2]. Intelligent devices provide essential moral subjects besides the numerous potential benefits. The security and privacy of information is a concern. The issue of health privacy should be given the first priority as the health-related information is one of the most sensitive categories of privacy. Wearable technology can collect very large amounts of personal data that can either be breached through malicious hacking or negligence and find itself in an unsecured place. The other important issue is the security of devices [3]. Human-centered design (HCD) performs better than traditional wearable sensors because it prioritizes user comfort, usability, trust and engagements, leading to continued adoption and successful management of chronic diseases, as opposed to technical monitoring only [4]. Object detection simplifies automation in various ways in terms of identifying objects, classifying item tracking, and classifying characteristics such as color, characteristics, and small details of a particular object. Artificial intelligence (AI) is an omnipresent discipline, and in most of its applications, it makes use of automation, especially when detecting objects: image extraction, safety, monitoring, and computerized analysis of automotive systems and equipment [5]. By combining ergonomics, user experience, Internet of Medical Things (IoMT), and intelligent healthcare systems, this chapter aims to investigate how HCD principles improve the usability, acceptability, and efficacy of wearable sensor technologies for managing chronic diseases.

20.2 Medical and Chronic Condition Treatment Applicable Equipment

Wearable technology facilitates the tracking of an individual's activities, routines, and genetic markers. The majority of people assess their heart rate, pulse, thermal regulation, blood oxygen content, posture, and level of activity. This is accomplished using ballistocardiograms and electrocardiogram (ECGs). Scientific information can be gathered using wearable cameras or video recorders.

20.2.1 Wearable Devices and Data Collection Infrastructure

- Watches, shoes, glasses, clothes, earrings, and accessories can all incorporate electronic devices.
- For ongoing surveillance, additional wearable technology can also be surgically placed.
- Scientific and social information is gathered via wearable cameras and video recorders.
- Wearable data is sent to distant servers via smartphones, which function as gateways.
- Cloud-based technologies are used to maintain and analyze the data gathered.
- Real-time access to and analysis of health data is supported by smartphones and handset gadgets.

20.2.2 Wearable Technology for Motion and Functional Evaluation

Electromagnets, acceleration sensors, and angled video recordings are examples of wearable technology, which make it possible to analyze walking habits, equilibrium, posture, and movement in great detail. Through constant and real-time gait examination, these devices help doctors identify mobility problems, measure treatment progress, evaluate fall risk, and enhance health monitoring. The leading chronic illnesses are cancer, heart disease, cerebrovascular disease, strokes, asthma, diabetes, neurological disorders, chronic renal disease, obesity and joint pain. The rapidly growing technologies of AI, blockchain, and wearable gadgets can be used to monitor the health of people with chronic illnesses.

20.2.3 Advantages of Wearable Technology in chronic illness

Wearable gadgets are used in chronic care tracking, observing patients, and hospital stations. While smartwatches can report daily activities and weather, necklaces and monitors can track medical records and prescriptions. For tracking patients, wearable technology is essential [6]. Clinicians can better deal with their patients by using wearable technology to track patient actions and data. Monitors and bracelets offer vital data in hospitals and senior residences.

20.2.4 Passive Wearable Monitoring and Its Role in Chronic Disease Treatment

Numerous wearable gadgets have been created that can passively gather real-time medical, cognitive, and ecological information while people go about their daily lives (a process known as mobility monitoring). These cutting-edge wearable sensors have great potential for tracking patients for a longer duration of time and gathering data in real-time during a person's regular everyday activities. Additionally, through alerts, smartphone synchronization, and/or videoconferencing system connectivity, these devices can facilitate the direct transmission of behavioral and therapeutic aspects to patients, thereby enhancing therapy for chronic conditions that necessitate continuous monitoring and adherence. New wearable medical sensors are not limited to people who have been diagnosed with or are at high risk of developing a chronic illness.

20.2.5 The Wearable Health Emerging Technology User Acceptance and Adoption Challenges

Adoption Insecurity: There is uncertainty about why the wearable health devices may or may not be adopted due to the lack of knowledge about the attributes, perceived difficulty, and lack of trust. Consequently, the real-life deployment heavily depends on the user acceptance.

First Measurement: It is vital as it has an effect on the improvement of the product, deployment strategies, marketing, and client retention, which increases the likelihood of massive acceptance [7].

Acceptability by Users: When the wearable sensors seem to users disruptive, obtrusive, inaccurate, or incompatible with their routine habits and requirements, even highly accurate and clinically validated smart sensors will lack the intended impact.

20.3 Advanced Wearable and On-Body Sensing Technologies for Continuous Monitoring of Chronic Diseases

Wearable sensing tools enable the monitoring of long-term conditions easily with high levels of adherence and portability of records. They are added to soft and stretchable materials, which enhances practicality and convenience.

Aerosol Monitoring: There are a vast number of on-skin analysis methods, including patches, stickers, earpieces and socks, enhancing the monitoring of sweat biomarkers. Nonetheless, automated sweat collection is disadvantageous because of the sweat coagulation and low quantities, including spatial resolution as well as accuracy problems. To solve these issues, hydrogels are being considered in digital interfaces to enhance sampling. Sweating can be controlled by iontophoresis and long-term local activation can be performed, as well as sensor capture of diseases such as irritation, joint pain, and diabetes can be tracked in real time.

Tracking of Interstitial Fluid (ISF): ISF is the profile of the blood biomarkers that contain minimal cellular interference. Challenges in the extraction of ISF include limited volume and hydraulic restriction by the outside matrix. Studies of such methods as microneedles and microdialysis are being pursued. To measure glucose in ISF, continuous glucose meters (CGMs) require sensor instruments that are treated with enzymes, but they also have issues of sensor drift and frequent replacement. Though microneedles provide a low invasive option, the penetration of the skin is not always uniform meaning that it may affect accuracy. Reverse iontophoresis can also be used to promote the flow of ISF using electric current to extract certain small compounds selectively, which may be diluted and thus harder to analyze.

Observation of Wound Exudate: Wound exudate in chronic wounds with related disorders such as diabetes, cardiovascular, and stress injuries may be observed with implanted devices like bandages. On-site measurements of metabolites in the wound fluid can be done using soft microfluidic molecular sensors to aid clinical testing and therapy.

Respiratory and Mouth Surveillance: Dental protection and tooth stickers are examples of devices that make it possible to directly monitor fluid biomarkers, markers of regional diseases like dental decay, and external chemicals like food and medication intake. Current inner-oral detectors ignore solid objects in favor of detecting liquids. Foods, hence, solid-body sensor development should be the main goal of future research. Stability issues brought on by the regular deterioration and salivary contaminants must also be addressed.

Vision Examination: The constant monitoring of retinal biomarkers for conditions like glaucoma and metabolites for diabetes is made possible by tear analysis, which was significantly enhanced through projects like Google's smart contact lens (SCLs). While SCLs use biomolecular sensors for ongoing monitoring, eyeglasses can gather tears using fluid sample units Moreover, damaged sensors that can be worn can be used together with therapeutic features such as drug therapy and mechanical or electric stimulation to accelerate healing. It is still a challenge of miniaturizing power supply and circuitry to the lenses [8]. These are examples such as wireless data transmission and power. Even though SCLs have tremendous potential to monitor diabetes, they can only be practically applied under the condition of variation of blood glucose and tear concentration, which demand precise sampling and individual adjustments to ensure the accuracy of results.

As noted in Table 20.1, privacy, reliability, and system integration are seen as challenges, whereas wearable gadgets and their usefulness in continuous tracking, chronic disease management, caring seniors, and barriers are summarized [9]. However, wound tissue testing in situ is challenging because the reliability of data can be impacted by interference, a small sample quantity, and a complicated makeup that includes debris and cell death.

TABLE 20.1 Applications, benefits, and challenges of wearable strategies in long-lasting illness and elderly care [9]

Components	Summary	Essential devices	Benefits
Continuous monitoring	Wearables improve diagnostic precision and offer forecast health data through tracking vital signs in real time.	Healthcare monitors and smartwatches	Provide instant data gathering and early health problem detection.
Chronic disease monitoring	Devices are essential for controlling chronic illnesses like diabetes and stroke because they monitor energetic signs like heart rate and glucose.	Glucose sensors and heart rate sensors	Effective management of chronic illnesses and fewer issues
Elderly care	Wearables provide remote oversight, which enhances the elderly's standard of living by lowering hospital visits and facilitating prompt interventions.	Devices for remote wellness tracking	Improved the level of life and lowered medical expenses
Challenges	Acceptance is still significantly hampered by privacy issues, device accuracy, and interaction with medical facilities.	Different wearables	Possibility of increased healthcare effectiveness

20.3.1 Safety and Secrecy of Information Subjects

One of the most pressing concerns of wearable health devices is the maintenance of the safekeeping of the personal information of patients. Accessing it illegally or stealing information is a high risk as such gadgets always record the parameters of physiological activity, such as the heart rate, glucose levels, and movements. Cyber-attacks targeting medical records have been on the rise, and wearable devices with poor authentication and verification protocols can become a vulnerability of hackers.

20.3.2 Gadget Accuracy and Reliability

In managing chronic patient diseases with the ability to influence drug changes or treatment, health data evaluation must be accurate. Nonetheless, connected devices can be affected by environmental factors (humidity and temperature) or improper placement of the gadgets, as well as the shortcomings of sensor technology. Even consumer-level fitness devices, such as those, may not be as accurate as clinical devices, which may cause missed notifications or false alarms. Although both clinical validation and calibration are important processes, not every equipment available in the market is subjected to thorough testing before/before it is actually utilized. Healthcare professionals can have trouble combining patient data from many sources in the absence of defined data formats and ways to interact. This lack of compatibility restricts the possibilities for sophisticated analytics and AI-driven insights, in addition to impeding the effectiveness of monitoring patients online.

20.3.3 Next-Generation Wearables for Personalized Health and Therapy Management

Precision medicine, in which treatments are tailored based on a patient's biological composition, lifestyle, and regular therapeutic data, promises to help the next generation of wearable technology. For instance, CGM readings are already used by smart insulin pumps to modify insulin supply, and similar ideas are being developed for asthma treatment, Parkinson's disease relief, and cardiac rehabilitation [10]. The importance of wearables in the organization of enduring sicknesses is further strengthened by reinforcing the developments in cloud computing, wireless networking, and AI. However, there are certain difficulties in integrating them into the traditional system of healthcare. Data security, patient confidentiality, device precision, and user acceptance are still major obstacles. Wearable-enabled customized therapies can include continuous input loops, in which the gadget assesses patient compliance, modifies therapeutic guidance, and monitors the efficacy of treatment.

20.4 Integrating HCD, Ergonomics, and Socio-technical Principles in System Design

An example of the design theory and design approach that places human requirements, capabilities, and constraints at the core of the system development is called HCD. In the lifespan of a technology, it puts a lot of emphasis into contextual modification, communication with stakeholders, and trial with design. HCD includes more socio-technical integration besides usability and appearance. In the design and implementation, HCD, historically based on system theory, ergonomics and data systems research, puts primary emphasis on the demands, abilities, and limitations of different classes of users. This is tantamount to designing toward healthcare professionals, consumers, family members, managers, and sometimes even regulators who have some level of responsibilities, limitations and expectations.

20.4.1 Human Relations and Ergonomics

- The HCD of digital health, based on the ergonomics perspective, planning of human capabilities and constraints to improve efficiency, safety, and wellness.
- The concepts of human factors engineering point out that, with constant testing and refinements, instruments ought to be customized to the users other than vice versa. An example is that the social factors approach will need to evolve to consider a host of opinions and enable the real user participation of the design as care becomes more dispersed and complicated.
- This can involve applying user-centered design techniques of task evaluation, user testing and modeling in a real healthcare setting. Notably, ergonomics is a prompt that the effectiveness or ineffectiveness of a technology lies in macro-factors such as team coordination, the environment, and organizational culture almost as much as in minute details such as the design of the screen and cognitive fatigue.
- Daily activities or mental stress on customers cause medical equipment failures more often as opposed to the technology itself. Ergonomically sensitive design may help to reduce errors and enhance the level of acceptance by looking at the human, technical and process elements as a system.

20.4.2 Mobile Health (mHealth): Usability and User Engagement

Wearable technologies and mHealth apps present the most direct connection between the digital devices and healthy consumers. These technologies, covering wearable activity and vital indicator trackers, to smartphone applications to manage chronic diseases have a massive potential of empowering patients and promoting preventive treatment.

20.4.3 HCD to Effective and Culturally Sensitive mHealth

To be effective, digital health design needs mHealth tools to be integrated in everyday lives and healthcare systems. To enhance patient interactions, such as a heart rate monitoring application should allow reporting data to healthcare professionals. HCD is attempting to bring together the private medical technologies and official healthcare systems in order to ensure that clinicians are not overly overwhelmed by the applications in the context of mHealth. Good designs make use of user information as part of existing processes and prevent clinicians from being alerted unless necessary. Moreover, the emphasis on the importance of shared tasks may also increase the adherence of patients by using mechanisms, like alerts about the family members [11]. The user research stresses that to make the material and the layout effective; one should be able to consider the real life and be flexible. Figure 20.1 shows how to facilitate individualized monitoring, therapies, and enhanced everyday health through involvement of patients, caregivers, clinicians, and developers through process of iterative user testing, codesign, and real-world feedback.

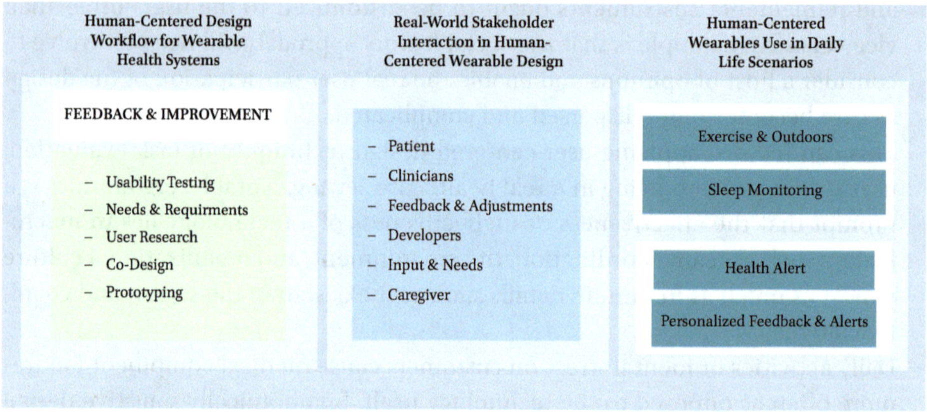

FIGURE 20.1 Human-Centered Wearable Health System Design Workflow. (Source: Own elaboration)

20.4.4 The Use of Human-Centered Design to mHealth: Definition, Standards, and Adoption

HCD Effectiveness: HCD-based treatments are more usable, suitable, and effective compared to treatments developed under the design techniques of conventional design.

Different Definitions of HCD: The adaptability and flexibility of HCD in other areas are manifested in the multiple definitions of HCD which often differ with the area of application.

ISO Standard Perspective: The standard lays a lot of emphasis on improving system usability by focusing on real system use and applying human factors, ergonomics, and usability engineering methods.

Empathy-Based Design HCD: It highly values the importance of understanding and empathizing with target users prior to implementing technology or design interventions to change or improve their behaviors [12].

20.5 Human-Centered IoT and Machine Learning (ML)-Based Smart Healthcare System for Diabetes Management

IoT-Based Customized Care: Internet of Things (IoT) devices enhance clinical outcomes and patient engagement by utilizing distinctive patient health data to tailor warnings and advice and treatment plans.

ML's Role: Through examining massive quantities of health information to create accurate patient-specific models, machine learning improves personalization. This is especially useful for managing chronic illnesses like diabetes.

Human-Centered IoT Systems: When abnormal health occurrences or emergencies occur, human-centered IoT devices incorporate decision-support processes that produce real-time notifications for healthcare providers or caregivers.

Automatic Clinical Decision-Making: ML algorithms examine the entered physiologic data and recommend suitable courses of treatment based on clinical guidance and learning patterns, particularly for controlling regular, high, and low blood sugar concentrations. Five sensors are included in the smart healthcare organization for tracking diabetes patients' health: humidity, body temperature, blood vessel pressure, respiration level, and glucose concentration sensors are among the sensors used. Worn on the hand, these human-centered IoT devices send information signals to the Arduino Nano microcontroller.

20.5.1 Smart IoT Architecture for Continuous Diabetic Health Monitoring and Clinical Decision Support

Data gathering from sensors affixed to the body of the diabetic patient. When data and data fluctuation patterns are continuously monitored, temporal variations in health conditions can be missed. On the server, the physician can see graphs and tendencies that show progressive shifts in health and past medical conditions that the algorithm has forecast. This can help with decisions about medicine and dose adjustments and lessen the impact of changes in time. The microcontroller that collects information from the sensor is the Arduino Nano. A smart importance system is applied to data gathered to prioritize extra information of an enduring nature, allowing essential and vital affected roles to get the health center suggestions more efficiently. Since there is a lot of data available for ongoing monitoring, it is uploaded to the IoT server [13]. To reduce problems, data preparation and the elimination of inconsistent and inaccurate information from stored data are carried out.

20.5.2 The Concern: Chronic Conditions' Effects and Administration

Chronic illnesses like diabetes, heart disease, stroke, and persistent obstructive pulmonary disease are becoming more common, as is the percentage of individuals with many health disorders, or dual diagnoses. Healthcare organizations that offer compartmentalized services do not help individuals through long-lasting diseases, who have complicated desires and need continuing treatment and preventative provision. In a clinical setting, personal health information (such as blood pressure and heart rate) that is needed to guide therapy is typically recorded episodically, creating a disjointed picture of a person's well-being position. Participating in the upkeep, observation, and control of daily actions to promote health is known as self-care. On a personal level, symptoms, pharmaceutical adverse effects, and a lack of expertise and abilities are some of the identified obstacles to self-care. However, accountability does not end here. Effective management of chronic illnesses requires a plan that considers behavioral, social, personal, and infrastructure factors in order to promote health and improve the efficacy of subsequent medical institutions.

Moving Toward a Resolution: Digital Health: The term "digital health" describes the application of technologies, such as the web, widespread sensing, and computer technology, to change healthcare delivery and health systems. Electronic medical records save patient information, telemedicine and telecommuting offer remote diagnosis and therapies, and analytics help practitioners make informed choices [14].

Figure 20.2 visualizes a diabetic patient engaging with wearable sensor equipment in the center of the illustration. In addition to emphasizing comfort, availability, simple interfaces, alarms, safe information, and ethical AI, surrounding factors also highlight user demands, usability, and privacy.

20.6 IoT's Place in Contemporary Health Monitoring Systems

The IoT has revolutionized strength nursing by providing actual surveillance of enduring health. By examining individual health trends, IoT also facilitates tailored healthcare. Patients get tailored information and recommendations using applications or other platforms that are readily available. IoT devices will be able to keep track of regular monitoring, reminders to take medication, and useful tips to ease the burden of individuals with chronic conditions such as diabetes or high blood

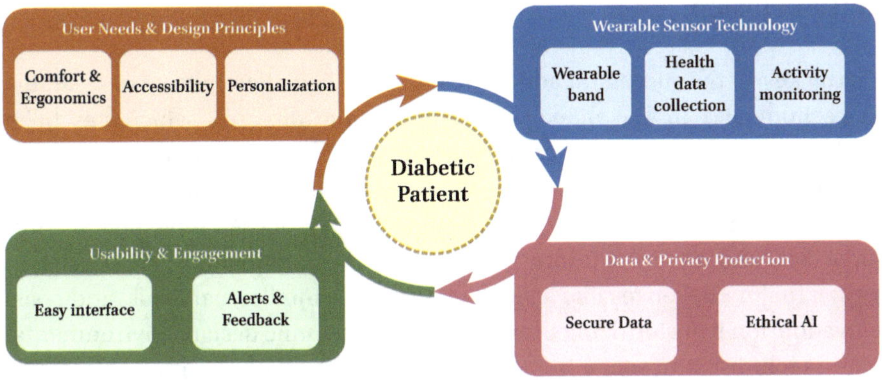

FIGURE 20.2 Framework for diabetic patient-centerpatient-centered wearable health monitoring integrating user-centered design, wearable sensors, usability, engagement, and data privacy for effective, secure, and personalized health management (Source: Own elaboration)

pressure. Examples of IoT applications are wearable devices that measure vital signs and provide health data, including health monitors and sensors.

20.6.1 IoMT Systems in Healthcare Monitoring

Components and Functionality: The IoMT is a system of interconnected gadgets and devices that folds and assesses well-being data in an effort to improve patient care. Such devices consist of wearables such as fitness trackers and smartwatches, and specialized medical devices such as heart rate cuffs, diabetes meters, and ECG monitors.

20.6.2 The Essential Aspects of IoMT Systems

- **Sensors:** These can be used in medical devices or can be worn by individuals as continuous health tracking devices that collect biometric data, including the level of the heart, pulse, and glucose.
- **Data Transmission Modules:** The data transfer between detectors and the handling units is to be ensured through the reliability data transfer in real time through the transfer of data using technologies such as Wi-Fi, Bluetooth, Zigbee, or cellular connection.
- **Preprocessing Units:** Process sensor information to be used in advanced applications by computing it on clouds or in a device.
- **Databases:** Clinical information is stored safely and often using the internet due to flexibility and easy accessibility which enables medical professionals to look and utilize the data [15].

20.7 Conclusion

HCD is the key to the successful adoption and sustained effectiveness of wearable sensor technologies in the management of chronic diseases, which has been revealed in this chapter. Despite the fact that wearable sensing, IoMT, and ML innovations provide the opportunity to monitor continuously and offer personalized treatment, their real effect largely relies on the ease of use, convenience, trust, ease of access, and integration into the routine of life and practice. HCD helps in addressing major challenges like a lack of adherence, intellectual loads, and resistance to adoption by prioritizing user demands, ergonomic design, environmental usage and technical and social fit. Altogether, HCD wearable health systems provide the groundbreaking approach to the management of chronic diseases that will facilitate more reliable, ethical, and patient-centered digital health services, which will lead to improved health outcomes and life quality.

References

[1] Mbata AO, et al. Preventive medicine and chronic disease management: Reducing healthcare costs and improving long-term public health. International Journal of Multidisciplinary Research and Growth Evaluation, 2024;5(06):1584–1600. https://doi.org/10.54660/.IJMRGE.2024.5.6.1584-1600.

[2] Adepoju VA, et al. Wearable Technology in the Management of Chronic Diseases: A Growing Concern. Chronic Diseases and Translational Medicine, 2025;11(02):117–121. https://doi.org/10.1002/cdt3.156.

[3] Lewczak Z, Mitchell M, Mitchell MG. Wearable technology and chronic illness: Balancing justice and care ethics. Cureus. 2024;16(11). https://doi.org/10.7759/cureus.73686.

[4] Rani P, et al. Simulation of the lightweight blockchain technique based on privacy and security for healthcare data for the cloud system. International Journal of E-Health and Medical Communications (IJEHMC), 2022;13(4):1–15. https://doi.org/10.4018/IJEHMC.309436.

[5] Yadav SP, et al. An improved deep learning-based optimal object detection system from images. Multimedia Tools and Applications, 2024;83(10):30045–30072. https://doi.org/10.1007/s11042-023-16736-5.

[6] Murala DK, Panda SK, Dash SP. MedMetaverse: Medical care of chronic disease patients and managing data using artificial intelligence, blockchain, and wearable devices state-of-the-art methodology. IEEE Access. 2023;11:138954–138985. https://doi.org/10.1109/ACCESS.2023.3340791.

[7] Materia FT, Smyth JM. Acceptability and concerns about innovative wearable health sensors in persons with and without chronic disease diagnosis. Internet Interventions. 2024;35:100702. https://doi.org/10.1016/j.invent.2023.100702.
[8] Tu J, et al. Wearable biomolecular sensing nanotechnologies in chronic disease management. Nature Nanotechnology. 2025;20:1–17. https://doi.org/10.1038/s41565-025-02010-2.
[9] Obianyo C, et al. The future of wearable health technology: From monitoring to preventive healthcare. World Journal of Biological, Pharmaceutical and Health Sciences. 2024;20:36–55. https://doi.org/10.30574/wjbphs.2024.20.1.0709.
[10] Thompson E. Wearable Health Devices in Chronic Disease Management: Opportunities, Challenges, and Future Directions. Frontiers in Healthcare Technology. 2025;2(1):27–33. https://doi.org/10.71465/fht298.
[11] Tzimourta KD. Human-Centered Design and Development in Digital Health: Approaches, Challenges, and Emerging Trends. Cureus. 2025;17(6). https://doi.org/10.7759/cureus.85897.
[12] An Q, et al. Sustainable development for mobile health apps using the human-centered design process. JMIR Formative Research. 2023;7:e45694. https://doi.org/10.2196/45694.
[13] Maqbool S, et al. A smart sensing technologies-based intelligent healthcare system for diabetes patients. Sensors, 2023;23(23):9558. https://doi.org/10.3390/s23239558.
[14] Nourse R, et al. The role of a smart health ecosystem in transforming the management of chronic health conditions. Journal of Medical Internet Research. 2023;25:e44265. https://doi.org/10.2196/44265.
[15] Mohammed SY, Aljanabi M. Human-Centric IoT for Health Monitoring in the Healthcare 5.0 FrameworkDescriptive Analysis and Directions for Future Research. Edraak. 2023;2023:21–26. https://doi.org/10.70470/EDRAAK/2023/005.

CHAPTER 21

Validation and Accuracy of Wearable Sensor Data in Chronic Disease Monitoring and Diagnosis

Krishnamoorthy N. R., Mahendra Kumar Verma, Ipsita Dash

Abstract Wearable sensor technologies are increasingly being used for continuous monitoring and diagnostic support in chronic disease management, providing ongoing insights beyond the traditional clinical environment. Validity, accuracy, and reliability of the data produced in various physiological and environmental settings are critical to how well these systems are of clinical use. This chapter will describe the principles, methodologies, and clinical factors of validating the wearable sensor data and evaluating the accuracy of measurement of chronic diseases monitoring and diagnosis. The chapter presents fundamental ideas of accuracy, precision, reliability, and validity in wearable sensing. Comparison to clinical reference standards description is described, benchmarking measures of error analysis, assessment of agreement, and diagnostic performance measures are described. Conditional-specific validation methods are discussed in cardiovascular, metabolic, respiratory, and movement related chronic conditions. Factors influencing data quality, such as sensor placement, motion artifacts, calibration drift, and population variability, are analyzed. Robust validation and accuracy assessment are essential for establishing confidence in wearable sensor data, enabling safe clinical integration, reliable diagnosis, and effective long-term chronic disease management.

Keywords Wearable sensors – data validation – measurement accuracy – chronic disease monitoring – digital health – clinical diagnostics

21.1 Introduction

Chronic illnesses are defined as a collection of long-lasting medical conditions with progressively lengthy clinical features, treatment, and outcomes. They can also be classified in terms of their duration (generally over 3 months, or longer). Due to the lack of widely used outcome-measuring scales, chronic diseases are complicated, multifaceted problems that are challenging to quantify and categorize. Early detection, therapy, and restoration are long-term procedures that are influenced by a variety of psychological and societal risk factors. Over the course of days and months, wearable technology provides a way to covertly gather lifestyle information and physiological signs [1]. Electrical gadgets carried on human bodies that are

 | HTTPS://DOI.ORG/10.1515/9783112237083-021

capable of gathering, processing, and transmitting a variety of physiological data are called smart wearable sensors. These new wearable technologies are always getting better in terms of mobility, convenience, and detection accuracy when compared to conventional monitoring systems [2]. Wearable technology makes it possible to continuously monitor health parameters in people with long-term illnesses, which can enhance quality of life (QoL) by promoting healthy behaviors and actively engaged patients. Some of the barriers to its introduction into medical treatment are data privacy, sensor accuracy, expensive costs, and the necessity to maintain data collection and assessment regularly and ensure that patients trust the implementation [3]. The exchange of clinical data improves the performance of healthcare infrastructure and increases the quality of care. Patients must also have and control their own medical information, as one of the greatest assets instead of letting it be scattered among numerous healthcare providers [4]. Task plans were being made by top managers and executed by the juniors. Object localization is a very important role in the current automation, including in areas such as item recognition, classification, tracking, and accurate attribute detection [5].

21.1.1 Key Features

- **Comprehensive Frameworks of Validation:** The chapter applies accuracy, reliability, and consistency criteria to compare wearable sensor validation methods with clinical reference standards.
- **Condition-Specific Accuracy Analysis:** It is an evaluation of wearable sensor accuracy in many chronic diseases, including respiratory, physiological, cardiac, and movement-related diseases.
- **Assessment of Clinical Readiness and Data Quality:** This chapter also explores the variables that affect the accuracy of data in addition to the discussion of the practical implementation of the statistic, legal, and clinical issues.

Chapter Objective: The chapter investigates the validation of rules and methods of accuracy assessment of wearable e-sensor data used in the context of chronic disease tracking to guarantee accurate measurements, clinical availability, and safe integration with medical institutions used to make effective diagnostic decisions and treat a patient on a long-term basis.

21.2 Potential Benefits and Clinical Impact of Wearable Sensors in Healthcare

Mobile devices and the continuous well-being information provided by them can transform treatment and health care in many ways. The gadgets open a number of important benefits, which are as follows:

- **Constant, Actual Surveillance:** Wearable detectors enable inactive surveillance of comprehensive functional and physical information around-the-clock or during important activities. Minor changes over time may manifest as novel diseases or the impact of treatment and changes in lifestyle.
- **Early Disease Detection and Intervention:** Machine learning (ML) models can be used to identify the presence of future health problems based on continuous sensor data on wearables. According to the violation of individual norms, it may make it possible to initially discover and predict illnesses such as core illness, illnesses of Alzheimer, etc.
- **Personalized and Optimized Healthcare:** Through a thorough understanding of a person's composition and health trends, detectors combined with heredity, intestinal tract, and other health information can power personalized medication.
- **Patient Chronic Illness Management:** The passive monitoring that is available in wearables simplifies tracking the key metrics between appointments with a doctor, which helps in the treatment of chronic diseases. This is especially beneficial in such conditions as diabetes that should be meticulously handled. Patients and medical workers can more easily assess the impact of medication changes.
- **Reduced Hospitalization:** Wearable health tracking reduces unnecessary doctor visits and hospitalization through providing frequent, objective check-in examinations and support prompt interventions where necessary. This reduces medical costs and enhances efficacies to the patients.
- **Population Health Measurement:** Anonymized data that is collected by wearable users provides an abundance of population health data. This helps in developing orientation well-being and description of temporal, and regional strength designs and differences.
- **Clinical Trial Endpoints and Digital Biomarkers:** The sensor-generated data that can be utilized as innovative digital biomarkers include activity, heart rate variability (HRV), and sleep duration [6]. They can provide objective measurements of patient trials and studies in the field of medicine besides patient questionnaires and laboratory tests.

21.2.1 Continuous Telemonitoring of Vital Signs Using Wearable Sensors

An increasing number of wearable sensors are being used in continuous telemonitoring to observe physical indicators and vital signs throughout the daily lives of an individual, such as blood oxygen level, heart rate (HR), respiration motion, and other cardiovascular signals. The approach enables home-based care, continues monitoring of chronic illnesses, and enables identifying health deterioration at an early stage, which may prevent significant effects. Wearable technology differs in terms of their location on the body, feasibleness or disposability, battery life, and how the data can be relayed and all this may impact the reliability and usefulness of the wearable technology in real life scenarios. These devices need to be accurate, reliable, be certified as medical equipment, and address clinical and regulatory needs so as to be effectively and successfully incorporated into the healthcare processes. Precision of sensors may be affected, though, by such aspects as daily activities, patient behavior, and less supervision at home. Consequently, there was no cross-over between performance in controlled or clinical environments and performance in the home or free-living environments. Hence, to ensure that the collected data is clinically useful and important, continuous telemonitoring requires extensive validation and contextual evaluation [7].

Wearable expertise with artificial intelligence (AI) has since become a powerful aid to the current well-being nursing, facilitating the initial detection of such diseases as mental health issues, respiratory diseases, and cardiovascular anomalies in the first place. An example of such strategies that employ AI to identify early signs of diseases includes smartwatches, biosensors, and suitability trailers, which are devices that acquire a constant stream of physical data. This section discusses the technological basis, applications, and practical implications of AI-based wearables and provides a discussion of the implementation challenges, as well as highlights the role of the wearables in preventive healthcare.

21.2.2 Advanced Sensor Integration and Data Intelligence in AI-Driven Wearable Healthcare Systems

Examples of advanced sensors that enhance accuracy of health measurements include photoplethysmography (PPG), electrocardiogram (ECG), accelerators, and pulse oximeters. AI algorithms process these huge datasets to detect patterns, eliminate noise, and differentiate between healthy and diseased cases. Cloud-based computing also makes scalable assessment and easy access to health data by the patients and physicians possible.

Mobile applications are easy to use and provide immediate feedback and health warnings, thus improving patient engagement. However, challenges remain, including data privacy, energy efficiency, regulatory compliance, and limited integration with digital medical records. These issues need to be addressed by safe standards of data and interoperable platforms to achieve the maximum clinical value of AI-based wearable solutions.

Circulatory tracking: The possibility to constantly measure heart activity in any real-life situation has revolutionized the sphere of cardiac nursing and been implemented through AI-based wearables. These devices are able to assess autonomic nerve activity, diagnose the first signs of cardiac deterioration, and abnormal heart rhythms through the analysis of ECG, PPG, and HRV signals. These parameters improve the disease management in the long term, reduce emergency cases, and facilitate early interventions. Wearable patches are used to detect intermittent cardiac abnormalities, as they enable longer monitoring periods than traditional methods. Nevertheless, to be used on a large scale, the challenge of cost of devices, their long-term convenience, and accessibility among different demographics should be addressed.

Respiratory and Mental Health Measurement: Wearable technologies can detect respiratory and mental health issues by continuously tracking oxygen levels, respiratory habits and activity levels, as well as indicators of physiological stresses with the help of AI. The devices are a noninvasive and scalable alternative to the conventional methods of diagnosis since they allow early detection of respiratory dysfunction, sleep-related breathing disorders, and stressful mental health abnormalities [8]. Immediate input is facilitated in clinical decision-making as well as in aggressive self-management. The AI-driven wearables in diagnosis and monitoring and preventative healthcare summarizes wearable technology, applied AI methods, performance outcomes, and their implementation implications through the data presented in Table 21.1.

TABLE 21.1 AI-enabled wearable devices for cardiovascular, respiratory, and mental health monitoring.

Device/sensor	AI technique	Performance metrics	Clinical impact
Fitbit (pulse oximetry sensor)	ML	Notices breathing	Distress: 20% reduction in hospitalizations
Speedometer and SpO_2 wristwatch	Convolutional neural network (CNN)	Accurateness in sensing	30% lower diagnostic expenses
Monitor (activation, HRV)	ML	Associates with psychological well-being conditions	Better handling of stress
Apple Watch (ECG, PPG)	Deep learning (DL)	Sensitivity: 87%	Initial analysis, condensed injury danger

TABLE 21.1 AI-enabled wearable devices (*cont.*)

Device/sensor	AI technique	Performance metrics	Clinical impact
Zio patch (ECG)	AI algorithms	Effectiveness for arrhythmias	Decreased alternative calls
Wearable (accelerometer)	ML	Accuracy: 85% for risk profiling	Improved defensive maintenance

Source: Reference [8].

21.3 Instrumentation for Physiological and Activity Monitoring

Physical Activity (Dynaport Move Monitor): It is a triaxial accelerometer which is a validated instrument to measure the amount of physical activity in healthy participants and chronic obstructive pulmonary disease (COPD) patients. It was worn at the lower back for 7 days in a row while awake; days with fewer than 8 h of wear time were not included. Daily step counts and wearing times were examined.

Cardiac Autonomic Function Polar H10: HR and HRV**Error! Bookmark not defined.** can be observed with the heart-autonomic purpose (Polar H10), a certified chest-strap sensor. Data was recorded at 1 s period while individuals wore the gadget nonstop for 5 days. HRV (RMSSD) was computed using Kubios HRV software, and idle HR was obtained from resting data.

Oxygen Saturation (Nonin WristOX$_2$ 3150): The reference tool for measuring peripheral capillary oxygen saturation (SpO_2) in COPD patients is a clinically recognized finger pulse oximeter. After removing incorrect information and short sleep intervals, the average overnight SpO_2 was computed [9]. Accuracy, reliability, mean absolute error (MAE), and Bland-Altman analysis are used to verify wearable devices that monitor HR, HRV, and activity, as shown in the image. In the end, it improves QoL by highlighting advantages such as early detection and individualized care, as well as difficulties such as data privacy, sensor accuracy, and high prices, which were depicted in Figure 21.1.

21.3.1 HRV Assessment

To guarantee reliable HRV assessment, individuals were asked to refrain from eating, drinking, smoking, and engaging in severe physical activity for 12 h before testing. Participants were seated comfortably and allowed to breathe regularly in a quiet, temperature-controlled room during the morning measurements. The

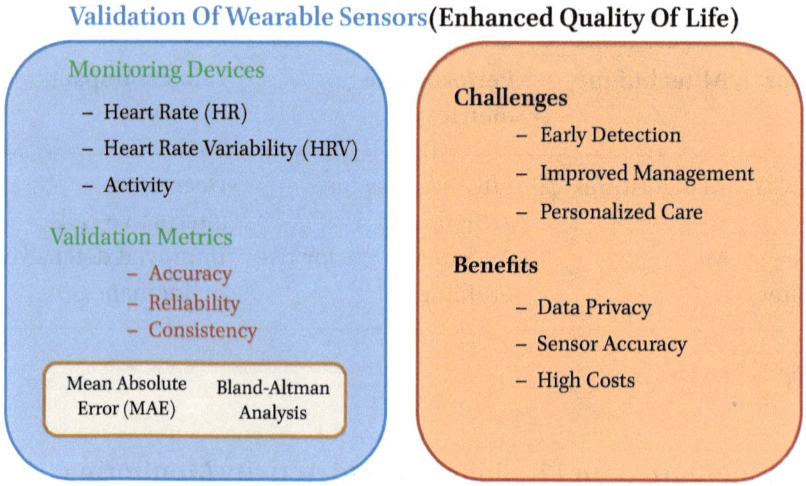

FIGURE 21.1 Validation framework for wearable health sensors. (Source: Own elaboration)

Zephyr BioHarness-3, a lightweight wearable sensor placed on the chest that collects cardiac electrical activity and computes beat-to-beat (R-R) durations at a high processing rate, was used to gather HRV data. HR data were wirelessly transferred to a mobile device and constantly monitored for several minutes following a brief resting time.

Thereafter, the data were collected and then cleaned and analyzed with specialized HRV software. Any signal artifacts or irregular beats were located and corrective measures were taken to ensure that the data was of high quality. Time and frequency of HRV were calculated based on the established clinical standards. All in all, time domain analysis was used to measure the HRV, and the frequency domain analysis to measure the activity of the sympathetic and parasympathetic nervous system.

21.3.2 Outcome: Reduced HRV in COPD and Feasibility of Wearable-Based Monitoring

The study [10] demonstrated that HRV is decreased in patients with slight to extreme COPD and that HRV is correlated with a number of performance metrics and health status indicators but is unaffected by the disease's development as indicated by deteriorating lung function. According to available research, decreased HRV in COPD is most likely caused by the impacts of gasped inhalers on the self-regulating circuitry. Furthermore, the assessment supports the viability of measuring HRV in COPD patients using a wearable biosensor worn on the chest. This is because it can be essential to continuously monitor physiological parameters (such as HRV and activity level) to evaluate patients' functional status, identify improper metrics early, and direct targeted treatment. Wearables can provide important information about a person's health and well-being by tracking important biomarkers such as HRV, HR, breathing, humidity, and psychological restrictions.

Data Alignment and Temporal Aggregation of Wearable Sensor Validation: Three sets of data were constructed through standardization of data collected at varying rates (1–30 s): per-second, 10 s, and 60 s. The signals were temporally aligned with the reference with per-second-data directly paired and 10 s averaged data over the last 10 s and 60 s averaged data over synchronized 60 s windows. The missing measurements were deleted through pairwise deletion and no additional preprocessing was performed to allow outliers to indicate realistic situations.

Validation of HR Accuracy Across Devices and Dynamic Conditions: Using measures such as MAE, mean absolute percentage error (MAPE), Spearman association, and Bland-Altman analysis, the accuracy and agreement of the device were assessed. Each participant's metrics were computed and compiled at the group level. Analyses contrasted the performance of wrist-worn devices, such as the Fitbit Charge 5, WHOOP 4.0, Withings Scan watch, and the research-grade Embrace Plus, with the chest-worn Zephyr BioHarness 3.0 and 12-lead ECG during steady states and transitions. Zephyr's high-resolution validation allowed for the evaluation of fast HR dynamics without smoothing or averaging. Variations in device accuracy under dynamic situations were highlighted by applying statistical bounds and correlation requirements to judge performance. Strong HR accuracy was maintained by the Zephyr BioHarness 3.0 in every situation. The Garmin Vivosmart 4 was more stable during transitions, while the Fitbit Charge 5 and Sense 2 displayed the finest general correctness among wrist-worn devices. While bigger averaging windows enhanced performance, accuracy decreased for all wrist-worn devices during abrupt changes in HR [11].

21.3.3 Wearable Diagnostic Assessment System for Attention-Deficit/Hyperactivity Disorder Evaluation

Six wearable motion sensors, interactive exercises, and a touchscreen interface are all integrated into the wearable diagnostic assessment (WeDA) system, a diagnostic platform built on the *Indicative and Numerical Physical of Psychological Illnesses*, 5th ed (DSM-5). It uses postural data and timed exercises to record cognitive and motor performance. These signals are analyzed by ML models in conjunction with SNAP-IV scores to facilitate objective evaluation of attention-deficit/hyperactivity disorder (ADHD).

Components and Architecture of the WeDA System: In accordance with the DSM-5 recommendations, the WeDA scheme is a sensor-based diagnostic tool. It incorporates six wearable gesture devices on the skull, pointers, bases, and stomach, as well as interactive 3D-printed devices and a touchscreen computer. These components enable simultaneous collection of physical movement and cognitive functioning data to be collected during assessment.

Design of Tasks and Acquisition of Diagnostic Data: Ten mental-motor activities of different durations (timing tasks) are taken to assess coordination, balance, attention, and reaction. Wearable sensors are used to measure performance using task accuracy, mistake rate, time, and posture data. This data collection gives a detailed description of both cognitive and cognitive characteristics relevant to diagnosis.

Clinical Linkage and Diagnostic Modeling (SNAP-IV): Parent-reported SNAP-IV, the clinical baseline is rated on a four-point scale using 26 items, in measuring symptoms associated with ADHD. A mean subgroup score of above 1.2 indicates a probable analysis of ADHD.

Processing and Interpretation: To ensure that the data collection was done in a more comfortable setting, it was carried out in a child-appropriate environment. During the test, parents filled out the SNAP-IV questionnaire without speaking to the child. After being given instructions, kids used six motion sensors to accomplish WeDA game-based tasks. Children were given the option to select the task sequence and mobility, and task-performance data were continually recorded and stored during task execution.

Result: With greater sensitivity, specificity, and overall accuracy than SNAP-IV, the WeDA system demonstrated superior diagnostic performance. Its improved impartiality and dependability were demonstrated by its continuously higher diagnostic efficacy across a number of validity measures. High diagnostic accuracy was attained by both approaches, confirming the WeDA system's efficacy as a supplementary diagnostic tool for ADHD [12]. Table 21.2 evaluates wearable sensor performance using accuracy, validity, reliability, signal quality, and clinical outcome metrics.

21.4 Comprehensive Evaluation of Nutritional Status and Diagnostic Performance Using Global Leadership Initiative on Malnutrition Standards

The study by Kinugasa et al. [13] was based on Global Leadership Initiative on Malnutrition (GLIM) criteria and it assessed the analytic expression of nourishing pointers in subjects, including the body weight, muscle mass, intake of diet and inflammation. The physical functioning was measured with the help of a 6-min walk test, and the arm and calf measurements that estimated the muscle mass. Clinical outcomes in combination with these measurements provide a comprehensive evaluation of nutritional condition and its health outcomes.

TABLE 21.2 Summary of key evaluation metrics assessing accuracy, validity, reliability, signal quality, and clinical usefulness of wearable sensors.

Category	Metric	Purpose
Accuracy metrics	MAE	Average absolute difference between wearable sensor values and reference measurements
	MAPE	Percentage-based error, useful for comparing across subjects
Validity metrics	Sensitivity	Ability to correctly identify true positives
	Specificity	Ability to correctly identify true negatives
	Accuracy	Overall correctness of classification
	Area under the curve (AUC)	Overall diagnostic performance across thresholds
	Bland-Altman analysis	Agreement and limits of agreement (LoA)
Reliability metrics	Intraclass correlation coefficient (ICC)	Repeatability and consistency of measurements
	Coefficient variation (CV)	Relative variability of repeated measurements
Temporal and signal quality metrics	Data availability	Percentage of usable data
	Signal-to-noise ratio (SNR)	Quality of recorded signals
	Artifact rate	Motion or noise contamination
Clinical outcome metrics	HRV indices (RMSSD, LF/HF)	Autonomic nervous system assessment
	Event detection rate	Detection of abnormal episodes

Source: Own elaboration

21.4.1 Evaluation of Diagnostic Performance

- GLIM standards were employed to assess three dietary indicators: low body weight, low muscle mass, decreased dietary intake, and inflammation.
- Incorporated medical results (heart failure rehospitalization, all-cause death) and exercise performance to evaluate the practical effect of malnutrition.

21.4.2 Inflammation, Weight, and Diet

- Inflammation is indicated by blood albumin levels (<3.2 mg/dL).
- Mini nutritional assessment (MNA) was used to track food consumption and weight loss; eating less or losing 1 kg or more in three months suggested danger.
- Asian BMI criteria were used to determine underweight

21.4.3 Physical Fitness and Strength Form

- Arm circumference, biceps skinfold, and arm muscle length (%AMC < 90% indicates low muscle) were used to measure upper-body muscle.
- Calf circumference was used to measure lower-body muscle (<34 cm for men, < 33 cm for women = low muscle).
- The 6-min walk was used to measure fitness; walking less than 300 m demonstrated poor physical function.

Through extensive nutritional and functional evaluation, the figure illustrates how nutritional signs, inflammation, dietary intake, muscle mass, and physical condition all contribute to diagnostic efficacy, which is connected to clinical health consequences such as heart failure, rehospitalization, and all-cause mortality, as shown in Figure 21.2.

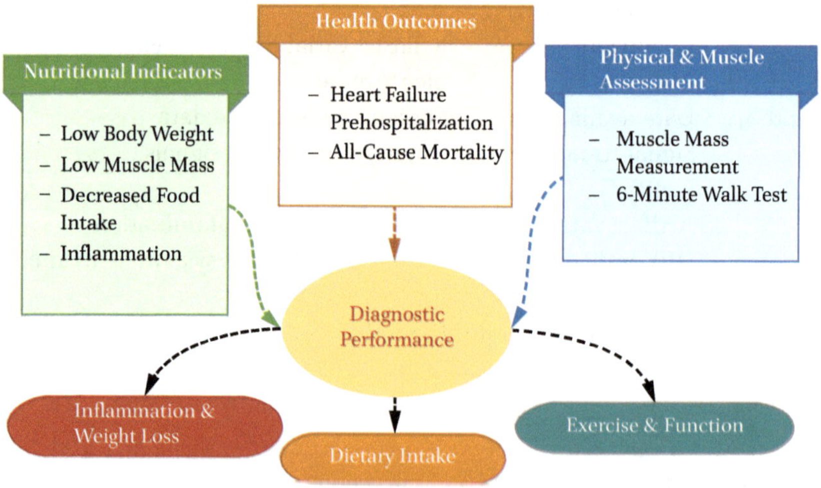

FIGURE 21.2 Diagnostic performance framework integrating nutrition, muscle status, and outcomes. (Source: Own elaboration)

21.5 Key Design and Adoption Challenges in Wearable Healthcare Systems

Vital human indicators can be continuously and noninvasively monitored by wearable technology with integrated transmission sensors. The data is then transferred to an electronic device for storage and viewing. The combination of several methodologies, including data analysis, programming, microelectromechanical technologies, and sensing methods, has made this feasible. Sensor accuracy, determining demands pertaining to consumers and commercial representations for financial utilization, information secrecy, and safety are among the design elements. Additionally, it is difficult to protect patients using imaging devices and isolated innovations in fatherlands since it affects individuals' behavior and trust in Internet of Things (IoT) devices. The global pandemic, the aging population, and the increasing prevalence of chronic diseases have all encouraged people to lead healthier lifestyles. Technology led to the development of state-of-the-art healthcare devices and personalized applications for diagnosing common illnesses, monitoring physiological processes, supporting patients throughout treatment, and giving daily medications. Though these devices get smaller, their development complexity has increased due to different device incorporation methods, different associativity possibilities, and little electric control needs for a portable form factor.

Sensor accuracy, determining demands pertaining to people and commercial replicas for marketplace utilization, information secrecy, and safety are among the design elements. The most well-known topics that come up when designing wearable devices with sensors are lack of trust, moral dilemmas, and privacy concerns. This concept is challenged by the use of vision sensors and remote innovations in homes for patient protection, which affects patients' trust and behavior when using IoT devices. Researchers are continuously examining smart sensing systems due to the declining health of older adults in general [14].

21.6 Future Directions and Emerging Opportunities in Wearable Sensor Technologies for Healthcare

- **Advanced Materials with Several Uses:** Creation of skin-like, multifunctional, highly sensitive composites that combine comfort, defense, and sensing for persistent wearable health monitoring.
- **Tools for Noninvasive Biosensors:** Expansion of biosensors that utilize sweat and tears (such as smart eyeglasses) for painless, continual monitoring of a variety of biomarkers, backed by increased precision and clinical testing.

- **AI-Powered Customized Medical Treatment:** IoT and AI integration to provide personalized health estimation, adaptive health instruction, and immediate analysis of information from a variety of electronic devices.
- **Wearables that Do Not Require Batteries:** Expansion of battery-free, energy-efficient systems that use technologies such as near-field communication to monitor continuously and in real-time without the need for large power sources.
- **Scalable Production and Marketing:** Development of large-scale, low-cost manufacturing methods to facilitate wearable biosensor manufacturing and broader consumer and clinical usage.
- **Enhanced Sensor Dependability and Performance:** Ongoing research is being conducted to address issues such as power usage, hysteresis, movement distortions, and long-term stability in order to ensure high-quality reliability.
- **Integration and Management of Big Data:** Creation of reliable platforms and systems for gathering, organizing, analyzing, and integrating massive amounts of wearable data into medical facilities to provide useful clinical insights [15].

21.7 Conclusion

The growing importance of wearable medical devices in promoting continuous, real-time, noninvasive health monitoring in the treatment of chronic diseases and early diagnosis has been emphasized in this chapter. The invention of sensor technology, the science of materials and AI has significantly increased the accuracy of data, clinical relevance and the participation of users. Other problems that still impede the large-scale adoption include validation concerns, data accuracy, privacy, power management, interoperability with healthcare systems, and acceptance by the user. To overcome these constraints, turning wearable technology into reliable, clinically integrated health care systems will demand solving these constraints by robust, strong validation, secure data frameworks, cost-effective designs and interdisciplinary teamwork.

References

[1] Chakilam C. Integrating machine learning and big data analytics to transform patient outcomes in chronic disease management. Journal of Survey in Fisheries Sciences. 2022;9(3):118–30. https://doi.org/10.53555/sfs.v9i3.3568.

[2] Tao Q, et al. Clinical applications of smart wearable sensors. Iscience. 2023;26 (9):107485. https://doi.org/10.1016/j.isci.2023.107485.

[3] Adepoju VA, et al. Wearable technology in the management of chronic diseases: a growing concern. Chronic Diseases and Translational Medicine. 2025;11(02):117–21. https://doi.org/10.1002/cdt3.156.

[4] Rani P, et al. Simulation of the lightweight blockchain technique based on privacy and security for healthcare data for the cloud system. International Journal of E-Health and Medical Communications (IJEHMC). 2022;13(4):1–15. https://doi.org/10.4018/IJEHMC.309436.

[5] Yadav SP, et al. An improved deep learning-based optimal object detection system from images. Multimedia Tools and Applications. 2024;83(10):30045–72. https://doi.org/10.1007/s11042-023-16736-5.

[6] George ASH, Shahul A, George AS. Wearable sensors: a new way to track health and wellness. Partners Universal International Innovation Journal. 2023;1(4):15–34. https://doi.org/10.5281/zenodo.8260879.

[7] Haveman ME, et al. Continuous monitoring of vital signs with wearable sensors during daily life activities: validation study. JMIR Formative Research. 2022;6(1):e30863. https://doi.org/10.2196/30863.

[8] Egbuna I, et al. Advancing early disease detection with AI: innovations in medical imaging, EHR analytics, and wearable technologies. International Journal of Life Science Research Archive. 2025;9:120–40. https://doi.org/10.53771/ijlsra.2025.9.1.0049.

[9] Hermans F, et al. Validity of a consumer-based wearable to measure clinical parameters in patients with chronic obstructive pulmonary disease and healthy controls: observational study. JMIR mHealth and uHealth. 2024;12(1):e56027. https://doi.org/10.2196/56027.

[10] Park S-C, et al. Use of a wearable biosensor to study heart rate variability in chronic obstructive pulmonary disease and its relationship to disease severity. Sensors. 2022;22(6):2264. https://doi.org/10.3390/s22062264.

[11] Oost V, Nina C, et al. Accuracy of heart rate measurement under transient states: a validation study of wearables for real-life monitoring. Sensors. 2025;25(20):6319. https://doi.org/10.3390/s25206319.

[12] Luo J, et al. A wearable diagnostic assessment system vs. SNAP-IV for the auxiliary diagnosis of ADHD: a diagnostic test. BMC Psychiatry. 2022;22(1):415. https://doi.org/10.1186/s12888-022-04038-3.

[13] Kinugasa Y, et al. Diagnostic performance of nutritional indicators in patients with heart failure. ESC Heart Failure. 2022;9(4):2096–106. https://doi.org/10.1002/ehf2.13886.

[14] Mukhopadhyay SC, Suryadevara NK, Nag A. Wearable sensors for healthcare: fabrication to application. Sensors. 2022;22(14):5137. https://doi.org/10.3390/s22145137.

[15] Sharma A, et al. Advancements and future prospects of wearable sensing technology for healthcare applications. Sensors & Diagnostics. 2022;1(3):387–404. https://doi.org/10.1039/d2sd00005a.

Index

 | HTTPS://DOI.ORG/10.1515/9783112237083-022

Mathematical Methods in the Digital Age

Volume 1
Satya Prakash Yadav, Abhishek Kumar, Ashutosh Kumar Dubey, Victor Hugo C. de Albuquerque, Adrian Rodriguez-Aguiñaga, Angeles Quezada
Medical Image Processing – Multimodal Fusion Techniques, 2024
ISBN 978-3-11-139549-4, e-ISBN (PDF) 978-3-11-143597-8, e-ISBN (EPUB) 978-3-11-143599-2

Volume 2
Umesh Kumar Lilhore, Vishal Dutt, T. Ananth Kumar, Martin Margala, Kaamran Raahemifar
Math Optimization for Artificial Intelligence – Heuristic and Metaheuristic Methods for Robotics and Machine Learning
ISBN 978-3-11-143605-0, e-ISBN (PDF) 978-3-11-143618-0, e-ISBN (EPUB) 978-3-11-143624-1

Volume 3
Hemant Kumar Saini, S.B. Goyal, Sardar M.N. Islam and T. Ananth Kumar
Quantum Unmanned Aerial Vehicle – Achieving Cybersecurity and AI Strategies
ISBN 978-3-11-161930-9; e-ISBN (PDF) 978-3-11-161932-3; e-ISBN (EPUB) 978-3-11-161935-4

Volume 4
Shalli Rani, Syed Hassan Sha, Ayush Dogra
Mathematical Computing and Sustainability – Predictive Modeling and Ramifications of Intelligent Systems
ISBN 978-3-11-161111-2, e-ISBN (PDF) 978-3-11-161203-4; e-ISBN (EPUB) 978-3-11-161232-4

Volume 5
Satya Prakash Yadav, Abhay Bansal, Victor Hugo C. de Albuquerque, Fadi Al-Turjman, Houbing Herbert Song
Octonion-Sparse Based Image Processing – Multispectral Imaging and Feature Extraction
ISBN 978-3-11-914666-1, e-ISBN (PDF) 978-3-11-220807-6; e-ISBN (EPUB) 978-3-11-220835-9

Volume 6
Ganesh Gopal Devarajan, Daniel Arockiam, Senthil Murugan Nagarajan, Rashmi Agrawal
Healthcare 5.0 with Fuzzy Logic – Artificial Intelligence, Cyber-Physical Systems, and Medical Internet of Things
ISBN 978-3-11-914504-6, e-ISBN (PDF) 978-3-11-221572-2; e-ISBN (EPUB) 978-3-11-221592-0

Volume 7
Narinder Kaur, Bobbinpreet Kaur, Yonis Gulzar, Dac-Nhuong Le
Computational Optimization – Machine Learning and Fuzzy Systems
ISBN 978-3-11-914376-9, e-ISBN (PDF) 978-3-11-221943-0, e-ISBN (EPUB) 978-3-11-221944-7

Volume 8
Suresh Kumar Krishnadhas, Ananth Kumar Tamilarasan, Pramod Singh Rathore, Abhishek Kumar
Artificial Intelligence and Robotics – Transformative and Computational Algorithms for Enhanced Problem Solving
ISBN 978-3-11-914520-6, e-ISBN (PDF) 978-3-11-221437-4, e-ISBN (EPUB) 978-3-11-221466-4

Volume 9
Monika Sethi, Shivani Sood, Saravjeet Singh, Arfat Ahmad Khan
Applied Mathematics for Healthcare Intelligent Systems – Data Representation, Smart Healthcare, Deep Learning and Medical Imaging
ISBN 978-3-11-222621-6, e-ISBN (PDF) 978-3-11-222622-3, e-ISBN (EPUB) 978-3-11-222623-0